Genotoxicology of
N-Nitroso Compounds

TOPICS IN CHEMICAL MUTAGENESIS
Series Editor: **Frederick J. de Serres**
National Institute of Environmental Health Sciences
Research Triangle Park, North Carolina

Volume 1   Genotoxicology of N-Nitroso Compounds
            Edited by T. K. Rao, W. Lijinsky, and J. L. Epler

Volume 2   Single-Cell Mutation Monitoring Systems: Methodologies and Applications
            Edited by Aftab A. Ansari and Frederick J. de Serres

A Continuation Order Plan is available for this series. A continuation order will bring delivery of each new volume immediately upon publication. Volumes are billed only upon actual shipment. For further information please contact the publisher.

Genotoxicology of N-Nitroso Compounds

Edited by

T. K. RAO
*Environmental Health Research and Testing, Inc.
Research Triangle Park, North Carolina*

W. LIJINSKY
*NCI-Frederick Cancer Research Facility
Frederick, Maryland*

and

J. L. EPLER
*Oak Ridge National Laboratory
Oak Ridge, Tennessee*

PLENUM PRESS • NEW YORK AND LONDON

Library of Congress Cataloging in Publication Data

Main entry under title:

Genotoxicology of N-nitroso compounds.

(Topics in chemical mutagenesis; v. 1)
Includes index.
1. Nitroso-compounds—Toxicology—Addresses, essays, lectures. 2. Carcinogenicity testing—Addresses, essays, lectures. 3. Genetic toxicology—Addresses, essays, lectures. 4. Mutagenicity testing—Addresses, essays, lectures. 5. Chemical mutagenesis—Addresses, essays, lectures. I. Rao, T. K., 1946-   . II. Lijinsky, W., 1928-   III. Epler, J. L. IV. Series. [DNLM: 1. Nitroso compounds—Toxicity. 2. Mutagens. 3. Carcinogenesis. QZ 202 G335]
RC268.7.N58G46 1984   615.9′5141   83-23716
ISBN 0-306-41445-7

©1984 Plenum Press, New York
A Division of Plenum Publishing Corporation
233 Spring Street, New York, N.Y. 10013

All rights reserved

No part of this book may be reproduced, stored in a retrieval system, or transmitted in any form or by any means, electronic, mechanical, photocopying, microfilming, recording, or otherwise, without written permission from the Publisher

Printed in the United States of America

Contributors

A. W. ANDREWS, Program Resources, Inc., and Chemical Carcinogenesis Program, LBI-Basic Research Program, NCI-Frederick Cancer Research Facility, Frederick, Maryland 21701

R. K. ELESPURU, Biological Carcinogenesis Program, NCI-Frederick Cancer Research Facility, Frederick, Maryland 21701. *Present address*: Fermentation Program, NCI-Frederick Cancer Research Facility, Frederick Maryland 21701.

J. L. EPLER, Biology Division, Oak Ridge National Laboratory, Oak Ridge, Tennessee 37830

J. G. FARRELLY, Chemical Carcinogenesis Program, NCI-Frederick Cancer Research Facility, Frederick, Maryland 21701

J. B. GUTTENPLAN, Department of Biochemistry, New York University Dental Center, New York, New York 10010

L. I. HECKER, Chemical Carcinogenesis Program, NCI-Frederick Cancer Research Facility, Frederick, Maryland 21701

T. HO, Biology Division, Oak Ridge National Laboratory, Oak Ridge, Tennessee 37830

A. W. HSIE, Biology Division, Oak Ridge National Laboratory, Oak Ridge, Tennessee 37830

E. HUBERMAN, Argonne National Laboratory, Division of Biological and Medical Research, Argonne, Illinois 60439

C. A. JONES, Argonne National Laboratory, Division of Biological and Medical Research, Argonne, Illinois 60439

W. LIJINSKY, Chemical Carcinogenesis Program, NCI-Frederick Cancer Research Facility, Frederick, Maryland 21701

V. D. MITCHELL, Genetic Toxicology Branch, Food and Drug Administration, Washington, D.C. 20204

M. J. PRIVAL, Genetic Toxicology Branch, Food and Drug Administration, Washington, D.C. 20204

T. K. RAO. Biology Division, Oak Ridge National Laboratory, Oak Ridge, Tennessee 37830. *Present address*: Environmental Health Research and Testing, Inc., Research Triangle Park, North Carolina 27709

J. R. SAN SEBASTIAN, Biology Division, Oak Ridge National Laboratory, Oak Ridge, Tennessee 37830. *Present address*: Pharmakon Research International Inc., Waverly, Pennsylvania 18471

Foreword

Topics in Chemical Mutagenesis is a new series dedicated to studies in the areas of environmental chemical mutagenesis and genetic toxicology. In this series we will explore some of many topics that are emerging in these rapidly developing fields.

The purpose of the present volume is to attempt to organize and compare the genotoxic properties of the N-nitroso compounds. This is a particularly interesting class of compounds because of the problems encountered with the *Salmonella* assay of Ames in generating both false positive and false negative results. The battery approach using a number of assay systems seems more appropriate to evaluate chemicals in this class.

Topics to be discussed in other volumes in this series include single-cell mutation monitoring systems, the detection of genetic damage in mammalian germ cells, the mutagenicity of pesticides, problems in monitoring human populations in genetic toxicology, and a glossary of terms in genetic toxicology. All of these books are in various stages of development and should appear within the next few years.

Frederick J. de Serres
Series Editor

Preface

During the past ten years there has been an explosive development in the number of short-term tests to predict the biological risks, especially risks of cancer, in exposure to xenobiotic chemicals. The number of published articles in this area has reached many thousands a year and there are several new journals devoted almost entirely to the presentation of the results obtained in these tests. The developers and large-scale users of these tests often rival one another in their claims of validity as predictors of carcinogenicity. Many of the test systems have mutagenesis as the measured end point, and it is frequently forgotten that mutagenesis itself is a biological hazard and that the measurement of a mutagenic risk to man is of equal importance with the estimation of carcinogenic risk. A number of books and review articles have been written about the development and application of short-term assays for mutagenesis–carcinogenesis, but none has focused on the applications of these assays to a single group of well-tested carcinogens sufficiently large and important to be a guide to understanding the complex processes of mutagenesis and carcinogenesis and the relation between them.

Almost three years ago, Dr. F. J. de Serres organized a meeting at the National Institutes of Health that was attended by a number of scientists who had worked, sometimes cooperatively, sometimes alone, in examining the behavior of a single group of compounds, the N-nitroso compounds, in the assay with which they were most familiar. From this group of scientists, all of whom presented interesting data, several volunteered to pool their results in the compilation of a single volume that would be devoted to the comparison of the results obtained with N-nitroso compounds from mainly a single source of consistent chemical quality, and an evaluation of those results as a predictor of one assay system by another, and that would be a means of illuminating the complexity of the mechanisms of biological action of these toxicants. It is this volume that we present as an effort to understand the meaning of all of the results, often disparate, that appear in the literature describing the application of all of these assay systems. The conclusions represent a consensus view of these combined efforts and are a beginning for the continuing process of unraveling the mechanisms of genetic toxicology through comparative studies of biological assays, based on a firm foundation of chemical structural relations among N-nitroso compounds.

T. K. Rao
W. Lijinsky
J. L. Epler

Contents

Chapter 1. Formation of N-Nitroso Compounds and
Their Significance 1
WILLIAM LIJINSKY

Chapter 2. N-Nitrosamine Mutagenicity Using the
Salmonella/Mammalian-Microsome Mutagenicity
Assay 13
A. W. ANDREWS AND W. LIJINSKY

1. Introduction 13
2. Materials and Methods 14
   2.1. Chemicals 14
   2.2. Bacterial Strains 14
   2.3. Experimental Procedures 14
   2.4. Preparation of Liver Homogenates 16
3. Mutagenic Activity and Chemical Structure 16
   3.1. Cyclic Nitrosamines 16
   3.2. Aromatic Nitrosamines and Related Compounds .. 24
   3.3. Acyclic Nitrosamines 24
   3.4. Nitrosamides 28
4. Conclusions/Discussion 34
5. References 43

Chapter 3. Structural Basis for Mutagenic Activity of N-Nitrosamines in the *Salmonella* Histidine Reversion Assay 45
T. K. RAO

1. Introduction 45
2. *Salmonella typhimurium*—Histidine Reversion Assay 46
   2.1. Bacterial Strains 46
   2.2. Mutagenesis Assay—Procedures 47
   2.3. Source of Nitrosamines 48
3. Mutagenesis Studies in *Salmonella typhimurium* 48
   3.1. Mechanism of Mutation Induction by Nitrosamines 48
   3.2. Relationship between Chemical Structure and Mutagenic Activity 48
   3.3. Nitrosamines—Model Compounds for Mutagenesis Studies 56
4. References 57

Chapter 4. Effects of pH and Structure on the Mutagenic Activity of N-Nitroso Compounds 59
JOSEPH B. GUTTENPLAN

1. Introduction 59
   1.1. Background 59
   1.2. Chemistry 60
   1.3. Alkylating Properties 60
   1.4. Organ Specificity 61
   1.5. *In Vitro* Mutagenesis 62
   1.6. Problems with Assays: Metabolism 62
   1.7. DNA Repair 63
   1.8. Permeation into Bacteria 63
   1.9. Optimizing Test Conditions 63
2. Methods 63
   2.1. Liver Extracts 63
   2.2. Mutagenesis Assays 63
   2.3. DMN Demethylase 63
   2.4. Isolation of DNA and Analysis of Alkylated Bases 63
3. Effects of pH on Mutagenesis and DNA Alkylation by N-Nitroso Compounds 66
   3.1. N-Nitrosamines 67
   3.2. Direct-Acting Compounds 67

3.3.	Alkylation of Isolated DNA	70
3.4.	Alkylation of DNA in *Salmonella*	71
3.5.	Possible Explanations: NMU, MNNG	71
3.6.	Possible Explanations: N-Nitrosamines	73
3.7.	Possible Relevance to Carcinogenesis	73
3.8.	Applications to *in Vitro* Mutagenesis Assays	73
4. Mutagenesis by N-Nitroso-N'-nitroso-N-alkylguanidines		74
4.1.	How to Compare Mutagenic Potencies	74
4.2.	MNNG	75
4.3.	ENNG	77
4.4.	PNNG	77
4.5.	Comparisons between Mutagens	78
5. Repair of O_6-Methylguanine and Its Effect on Mutagenesis		78
5.1.	O_6-MeG and 7-MeG in *Salmonella* DNA	78
5.2.	Threshold in O_6-MeG Formation	79
5.3.	Attempt to Observe Repair of O_6-MeG	80
5.4.	Dependence of Mutagenesis on O_6-MeG	80
6. Alternate Mechanisms of Mutagenesis by N-Nitroso Compounds		81
7. Mutagenesis by a Series of "Weakly Mutagenic" N-Nitrosamines: Effects of Structure		82
7.1.	Conditions	82
7.2.	Results: Di-N-alkylnitrosamines and Related N-Nitrosamines	83
7.3.	β-Substituted N-Nitrosamines	85
7.4.	α-Substituted N-Nitrosamines	86
8. Correlation between Mutagenic Potencies and Carcinogenic Activities of Some N-Nitroso Compounds		86
9. References		87

Chapter 5.	Induction of Bacteriophage Lambda by N-Nitroso Compounds	91
	ROSALIE K. ELESPURU	
1. Mechanism of Induction		91
2. Genetics and Methodology		91
3. Factors Affecting Induction by Nitroso Compounds		93
3.1.	Bacterial Strain	93
3.2.	Source and Amount of Activating Enzymes	95
3.3.	Media	97
3.4.	Kinetic Parameter	97

4. Structure–Activity Relationships	98
5. Relationships between Mutagenicity, Carcinogenicity, and Phage Induction	105
6. Mechanism of Action of Nitroso Compounds as Inducers of Bacteriophage Lambda	111
7. References	114

Chapter 6. The Relationship between the Carcinogenicity and Mutagenicity of Nitrosamines in a Hepatocyte-Mediated Mutagenicity Assay 119
CAROL A. JONES AND ELIEZER HUBERMAN

Chapter 7. Mutagenic Activity of Nitrosamines in Mammalian Cells: Study with the CHO/HGPRT and Human Leukocyte SCE Assays 129
TI HO, JUAN R. SAN SEBASTIAN, AND ABRAHAM W. HSIE

1. Introduction	129
2. Mutagenic Activity of Eight Nitrosamines in the CHO/HGPRT Assay	130
2.1. The CHO/HGPRT Assay	130
2.2. Mutagenic Activity of Nitrosodimethylamine as Studied in the CHO/HGPRT Assay	131
2.3. Mutagenicity of Eight Nitrosamines in the CHO/HGPRT Assay and Its Correlation with Carcinogenicity	133
3. Mutagenic Activity of 15 Nitrosamines in the Human Leukocyte SCE Assay	135
3.1. Sister-Chromatid Exchanges (SCE) in the Human Leukocyte Culture System	135
3.2. Induction of SCE by 15 Cyclic Nitrosamines in Human Leukocytes *in Vitro* and Its Correlation with Carcinogenicity	137
4. Summary and Concluding Remarks	142
5. References	143

Chapter 8.	Dimethylnitrosamine Demethylase and the Mutagenicity of Dimethylnitrosamine: Effects of Rodent Liver Fractions and Dimethylsulfoxide	149
	MICHAEL J. PRIVAL AND VALERIE D. MITCHELL	

1. Introduction	149
2. Materials and Methods	150
2.1. Chemicals and Media	150
2.2. Mutagenesis Assays	151
2.3. Preparation of S-9 Fractions and Microsomes	151
2.4. Enzyme Assays	152
3. Results	152
3.1. Kinetics of DMN Demethylase in Hamster Liver S-9	152
3.2. Correlation of Mutagenic and Enzyme Activities	152
3.3. Effects of Mixing Liver Fractions from Different Species	155
3.4. Cytosol Requirement for Mutagenicity of DMN	157
3.5. Effects of Dimethylsulfoxide	159
4. Discussion	161
4.1. Apparent K_m Values and Regulation of DMN Demethylase	161
4.2. Microsomal Inhibitor of DMN Mutagenesis	161
4.3. Cytosolic Activator of DMN Mutagenesis	162
4.4. Inhibition by Dimethylsulfoxide	163
4.5. Screening Chemicals with Hamster Liver S-9	163
5. References	164

Chapter 9.	The Relationship between Metabolism and Mutagenicity of Two Cyclic Nitrosamines	167
	JAMES G. FARRELLY AND LANNY I. HECKER	

1. Introduction	167
2. Materials and Methods	169
2.1. Chemicals	169
2.2. Preparation of Microsomes	169
2.3. Assay for Nitrosopyrrolidine Metabolism	170
2.4. Large-Scale Production of Metabolites	170
2.5. Mutagenesis Assay of NPYR with *E. coli* WU 3610 (*tyr, leu*)	171
2.6. Mutagenesis Assay of Nitrosohexamethyleneimine	171

3. Results and Discussion 172
   3.1. Nitrosopyrrolidine 173
   3.2. Nitrosohexamethyleneimine 180
4. Conclusion 184
5. References 185

Chapter 10. Structure–Activity Relations in Carcinogenesis by
             N-Nitroso Compounds 189
             WILLIAM LIJINSKY

1. Introduction 189
2. Nitrosoalkylamides 199
3. Alkylation by Nitrosamines 201
4. Nitrosomethylalkylamines (X) 204
5. Substituted Acyclic Nitrosamines 207
6. Cyclic Nitrosamines 212
7. Substituted Cyclic Nitrosamines 215
8. Derivatives of Nitrosopiperidine 221
9. Conclusion 225
10. References 226

Chapter 11. Comparison of Mutagenic and Carcinogenic Properties:
             A Critique 233
             W. LIJINSKY, J. L. EPLER, AND T. K. RAO

1. Introduction 233
2. Tabular Summation and Integration of Results 235
   2.1. Selection of Nitroso Compounds for the Comparative
        Study 235
   2.2. *In Vitro* Assays 235
   2.3. *In Vivo* Assay 236
3. Comparison of Mutagenic End Points 236
4. Relationship between Short-Term Test Results and
   Carcinogenicity 245
   4.1. Qualitative Relationships 245
   4.2. Quantitative Relationships 257
5. Precautions in Interpretation and Extrapolation 259
6. References 260

Index 265

Formation of N-Nitroso Compounds and Their Significance

William Lijinsky

Nitrosamines were first made more than 100 years ago by the simplest means, reaction of secondary amines with nitrous acid. They were not known to have adverse biological effects until 1937, when a case of poisoning by N-nitrosodimethylamine was reported but largely ignored.[1] A detailed investigation of the toxicity of nitrosodimethylamine began in England in the 1950s as a result of another accidental poisoning, the first results being reported by Barnes and Magee in 1954.[2]

The result of acute poisoning of rats by nitrosodimethylamine was extensive liver necrosis, which led rapidly to their death; liver necrosis was also the consequence of exposure of humans to large doses of this compound. With chronic administration to rats, a natural supplement to the acute study that produced such significant results, Magee and Barnes[3] began an increasingly intense examination of the toxicology of N-nitroso compounds, which is still in full flow.

Although it cannot be said that the mechanism of toxicity and carcinogenesis by nitrosodimethylamine is completely understood, the biochemical studies carried out by Magee and his collaborators have been of immense importance, following the discovery that chronic feeding of this compound to rats gives rise essentially to a 100% incidence of liver tumors.[3] A number of analogs and homologs of nitrosodimethylamine were tested within a short time after the reports of the carcinogenicity of this prototype, mainly for the purpose of establishing types of nitrosamine structure that were associated with carcinogenic activity. Information obtained by this approach, standard in toxicological

---

William Lijinsky • Chemical Carcinogenesis Program, NCI-Frederick Cancer Research Facility, Frederick, Maryland 21701.

studies, is important to an understanding of the mechanisms of carcinogenesis by this group of compounds. Some of this work was done by Magee and his collaborators in England, but the bulk of the important studies of the relationship between chemical structure and carcinogenic activity was carried out in Freiburg. Associates of H. Druckrey, and especially R. Preussmann, were the principal investigators, with the extensive studies of nitrosodiethylamine carcinogenesis in rats by D. Schmähl being the starting point. The dose–response study of nitrosodiethylamine in rats conducted by Schmähl, Preussmann, and Druckrey provided sufficient information for an attempt at quantification of carcinogenic response and risk extrapolation. The enormous amount of work carried out in Freiburg was summarized in a monumental paper in 1967.[4]

Among the N-nitroso compounds studied by the German group (many being tested in species other than the rat) were several nitrosoalkylamides, almost all of which were potent carcinogens. These were of special interest because several of them, including nitrosomethylurea and nitroso-N-methylurethane, had been commonly used in laboratories for decades in the preparation of the methylating agent diazomethane. It was tempting to believe that their action as carcinogens might be due to this ability to give rise to diazomethane, although this is not now thought to be true.[5] Another nitrosoalkylamide tested by Druckrey's group was nitroso-N-methyl-N'-nitroguanidine, originally prepared for use as a detonator, but now used as a standard mutagen, being a powerful mutagenic agent for bacteria and many higher organisms. Interest in the nitrosoalkylamides grew when it was found that they were not only mutagenic and carcinogenic by direct action (they produced tumors where they were applied, as distinct from nitrosamines, which are systemic carcinogens and are believed to require metabolic activation), but were directly acting alkylating agents for cellular macromolecules.

In the beginning it was not believed that nitrosamines represented a carcinogenic risk outside the laboratory, even though they are so easily formed from amines and nitrosating agents. It was not easy to determine nitrosamines analytically, and the sensitivity of methods available in the late 1960s was no greater than 1 to 10 parts per million, at which levels attempts to find them in the environment were not seriously made. An experiment of Druckrey's in the early 1960s was designed to test the obvious possibility that nitrosation of amines might occur in the acid milieu of the mammalian stomach and thereby result in exposure to this group of carcinogens. The amine chosen was diethylamine, which was fed to rats as the hydrochloride together with sodium nitrite for 2 years. It did not lead to the induction of tumors typical of treatment of rats with nitrosodiethylamine. It was concluded that formation of nitrosamines *in vivo* was unlikely.[6] It was later realized that nitrosation of diethylamine (and dimethylamine) is very slow because of the strong basicity of these amines, and therefore it was improbable that sufficient nitrosodiethylamine was formed to give rise to tumors within the lifetime of the rats. The nitrosation of secondary amines (formation of nitrosamines from secondary amines and nitrous acid) had been studied very thoroughly by Ridd,[7] and kinetic experiments showed that under most conditions the rate of reaction is proportional to the square of the nitrite concentration.

The reaction of tertiary amines with nitrous acid had been largely ignored because of the commonly held belief (based on inadequate experiments) that such a reaction did not take place. There had been sporadic reports in the literature of formation of nitrosamines by nitrosation of tertiary amines, but these findings could have been due to contamination of the tertiary amine with secondary amine. However, in 1967 Smith and Loeppky reported their studies of the reaction of tribenzylamine with nitrous acid, which showed that the reaction did take place,[8] but only at a weakly acid pH (above pH 3). Starting in the 1970s, this finding encouraged others, including this author, to undertake a more comprehensive study of the nitrosation of tertiary amines.[9] Among the tertiary amines of interest were the large number of drugs and agricultural chemicals that belong to this class.[10,11] There is, of course, a comparably large number of secondary amines that are used as drugs and agricultural chemicals. Although there have been some kinetic studies of nitrosation of tertiary amines,[12,13] the mechanisms are not well established and the stoichiometry of the reaction is not clear, although it is known that one or more nitrosamines and carbonyl compounds are the most prominent products.[14]

Comprehensive studies of amine nitrosation were very few until a decade ago. It was known that, in addition to nitrous acid, alkyl nitrites and nitrogen oxides (mixtures of NO_2 and NO) reacted with secondary amines to form nitrosamines, and several were prepared this way, e.g., nitrosoiminodiacetic acid.[15] Nitrogen oxides in flue gases are also the source of nitrosamines in beer.[16] It has been proposed that nitrogen oxides might lead to formation of nitrosamines *in vivo*, but this has been disputed. It is believed, now, that many types of oxidized nitrogen compounds can act as nitrosating agents, e.g., peroxyacyl nitrites and nitrates, which are components of the modern "smog" in cities (see review on nitrogen in AMBIO.[17]) Some N-nitroso compounds can themselves act as nitrosating agents for amines. Among them is the aromatic nitrosamine, nitrosodiphenylamine, which finds extensive use in the rubber industry and is itself a carcinogen,[18] but which can also be an effective nitrosating agent.[19] More recently it has been found that many aliphatic nitrosamines, as well as aromatic ones, are excellent nitrosating agents,[20] particularly in the presence of nucleophiles, such as thiocyanate, which act as catalysts. Thiocyanate has long been known as a good catalyst of nitrosation.[21] Nitrosamines usually considered noncarcinogenic, and therefore of little or no risk, such as nitrosoproline and nitrosohydroxyproline,[22-24] or others considered only weakly carcinogenic, such as nitroso-N-methylpiperazine,[4,25] react quite rapidly with easily nitrosated secondary amines, such as morpholine, at acid pH to form nitrosamines that can be carcinogenic, in this case nitrosomorpholine.[26] Nitrosoalkylureas are also good nitrosating agents, in some cases nitrosating themselves.[27]

In the course of repeating a study of the nitrosation of oxytetracycline by nitrous acid to form nitrosodimethylamine,[28] Mirvish and co-workers discovered that ascorbic acid was an effective inhibitor of nitrosation of amines[29] by competing with the amine for the nitrous acid, as demonstrated earlier by Dahn.[30] This finding, with its obvious implication for the prevention of formation of carcinogenic N-nitroso compounds, led to a large number of studies of the role of inhibitors and accelerators of nitrosation, in conjunction with studies of nitrosation itself. In addition to ascorbic acid, alpha tocopherol, glutathione, some phenols, and tannins are inhibitors of nitrosation, also through competition for nitrous acid.[31] Among accelerators of nitrosation, in addition to nucleophilic anions, are some phenols[32] and metal ions.[33] Among the more provocative discoveries was that of Keefer and Roller,[34] who showed that certain carbonyl compounds, including formaldehyde and chloral, were able to facilitate reaction of nitrites with secondary amines (but not with tertiary amines), even at alkaline pH. This could greatly broaden our awareness of the range of sources of exposure of humans to N-nitroso compounds. It goes far to explain the presence of nitrosodiethanolamine in synthetic cutting oils, which contain diethanolamine as an impurity in a main component, triethanolamine, in addition to nitrite.[35,36] Nitrosodiethanolamine is also present, in much lower concentrations, in many types of cosmetic preparation.[37] It is becoming increasingly obvious that nitrosodiisopropanolamine [nitroso-bis(2-hydroxypropyl)amine], a homolog of nitrosodiethanolamine, and seemingly a more potent carcinogen than the latter, is also quite widely distributed. The concentrations of the homolog are usually lower than those of nitrosodiethanolamine, and they are derived from nitrosation of the secondary and tertiary aromatic bases di-isopropanolamine and tri-isopropanolamine.

Another type of nitrosamine found in cutting oils, nitroso-5-methyloxazolidine,[38] illustrates the variety of N-nitroso compounds that can be formed in mixtures. It has been known for some time that alkanolamines react with aldehydes in the presence of nitrites to form nitrosooxazolidines and nitrosotetrahydrooxazines.[39] Many of these compounds are potent carcinogens, although some are not, and considerable work is in progress on their formation and occurrence.

There have been reports of nitrosamines in air of industrial areas, especially nitrosodimethylamine in the vicinity of factories in which dimethylamine is manufactured or used.[40] There have also been reports of the presence of nitrosamines in some foods in which their presence would be far from obvious, such as dried milk and vegetable oils. Some of these reports might be spurious, and the curious identification of a nitrosamine in dried milk as nitrosodiisopropylamine suggests that it might have arisen from a corrosion inhibitor di-isopropylamine nitrite.

The presence of nitrosamines has been reported in cheese, and this is understandable because of the use of nitrate to inhibit gas formation. It is probable that nitrite is formed by reduction of the nitrate by microorganisms and that the nitrite then nitrosates amines naturally present. In fish and meat cured with nitrite, the nitrosating agent is already present and it is not surprising that samples of cured foods often contain parts per billion of nitrosamines. Analytical methods for volatile nitrosamines are well worked out, but methods for nonvolatile nitrosamines, including nitrosoalkylamides, are at present not very good and rather insensitive. This makes the evaluation of the nitrosamine content of foods and environmental materials unsatisfactory and suggests that human exposure to N-nitroso compounds could be much greater than is presently suspected. On the other hand, many nonvolatile nitrosamines seem to be much less carcinogenic than their volatile counterparts, although some, such as nitrosoproline, have the potential for formation of carcinogenic nitrosamines through transnitrosation.[41] It is also important that nonvolatile nitrosamines can be converted by heat into volatile ones; for example, nitrosoproline in bacon could conceivably be the source of the nitrosopyrrolidine found in fried bacon,[42,43] although other mechanisms have been proposed for the formation of nitrosopyrrolidine by heating nitrite-treated fatty foods.[44]

The principal source of exposure of humans to N-nitroso compounds (other than nitrosodialkanolamines) seems to be their formation in the gastrointestinal tract, most notably the stomach. The nitrite can come from the foods in which it is present, or from saliva in which nitrite is formed by bacterial reduction of nitrate in the mouth. In either case it is likely that the nitrite enters the stomach and there reacts with secondary or tertiary amines to form nitroso compounds, as would be predicted from their chemical reactivity. Primary amines have also been suggested as sources of nitrosoalkylcyanamides, but without convincing evidence.[45]

Many studies have been carried out with the objective of determining the extent to which the combination of nitrite with secondary amines or tertiary amines in the stomach could present a carcinogenic risk to humans. Nitrite, as the commonest nitrosating agent to which people are exposed, will be of greater significance the higher the concentration, because of the square term in the kinetic equation.[7] Therefore, the nitrite obtained from foods, such as cured meats and fish, will be more important quantitatively than the nitrite in saliva, since nitrite concentrations in the saliva are normally low.[46] Early experiments showed that feeding of nitrite with high concentrations of dimethylamine,[47] pyrrolidine,[48] or phenmetrazine[49] to various animals with tied-off stomachs led to formation of detectable quantities of the corresponding nitrosamines in a comparatively short time. Further studies showed that sufficient nitrosamine could be formed in intact animals by feeding amines with nitrite to cause severe injury or death from toxicity.[50-52] The amines included a tertiary amine, aminopyrine,[51] which was long used as an analgesic. The formation of biologically active N-nitroso derivatives of a large number of amines used as drugs and agricultural chemicals has been demonstrated using bacterial mutagenicity tests of the products of nitrosation.[53] None of these results can be used by itself to quantify the risk in exposure to an amine when nitrite is present, even if the carcinogenic potency of the nitrosated product in animals is known. The task of risk extrapolation of this kind is not easy even when the amine is fed chronically to experimental animals together with nitrite. A large number of such experiments have been carried out, although their number has been limited by the considerable expense of such studies.

The earliest experiments of this kind, following that of Druckrey with its negative result,[6] were carried out by Sander and his co-workers and were first reported in 1969.[54] At the same time the currency of this concept was increased by reviews[42,55] and by several experiments in other laboratories,[56-60] using a variety of secondary and tertiary amines. Subsequent studies with other amino compounds have included the transplacental effect of alkylureas plus nitrite fed to pregnant females during a significant stage of gestation,[61] chronic feeding of alkylureas and nitrite to adults,[62] and concurrent feeding of a large number of drugs to rats together with nitrite.[63,64] Not all of these experiments have resulted in the induction of significant incidences of tumors, from which the conclusion may be drawn that insufficient carcinogenic nitroso compound was formed during the treatment to give rise to tumors within the lifetime of the animals. For example, the feeding of the insecticide carbaryl, together with nitrite, to pregnant rats failed to induce a significant incidence of tumors in the offspring,[65] which might also have been due to failure of the nitrosocarbaryl formed to cross the placenta, an aspect of the problem that was not studied at the time.

A number of likely candidates for precursors of carcinogenic N-nitroso compounds formed *in vivo*, and thereby posing a possible risk to man, include arginine, methylguanidine, piperidine, piperine, piperazine, and morpholine—a secondary amine often found in foods[66] and known to give rise to tumors when fed to animals with nitrite.[50] Others were chlorpromazine, chlordiazepoxide, methapyrilene, tolazamide, tolbutamide, fenuron, monuron, dimethyldodecylamine, dimethyldodecylamine oxide, nitrilotriacetic acid, allantoin, oxytetracycline, lucanthone, hydrochlorothiazide, chlorpheniramine, diphenhydramine, aminopyrine, dipyrone, disulfiram, thiram, and quinacrine. Most of these compounds had not previously been tested for chronic toxicity and were assumed to be free of risk. However, some of them were found, surprisingly, to be carcinogenic alone, in the absence of nitrite, and so the formation of nitroso derivatives from them would be an added risk. Some of the feeding studies of combinations of amines with nitrite are still in progress and have produced no results of consequence yet. The results of most of the studies have been published, at least

Table I
Amines Fed with Nitrite to Rats

Compound	Result	Tumors induced
Arginine	−	
Dimethyldodecylamine	+	Bladder, forestomach
Dimethyldodecylamine-N-oxide	+	Liver
Dimethylphenylurea	−	
Methylguanidine	−	
Monuron	±	
Morpholine	++	Liver
Piperine	+	Forestomach
Thiram	+	Nasal cavity, forestomach
Trimethylamine-N-oxide	−	

in summary form, and indicate that the conjecture of many years ago that human exposure to N-nitroso compounds formed *in vivo* could present a carcinogenic risk was not without foundation. The information available currently on these experiments is shown in Tables I and II. Therefore, in addition to the academic interest in this most unusual group of carcinogens, able to evoke

Table II
Amines Fed with Nitrite to Rats
Nitrosation of Drugs *in Vivo*

Drug	Result	Tumors induced
Allantoin	+	Liver, forestomach
Aminopyrine	+++	Liver
Chlordiazepoxide	±	
Chlorpheniramine	+	Liver
Chlorpromazine	−	
Cyclizine	−	
Diphenhydramine	U[a]	
Dipyrone	−	
Disulfiram	+	Esophagus
Hexamethylenetetramine	−	
Hydrochlorothiazide	U	
Lucanthone	±	
Methapyrilene	−	
Oxytetracycline	+	Liver
Quinacrine	−	
Tolazamide	−	
Tolbutamide	+	Liver, forestomach

[a] U = Unfinished.

tumorigenic effects in so large a variety of species, there is reason to believe that they present a perhaps considerable carcinogenic risk to man, and exploration of their biological activity is a worthwhile pursuit.

REFERENCES

1. H. A. Freund, Clinical manifestations and studies in parenchymatous hepatitis, *Ann. Int. Med. 10*, 1144–1155 (1937).
2. J. M. Barnes and P. N. Magee, Some toxic properties of dimethylnitrosamine, *Brit. J. Ind. Med. 11*, 167–174 (1954).
3. P. N. Magee and J. M. Barnes, The production of malignant primary hepatic tumors in the rat by feeding dimethylnitrosamine, *Brit. J. Cancer 10*, 114–122 (1956).
4. H. Druckrey, R. Preussmann, S. Ivankovic, and D. Schmähl, Organotrope carcinogene Wirkungen bei 65 verschiedenen N-Nitroso-Verbindungen an BD-Ratten, *Z. Krebsforsch. 69*, 103–201 (1967).
5. W. Lijinsky, J. Loo, and A. Ross, Mechanism of alkylation of nucleic acids by nitrosodimethylamine, *Nature 218*, 1174–1175 (1968).
6. H. Druckrey, D. Steinhoff, H. Beuthner, H. Schneider, and P. Klärner, Prüfung von Nitrit auf toxische Wirkung an Ratten, *Arzneim.-Forsch. 13*, 320–323 (1963).
7. J. H. Ridd, Nitrosation, diazotisation and deamination, *Q. Rev. 15*, 418–441 (1961).
8. P. A. S. Smith and R. N. Loeppky, Nitrosative cleavage of tertiary amines, *J. Am. Chem. Soc. 89*, 1147–1157 (1967).
9. W. Lijinsky, L. Keefer, E. Conrad, and R. Van de Bogart, The nitrosation of tertiary amines and some biologic implications, *J. Natl. Cancer Inst. 49*, 1239–1249 (1972).
10. W. Lijinsky, Reaction of drugs with nitrous acid as a source of carcinogenic nitrosamines, *Cancer Res. 34*, 255–258 (1974).
11. R. Elespuru and W. Lijinsky, The formation of carcinogenic N-nitroso compounds from nitrite and some types of agricultural chemicals, *Food Cosmet. Toxicol. 11*, 807–817 (1973).
12. F. Schweinsberg and J. Sander, Cancerogene Nitrosamine aus einfachen aliphatischen tertiären Aminen und Nitrit, *Hoppe-Seyler's Z. Physiol. Chem. 353*, 1671–1676 (1972).
13. A. R. Jones, W. Lijinsky, and G. M. Singer, Steric effects in the nitrosation of piperidines, *Cancer Res. 34*, 1079–1081 (1974).
14. G. M. Singer, The mechanism of nitrosation of tertiary amines, in: *N-Nitroso Compounds: Analysis, Formation and Occurrence*, IARC Scientific Publications No. 31, pp. 139–151 (1980).
15. J. V. Dubsky and M. Spritzmann, Die Salzbildung der Nitroso-, Nitro- und Phenyl-iminodiessigsäure, *J. Prakt. Chem. 96*, 105–111 (1916).
16. B. Spiegelhalder, G. Eisenbrand, and R. Preussmann, Contamination of beer with trace quantities of N-nitrosodimethylamine, *Food Cosmet. Toxicol. 17*, 29–31 (1979).
17. AMBIO, Vol. 6, Nos. 2 and 3, 1977.
18. R. H. Cardy, W. Lijinsky, and P. Hildebrandt, Neoplastic and non-neoplastic urinary bladder lesions induced in Fischer 344 rats and $B_6C_3F_1$ hybrid mice by N-nitrosodiphenylamine, *Ecotoxicol. Environ. Safety 3*, 29–35 (1979).
19. B. C. Challis and M. R. Osborne, Chemistry of nitroso compounds. Part VI. Direct and indirect transnitrosation reactions of N-nitrosodiphenylamine, *J. Chem. Soc. Perkin II*, 1526–1533 (1973).
20. S. S. Singer, W. Lijinsky, and G. M. Singer, Transnitrosation. An important aspect of the chemistry of aliphatic nitrosamines, in: *Environmental Aspects of N-Nitroso Compounds*, IARC Scientific Publications No. 19, pp. 175–181, International Agency for Research on Cancer, Lyon, France (1978).

21. E. Boyland and S. A. Walker, Effect of thiocyanate on nitrosation of amines, *Nature 248*, 601–602 (1974).
22. C. Chu and P. N. Magee, Metabolic fate of nitrosoproline in the rat, *Cancer Res. 41*, 3653–3657 (1981).
23. H. Garcia and W. Lijinsky, Studies of the tumorigenic effect in feeding of nitrosamino acids and of low concentrations of amines and nitrite to rats, *Z. Krebsforsch. 79*, 141–144 (1973).
24. W. Lijinsky and M. D. Reuber, Transnitrosation by nitrosamines *in vivo*, in: *N-Nitroso Compounds: Occurrence and Biological Effects*, IARC Scientific Publications No. 41, pp. 625–631, International Agency for Research on Cancer, Lyon, France (1982).
25. W. Lijinsky and H. W. Taylor, Carcinogenesis tests of nitroso-N-methylpiperazine, 2,3,5,6-tetramethyldinitrosopiperazine, nitrosoisonipecotic acid and nitrosomethoxymethylamine in rats, *Z. Krebsforsch. 89*, 31–36 (1977).
26. W. Lijinsky, H. W. Taylor, and L. K. Keefer, Reduction of rat liver carcinogenicity of nitrosomorpholine by alpha deuterium substitution, *J. Natl. Cancer Inst. 57*, 1311–1313 (1976).
27. S. S. Singer and B. B. Cole, Reactions of nitrosoureas and related compounds in dilute aqueous acid: Transnitrosation to piperidine and sulfamic acid, *J. Org. Chem. 46*, 3461–3466 (1981).
28. W. Lijinsky, E. Conrad, and R. Van de Bogart, Carcinogenic nitrosamines formed by drug/nitrite interactions, *Nature 239*, 165–167 (1972).
29. S. S. Mirvish, L. Wallcave, M. Eagen, and P. Shubik, Ascorbate–nitrite reaction: Possible means of blocking the formation of carcinogenic N-nitroso compounds, *Science 177*, 65–68 (1972).
30. H. Dahn, L. Lbewe, and C. A. Bunton, Uber die Oxydation von Ascorbinsäure durch salpetrige Säure. Teil VI: Ubersicht und Diskussion der Ergebnisse, *Helv. Chim. Acta 43*, 320–333 (1960).
31. S. S. Mirvish, Formation of N-nitroso compounds: Chemistry, kinetics and *in vivo* occurrence, *Toxicol. Appl. Pharmacol. 31*, 325–351 (1975).
32. B. C. Challis and C. D. Bartlett, Possible co-carcinogenic effects of coffee substituents, *Nature 254*, 532–533 (1975).
33. L. K. Keefer, Promotion of N-nitrosation reactions by metal complexes, in: *Environmental N-Nitroso Compounds Analysis and Formation*, IARC Scientific Publications No. 14, pp. 153–159, International Agency for Research on Cancer, Lyon, France (1976).
34. L. K. Keefer and P. P. Roller, N-nitrosation by nitrite in neutral and basic medium, *Science 181*, 1245–1247 (1973).
35. P. A. Zingmark and C. Rappe, On the formation of N-nitrosodiethanolamine from a grinding fluid under simulated gastric conditions, *Ambio 5*, 80–81 (1976).
36. T. Y. Fan, J. Morrison, D. P. Rounbehler, R. Ross, D. H. Fine, W. Miles, and N. P. Sen, N-nitrosodiethanolamine in synthetic cutting fluids: A part-per-hundred impurity, *Science 196*, 70–71 (1977).
37. T. Y Fan, U. Goff, L. Song, D. H. Fine, K. Biemann, and G. P. Arsenault, N-nitrosodiethanolamine in cosmetics, lotions and shampoos, *Food Cosmet. Toxicol. 15*, 423–430 (1977).
38. R. W. Stephany, J. Freudenthal, and P. Schuller, N-nitroso-5-methyl-1,3-oxazolidine identified as an impurity in a commercial cutting fluid, *Recl. Trav. Chim. Pays-Bas 97*, 177–178 (1978).
39. K. Eiter, K. F. Hebenbrock, and H. J. Kabbe, Neue offenkettige und cyclische α-Nitrosaminoalkyl-äther, *Justus Liebigs Ann. Chem. 765*, 55–77 (1972).
40. D. H. Fine, D. P. Roundbehler, E. D. Pellizzari, J. E. Bunch, R. W. Berkley, J. McCrae, J. T. Bursey, E. Sawicki, K. Krost, and G. A. DeMarrais, N-nitrosodimethylamine in air, *Bull. Env. Contam. Tox. 15*, 739–746 (1976).
41. S. S. Singer, W. Lijinsky, and G. M. Singer, Transnitrosation by aliphatic nitrosamines, *Tet. Lett.* 1613–1617 (1977).
42. W. Lijinsky and S. S. Epstein, Nitrosamines as environmental carcinogens, *Nature 225*, 21-23 (1970).

43. N. P. Sen, B. Donaldson, J. R. Iyengar, and T. Panalaks, Nitrosopyrrolidine and dimethylnitrosamine in bacon, *Nature 241*, 473–474 (1973).
44. R. N. Loeppky, W. Tomasik, J. R. Outram, and A. Feicht, Nitrosamine formation from ternary nitrogen compounds, in: *N-Nitroso Compounds: Occurrence and Biological Effects*, IARC Scientific Publications No. 41, pp. 41–56, International Agency for Research on Cancer, Lyon, France (1982).
45. C. Green, T. J. Hansen, W. T. Iwaoka, and S. R. Tannenbaum, Specific detection systems for the chromatographic analysis of nitrosamines, in: *Proceedings of the 2nd International Symposium on Nitrite in Meat Products*, Wageningen, The Netherlands, pp. 145–153 (1977).
46. S. R. Tannenbaum, A. J. Sinskey, M. Weissman, and W. Bishop, Nitrite in human saliva: Its possible relationship to nitrosamine formation, *J. Natl. Cancer Inst. 53*, 79–84 (1974).
47. S. Asahina, M. Friedman, E. Arnold, G. Millar, M. Mishkin, Y. Bishop, and S.S. Epstein, Acute synergistic toxicity and hepatic necrosis following oral administration of sodium nitrite and secondary amines to mice, *Cancer Res. 31*, 1201–1205 (1971).
48. T. S. Mysliwy, E. L. Wick, M. C. Archer, R. C. Shank, and P. M. Newberne, Formation of N-nitrosopyrrolidine in a dog's stomach, *Br. J. Cancer 30*, 279–283 (1974).
49. M. Greenblatt, V. Kommineni, E. Conrad, L. Wallcave, and W. Lijinsky, *In vivo* conversion of phenmetrazine into its N-nitroso derivative, *Nature New Biol. 236*, 25–26 (1972).
50. R. C. Shank and P. M. Newberne, Dose-response study of the carcinogenicity of dietary sodium nitrite and morpholine in rats and hamsters, *Food Cosmet. Toxicol. 14*, 1–8 (1976).
51. W. Lijinsky and M. Greenblatt, Carcinogen dimethylnitrosamine produced *in vivo* from sodium nitrite and aminopyrine, *Nature New Biol. 236*, 177–178 (1972).
52. M. A. Friedman, G. Millar, M. Sengupta, and S. S. Epstein, Inhibition of mouse liver protein and nuclear RNA synthesis following combined oral treatment with sodium nitrite and dimethylamine or methylbenzylamine, *Experientia 28*, 21–22 (1972).
53. A. W. Andrews, J. A. Fornwald, and W. Lijinsky, Nitrosation and mutagenicity of some amine drugs, *Toxicol. Appl. Pharm. 52*, 237–244 (1980).
54. J. Sander and G. Bürkle, Induktion maligner Tumoren bei Ratten Durch gleichzeitige Verfütterung von Nitrit und sekundären Aminen, *Z. Krebsforsch. 73*, 54–66 (1969).
55. J. Sander, Untersuchung über die Entstehung cancerogener Nitrosoverbindungen im Magen von Versuchstieren und ihre Bedutung für den Menschen, *Arzneim.-Forsch. 21*, 1572–1580 (1971).
56. H. W. Taylor and W. Lijinsky, Tumor induction in rats by feeding heptamethyleneimine and nitrite in water, *Cancer Res. 35*, 812–815 (1975).
57. M. Greenblatt, S. S. Mirvish, and B. T. So, Nitrosamine studies: Induction of lung adenomas by concurrent administration of sodium nitrite and secondary amines in Swiss mice, *J. Natl. Cancer Inst. 46*, 1029–1034 (1971).
58. H. W. Taylor and W. Lijinsky, Tumor induction in rats by feeding aminopyrine or oxytetracycline with nitrite, *Int. J. Cancer 16*, 211–215 (1975).
59. W. Lijinsky and H. W. Taylor, Nitrosamines and their precursors in food, in: *Origins of Human Cancer*, Cold Spring Harbor Symposium, Book C, pp. 1579–1590, Cold Spring Harbor Laboratory, Cold Spring Harbor, New York (1977).
60. W. Lijinsky, N-Nitrosamines as environmental carcinogens, in: American Chemical Society Symposium Series 101 (J. P. Anselme, ed.), pp. 165–173, American Chemical Society, Washington, D.C. (1979).
61. S. Ivankovic and R. Preussmann, Transplazentare Erzeugung maligner Tumoren, *Naturwissenschaften 57*, 460 (1970).
62. J. Sander, Induktion maligner Tumoren bei Ratten durch orale Gabe von N,N'-Dimethylharnstoff und Nitrit, *Arzneim.-Forsch. 20*, 415–419 (1970).
63. W. Lijinsky and H. W. Taylor, Feeding tests in rats on mixtures of nitrite with secondary and tertiary amines of environmental importance, *Food Cosmet. Toxicol. 15*, 269–274 (1977).

64. W. Lijinsky, Significance of *in vivo* formation of N-nitroso compounds, *Oncology 37*, 223–226 (1980).
65. W. Lijinsky and H. W. Taylor, Transplacental chronic toxicity test of carbaryl with nitrite in rats, *Food Cosmet. Toxicol. 15*, 229–232 (1977).
66. G. M. Singer and W. Lijinsky, Naturally occurring nitrosatable amines. I. Secondary amines in food, *J. Agric. Food Chem. 24*, 550–553 (1976).

2

N-Nitrosamine Mutagenicity Using the *Salmonella*/Mammalian-Microsome Mutagenicity Assay

A. W. ANDREWS AND W. LIJINSKY

1. INTRODUCTION

The *Salmonella*/mammalian-microsome mutagenicity assay has been used extensively to test a variety of chemicals.[1] The results of many of the assays employed soon after the Ames test began to gain worldwide acceptance showed that nitrosamines, as a chemical class, did not demonstrate a consistent correlation between mutagenicity and carcinogenicity.[2] However, the more this testing was used, the more information was uncovered that showed the flexibility and adaptability of the assay.

In addition to the basic plate incorporation assay,[3] the liquid preincubation assay,[4] and the liquid suspension assay,[5] there were two technical variations that increased mutagenic expression among the nitrosamines. The induction of liver enzymes with Aroclor 1254 produced more P448, whereas stimulation with phenobarbital induced more P450. These latter preparations seemed better suited to activate some nitrosamines.[6] The addition of hamster liver homogenates rather than the traditionally used rat or mouse S-9 also expanded nitrosamine activation.[7,8]

When a chemical's structure was altered, the *Salmonella* assay frequently detected differences in mutagenic potency. During the past six years more than

---

A. W. ANDREWS • Program Resources Inc., and Chemical Carcinogenesis Program, LBI-Basic Research Program, NCI-Frederick Cancer Research Facility, Frederick, Maryland 21701. W. LIJINSKY • Chemical Carcinogenesis Program, NCI-Frederick Cancer Research Facility, Frederick, Maryland 21701.

151 nitrosamines have been tested in our laboratory and the comparative results for most of them are presented here.

2. MATERIALS AND METHODS

2.1. Chemicals

The nitrosamines were prepared at the Frederick Cancer Research Facility, Chemical Carcinogenesis Program, by Drs. W. Lijinsky, J. E. Saavedra, S. S. Singer, and G. M. Singer. All were of high purity. Included in this study were 52 cyclic nitrosamines, 16 aromatic nitrosamines, 42 acyclic (aliphatic) nitrosamines, and 39 nitrosamides. They were dissolved in either sterile saline or Baker-analyzed dimethylsulfoxide (Me_2SO). Triphosphonucleotide (TPN) was from Boehringer-Mannheim and glucose-6-phosphate (G-6-P) from Sigma Chemical Corporation.

2.2. Bacterial Strains

Salmonella tester strains were obtained from Dr. B. N. Ames. The compounds were first screened in the five primary tester strains with and without added S-9 to establish a mutagenic pattern. Most nitrosamines were mutagenic with the base-pair substitution strains TA 1535, TA 100, or TA 1530. One chemical was mutagenic only with TA 1537. That result and the results using TA 1535 and TA 1530 are reported here.

The stock cultures were maintained as frozen permanents at $-80°$ C, and fresh 14-hr cultures were grown in Oxoid #2 nutrient broth on a reciprocating shaker at $37°$ C.

2.3. Experimental Procedures

The plate incorporation assays were performed as recommended by Ames *et al.*[3] with the modifications of Andrews *et al.*[9] Liquid preincubation assays followed the protocol of Yahagi[4] with a preincubation time of 30 min. Plates were then incubated at $37°$ C for 48 hr and revertant colonies were counted using a hand-held tally.[10]

Dose-response curves were constructed using the appropriate strain with ten doses over a 1–1000 μg range. Controls were the bacterial cells alone and the cells plus solvent. The number of revertants above the background considered significant was established at above twice the value of the historical control mean (Table I) or above twice the value of the current mean (Tables II–XIII), whichever was greater. The data presented in the tables show the type of S-9 preparation, tester strain, and plating technique that gave optimum mutagenic expression. At least two independent tests were done using duplicate plates at each dose.

Table I
Historical Controls for the Tester Strains

Nitrosamine type		Mean number of background revertants ± standard deviation		
		TA 1535	TA 1530	TA 1537
Cyclic	N =	94	10	
	− S-9	16 ± 6	20 ± 11	
	N =	88	8	
	+ RA	15 ± 5	25 ± 7	
	N =	4	2	
	+ RPb	18 ± 1	25 ± 1	
	N =	40	4	
	+ HA	14 ± 6	29 ± 3	
Aromatic	N =	32		
	− S-9	14 ± 6		
	+ RA	11 ± 5		
	N =	28		
	+ HA	11 ± 4		
Acyclic	N =	74	6	
	− S-9	14 ± 6	22 ± 3	
	+ RA	15 ± 6	22 ± 4	
	N =	56	6	
	+ HA	15 ± 7	36 ± 3	
Nitrosamide	N =	68		2
	− S-9	23 ± 8		18 ± 4
	+ RA	23 ± 10		20 ± 3
	N =		8	
	− S-9		22 ± 7	
	N =		4	
	RPb		28 ± 4	
	+ HA		28 ± 6	

[a] Abbreviations used in this table and the following tables: L, liquid preincubation assay; R, rat; H, hamster; A, Aroclor 1254; Pb, phenobarbital; 3, TA 1530; 5, TA 1535; 7, TA 1537; −, no S-9 added; +, S-9 added; T, toxic; NT, not tested; M, thousand; N, number of samples; No., see Chapter 9 for chemical structure. Underline indicates significant mutagenesis.

2.4. Preparation of Liver Homogenates

Male Sprague–Dawley rats (Charles River Breeding Laboratories, Wilmington, Massachusetts) were induced with either 500 mg/kg of Aroclor 1254 (one dose suspended in corn oil injected I.P. 5 days before sacrifice by decapitation) or sodium phenobarbital 100 mg/kg dissolved in saline and injected I.P. for 3 consecutive days and sacrifice was on day four. The male Syrian golden hamsters, bred in our laboratories, were stimulated with Aroclor 1254. The range of protein per plate for the Aroclor rat S-9s was 1.7–3.0 mg, for the phenobarbital rat S-9s 2.6–5.0 mg, and for the Aroclor hamster S-9s, 3.0 mg. The S-9 fractions were prepared according to the procedure described by Ames.[3]

3. MUTAGENIC ACTIVITY AND CHEMICAL STRUCTURE

3.1. Cyclic Nitrosamines

The cyclic nitrosamines used in this study comprise ring sizes of from 4 to 13 members and have one or two heteroatoms. In some instances they were dinitroso compounds. It is possible that humans are exposed to some of them. The dose responses of the 52 cyclic nitrosamines are presented in Tables II–VI.

3.1.1. Four- and Five-Membered Rings with One Heteroatom

N-nitrosopyrroline was not mutagenic with rat liver S-9 but was mutagenic when hamster S-9 was used. N-nitrosopyrrolidine was marginally active in these studies with rat liver S-9, but when hamster S-9 was used the compound was a very potent mutagen. N-nitrosoazetidine and -nornicotine, activated with rat S-9, were of low potency. See Table II.

3.1.2. Five-Membered Rings with More Than One Heteroatom

N-nitrosothiazolidine, -4,4-dimethyl-1,3-oxazolidine, and 2,4,4-trimethyl-1,3-oxazolidine were not mutagenic under any test conditions. N-nitroso-1,3-oxazolidine, -2-methyl- and -5-methyl-1,3-oxazolidine were direct-acting. The N-nitroso-2,4-dimethyl-, -2,2,5-trimethyl-, -2-ethyl-5-methyl-, and -2-isopropyl-4,4-dimethyl-1,3-oxazolidine were mutagenic only with hamster S-9. See Table III.

3.1.3. Six-Membered Rings with One Heteroatom

N-nitrosomethylphenidate was not mutagenic when tested with rat Aroclor or phenobarbital S-9 preparations and gave only an indication of mutagenicity at the 1000-μg dose with hamster S-9. It was not considered mutagenic

Table II

The Dose Responses for the Cyclic Nitrosamines with Four- and Five-Membered Rings with One Heteroatom

Compound	No.	S-9 Type	Strain	Control		10		25		50		100		250		500		1000	
				−	+	−	+	−	+	−	+	−	+	−	+	−	+	−	+
N-nitroso-																			
Azetidine	LI	RA	3	36	34	24	27	40	24	28	30	43	32	64	58	69	98	96	134
Pyrrolidine	LII	RA	3	15	17	16	20	12	16	14	24	13	16	12	27	11	45	18	66
		HA	3		27		71		101		204		269		501		1145		1269
Nornicotine	LXXXI	RA	3	27	27	22	19	27	23	13	25	15	21	33	27	34	59	20	87
Pyrroline	LXXXVI	RA	5	30	14	29	26	29	17	27	22	27	19	29	17	29	21	37	25
		HA	5		12		13		19		17		31		50		100		142

Table III
The Dose Responses for the Cyclic Nitrosamines with Five-Membered Rings with More Than One Heteroatom with Strain TA 1535

							μg/plate										
			Control		10		25		50		100		250		500		1000
Compound	No.	S-9 Type	−	+	−	+	−	+	−	+	−	+	−	+	−	+	−	+
N-nitroso-																		
1,3-Oxazolidine	LIX	RA	11	13	11	17	12	15	29	<u>37</u>	<u>41</u>	<u>62</u>	<u>74</u>	<u>111</u>	<u>137</u>	<u>181</u>	<u>234</u>	<u>293</u>
Thiazolidine		RA	13	10	18	8	25	13	17	8	16	11	15	21	20	19	19	26
		HA		10		16		13		5		13		17		18		<u>31</u>
2-Methyl-1,3-oxazolidine	LXIV	RA	10	13	12	12	10	14	14	23	16	20	27	<u>42</u>	<u>48</u>	<u>64</u>	<u>107</u>	<u>107</u>
5-Methyl-1,3-oxazolidine	LXIII	RA	10	15	6	22	18	30	15	<u>44</u>	17	<u>54</u>	<u>42</u>	<u>108</u>	<u>50</u>	<u>164</u>	<u>116</u>	<u>262</u>
2,4-Dimethyl-1,3-oxazolidine		RA	13	15	8	18	10	23	9	15	11	16	9	13	13	12	13	18
		HA		10		21		<u>35</u>		<u>53</u>		<u>87</u>		<u>164</u>		<u>212</u>		<u>305</u>
4,4-Dimethyl-1,3-oxazolidine		RA	14	15	14	9	17	19	16	20	21	23	13	20	15	14	12	17
		HA		22		19		18		17		19		22		26		21
2,2,5-Trimethyl-1,3-oxazolidine		RA	10	13	8	13	8	13	10	14	10	12	16	18	21	20	29	<u>37</u>
		HA		10		<u>37</u>		<u>70</u>		<u>94</u>		<u>125</u>		<u>168</u>		<u>208</u>		<u>204</u>
2,4,4-Trimethyl-1,3-oxazolidine		RA	17	20	15	27	16	29	17	28	21	28	19	27	19	22	17	19
		HA		28		20		36		34		42		38		34		50
2-Ethyl-5-methyl-1,3-oxazolidine		RA	12	8	8	5	5	8	7	13	7	14	8	19	13	25	18	<u>46</u>
		HA		5		8		19		<u>30</u>		<u>145</u>		<u>359</u>		<u>553</u>		<u>593</u>
2-Isopropyl-4,4-dimethyl-1,3-oxazolidine		RA	15	24	18	17	14	23	15	22	17	16	12	23	17	21	15	28
		HA		19		22		20		20		15		29		<u>48</u>		<u>50</u>

in these studies. N-nitroso-4-cyclohexylpiperidine was toxic to the bacteria and not mutagenic under the test conditions. N-nitroso-4-t-butyl- and N-nitrosoguvacoline were mutagenic only when activated by hamster S-9. N-nitroso-3,4-epoxypiperidine, -Δ^3- and -Δ^2-tetrahydropyridine, and -4-piperidone were direct-acting mutagens. The remainder, N-nitroso-piperidine, -3-methyl-4-piperidone, 4-phenylpiperidine, and -3-methyl-Δ^3-tetrahydropyridine were mutagenic with rat Aroclor S-9 mixes. There were no differences in the mutagenicity of N-nitroso-3,5-dimethylpiperidine or its *cis* and *trans* isomers. All were mutagenic with rat S-9 activation. See Table IV.

3.1.4. Six-Membered Rings with More Than One Heteroatom

N-nitrosothiomorpholine and -4-benzoyl-3,5-dimethylpiperazine were not considered mutagenic under any test conditions. N-nitroso-3,5-dimethylpiperazine and -2-ethyl-4-methyl-1,3-tetrahydrooxazine and 3,4,5-trimethylpiperazine needed activation by a hamster homogenate to be mutagenic. N-nitroso-2,6-dimethylmorpholine was marginally mutagenic when activated by rat S-9 but was a potent mutagen with the addition of hamster S-9, while the *cis* and *trans* isomers were essentially not mutagenic with added rat S-9. N-nitroso-2-methylmorpholine, 2-phenyl-3-methylmorpholine, tetrahydro-1,3-oxazine, and 4-acetyl-3,5-dimethylpiperazine were mutagenic with rat Aroclor S-9 but were of low potency. N-nitroso-2-ethyl-1,3-tetrahydrooxazine and nitrosomorpholine were mutagenic with both S-9 preparations, but the hamster S-9 gave many more revertant colonies than did the rat S-9. Dinitroso-piperazine, -2-methylpiperazine, -2,5-dimethylpiperazine, and -2,6-dimethylpiperazine were mutagenic with the rat liver homogenates. See Table V.

3.1.5. Seven-Membered Rings (or Larger) with One Heteroatom

N-nitrosododecamethyleneimine was mutagenic but toxic to the bacteria at the higher dosages; nitroso-hepta- and octa-methyleneimine when activated by rat S-9 were of high potency. The mutagenic potency of N-nitrosooctamethyleneimine was greatly enhanced when hamster S-9 was the activation system. N-nitrosohexamethyleneimine was a potent direct-acting mutagen. See Table VI.

3.1.6. Seven-Membered Rings with More Than One Heteroatom

Dinitrosohomopiperazine was a direct-acting mutagen. See Table VI. Of the 52 cyclic nitrosamines tested, 8 were nonmutagenic, 9 were direct-acting, 10 were mutagenic only with hamster liver S-9, and the remainder showed varying degrees of mutagenicity when activated by rat liver S-9.

Table IV

The Dose Responses for the Cyclic Nitrosamines with Six-Membered Rings with One Heteroatom with Strain TA 1535

Compound	No.	S-9 Type	Control		10		25		50		100		250		500		1000	
			−	+	−	+	−	+	−	+	−	+	−	+	−	+	−	+
N-nitroso-																		
Piperidine	LIII	RA	18	28	NT	NT	NT	NT	NT	NT	18	98	18	101	27	194	30	381
4-t-Butylpiperidine	LXXIX	RA	12	14	6	15	18	17	11	14	10	19	11	19	7	13	7	6
		HA		16		19		20		21		27		47		52		40
3,5-Dimethylpiperidine																		
-cis		RA	10	13	11	12	10	24	13	32	13	48	10	53	5	71	17	71
-trans		RA	10	13	17	26	11	29	10	29	12	42	7	51	14	57	5	72
		RA							8	40	8	56	8	74	5	79	3	74
4-Piperidone	CVIII	RA	17	22	22	31	25	25	17	18	18	30	40	42	50	55	74	156
3-Methyl-4-piperidone	CIX	RA	13	13	15	18	12	22	15	24	13	24	17	43	21	70	31	91
3,4-Epoxypiperidine		RA	11	10	14	13	29	16	40	25	58	42	64	91	203	108	464	214
4-Phenylpiperidine	LXXVIII	RA	13	12	13	26	10	31	9	40	9	51	5	67	10	96	9	133
4-Cyclohexylpiperidine	XCIV	RA	11	12	10	14	12	14	13	15	6	20	4	15	T	5	T	T
		HA		13		16		22		22		26		31		23		T
Methylphenidate	CI	RA	31	12	21	12	30	17	26	12	25	15	20	13	31	17	16	12
		HA		14		11		14		31		13		17		19		45
Δ³-tetrahydropyridine	XCV	RA	14	17	29	37	32	59	48	103	21	192	45	293	59	489	116	1262
Δ²-Tetrahydropyridine	XCVI	RA	11	14	26	21	52	26	82	35	106	42	285	85	595	176	813	296
3-Methyl-Δ³-tetrahydropyridine		RA	11	14	8	31	11	56	13	102	11	161	14	256	16	667	27	1283
Guvacoline	C	RA	9	14	11	14	11	10	9	18	9	14	12	17	11	24	14	43
		HA		12		16		17		30		47		104		230		436

Table V

The Dose Responses for the Cyclic Nitrosamines with Six-Membered Rings with More Than One Heteroatom

						μg/plate												
				Control		10		25		50		100		250		500		1000	
Compound	No.	S-9 Type	Strain	−	+	−	+	−	+	−	+	−	+	−	+	−	+	−	+
N-nitroso-																			
Morpholine		RA	3	10	21	15	20	5	25	15	22	9	25	15	42	14	62	17	87
		HA	3		31		129		225		210		682		1233		2230		2466
Thiomorpholine		RA	5L	28	14	25	14	27	18	20	20	19	13	27	14	38	15	23	10
		HA	5L		10		11		7		20		14		15		18		19
2-Methylmorpholine		RA	5	11	11	9	9	14	17	10	18	10	20	10	41	12	50	17	122
2,6-dimethylmorpholine		RA	5	11	10	NT	NT	NT	NT	9	14	9	26	NT	NT	9	71	29	96
		HA			6		NT		NT		1358		2253		NT		3410		3544
-cis		RA	5	14	13	13	11	14	18	10	15	16	18	14	23	14	36	19	33
-trans		RA	5	13	16	10	17	12	15	10	24	11	23	16	22	24	30	24	98
2-Phenyl-3-methylmorpholine		RPb	3	14	25	19	25	34	29	17	25	20	20	25	37	41	55	33	71
Tetrahydro-1,3-oxazine		RA	5	11	12	11	18	12	21	9	17	16	18	13	33	23	57	23	176
2-Ethyl-1,3-tetrahydrooxazine		RA	5	20	28	23	39	16	32	17	39	20	67	22	81	38	201	49	270
		HA	5		29		184		921		1197		1813		2233		2581		3625
2-Ethyl-4-methyl-1,3-tetrahydrooxazine		RA	5	15	10	9	10	14	16	10	15	10	13	13	19	12	22	13	45
		HA	5		10		22		32		102		296		522		719		902
3,5-Dimethylpiperazine	LXXIV	RA	5	30	11	31	14	23	18	32	18	28	17	31	20	27	26	28	29
		HA	5		11		78		168		287		674		1566		2538		3872
4-Acetyl-3,5-dimethylpiperazine	LXXVI	RA	5	25	22	20	28	35	37	13	39	17	42	15	80	15	140	18	248

(continued)

Table V (continued)

		S-9		Control		\[μg/plate\] 10		25		50		100		250		500		1000	
Compound	No.	Type	Strain	−	+	−	+	−	+	−	+	−	+	−	+	−	+	−	+
4-Benzoyl-3,5-dimethyl-piperazine	LXXVII	RA	5L	24	12	15	11	17	14	19	16	19	11	21	17	17	20	15	16
		HA	5L		13		15		17		24		15		27		32		37
3,4,5-Trimethyl piperazine	LXXV	RA	5	23	8	17	10	24	14	13	11	15	11	21	13	20	23	18	40
		HA	5		13		117		196		501		972		1327		1777		1987
Dinitroso-																				
Piperazine	LXVI	RA	5	19	25	22	20	9	19	16	16	18	25	23	76	23	92	13	89
2-Methylpiperazine	LXVII	RPb	5	20	17	12	22	19	30	16	36	12	43	11	96	9	151	31	464
2,5-Dimethylpiperazine	LXIX	RA	5	26	20	14	28	15	55	16	120	13	73	13	218	16	551	17	1189
2,6-Dimethylpiperazine	LXVIII	RPb	5	18	18	14	17	18	32	15	38	17	75	8	139	10	472	37	370

Table VI

The Dose Responses for the Cyclic Nitrosamines with Seven-Membered Rings or Larger with One or More Heteroatoms with Strain TA 1535

μg/plate

Compound	No.	S-9 Type	Control		10		25		50		100		250		500		1000	
			−	+	−	+	−	+	−	+	−	+	−	+	−	+	−	+
N-nitroso-																		
Hexamethyleneimine	LIV	RA	10	13	6	61	16	214	27	406	42	580	79	566	113	1030	121	1145
Heptamethyleneimine	LV	RA	21	30	31	73	19	115	20	141	18	493	17	928	27	1320	38	2248
Octamethyleneimine	LVI	RA	17	16	15	95	20	154	17	221	19	269	29	704	21	566	22	1015
		HA		13		486		754		1081		2857		4075		5699		6598
Dodecamethyleneimine	LVII	RA	20	17	9	52	18	78	18	138	19	168	3	163	T	87	T	40
Dinitroso-																		
Homopiperazine	LXXI	RPb	18	18	18	23	10	60	26	63	19	66	33	156	39	334	69	856

3.2. Aromatic Nitrosamines and Related Compounds

Table VII contains the results of the dose responses for 16 aromatic nitrosamines. N-nitroso-methylaniline, methyl-4-fluoroaniline, -diphenylamine, -methyl-4 carboxybenzylamine, and -phenylbenzylamine were not mutagenic. N-nitroso-methyl-2-phenylethylamine was the most potent mutagen among these aromatic compounds when activated with the rat liver preparation. N-nitroso-methylcyclohexylamine, -methylbenzylamine, and -methyl-4-methoxybenzylamine were not mutagenic when rat S-9 was added but were mutagenic when the hamster S-9 was incorporated. N-nitroso-methyl-4-methylbenzylamine was minimally mutagenic with rat S-9 and definitely mutagenic with the hamster preparation. N-nitroso-methyl-4-fluorobenzylamine was of about equal potency with either rat or hamster S-9, whereas with the methyl-4-cyano-, methyl-4-nitro-, and methyl-4-chloro-benzylamine the hamster S-9 gave a significantly greater number of revertants than the rat S-9. N-nitrosomethyl-4-nitroaniline was toxic at the higher dosages and showed no difference between activation by rat or hamster S-9. N-nitroso-α-ace-toxydibenzylamine was direct-acting but differed from other direct-acting compounds in this study in that the addition of the rat S-9 almost completely inhibited any mutagenic reaction.

In this group of 16 nitrosamines, 5 were not mutagenic, 3 were mutagenic only with hamster S-9, 1 was direct-acting, and the remainder were mutagenic with hamster and/or rat S-9.

3.3. Acyclic Nitrosamines

3.3.1. Symmetrical

N-nitroso-diethanolamine was not mutagenic within the dose range or with the rat or hamster S-9 preparations studied here. N-nitroso-dimethylamine and -diethylamine can be shown to be mutagenic if the liquid preincubation and phenobarbital induced rat S-9 mix are used, but both are mutagenic in the plate incorporation test with hamster liver S-9. N-nitroso-di-*n*-propylamine and di-isobutylamine were mutagenic with rat liver S-9. They were not tested with the hamster preparation. N-nitrosodiallylamine, -di-*n*-butylamine, and -di-*n*-octylamine were mutagenic with both rat and hamster liver S-9. N-nitroso-bis(2-hydroxypropyl)amine and -bis(2-oxopropyl)amine were not mutagenic when rat S-9 was added. Both were mutagenic with the addition of hamster S-9. See Table VIII.

Table VII
The Dose Responses for the Aromatic Nitrosamines with Strain TA 1535

							μg/plate												
			Control		10		25		50		100		250		500		1000	
Compound	No.	S-9 Type	−	+	−	+	−	+	−	+	−	+	−	+	−	+	−	+
N-nitroso-																			
Methylcyclohexylamine	XIII	RA	24	10	15	7	21	10	22	10	22	11	19	11	29	12	29	16
		HA		10		8		9		18		23		39		53		51
Methylaniline	XIIa	RA	7	5	4	5	4	6	2	2	5	5	3	11	4	6	10	9
		HA		4		3		6		7		6		5		6		7
Methyl-4-fluoroaniline	XIId	RA	13	8	22	15	20	10	15	8	18	15	17	12	11	11	12	8
		HA		10		8		12		15		11		13		9		11
Methyl-4-nitroaniline	XIIe	RA	22	9	20	9	23	15	24	23	40	46	74	39	21	18	T	T
		HA		17		9		18		22		40		37		1		T
Methyl-2-phenylethylamine	XXIV	RA	22	29	15	247	18	754	17	827	15	1827	20	2030	14	1987	27	2407
		HA																
Diphenylamine		RA	5	10	7	11	8	12	4	8	5	7	T	9	T	12	T	11
		HA		16		7		11		8		8		13		13		10
Methylbenzylamine	XXV	RA	12	11	12	15	18	21	24	17	17	22	23	24	15	16	15	33
		HA		18		12		24		32		44		58		136		134
Methyl-4-methylbenzylamine		RA	22	9	15	10	19	10	23	13	13	16	18	14	20	36	17	53
		HA		10		9		14		15		19		20		23		139
Methyl-4-methoxybenzylamine		RA	11	10	24	23	18	11	19	13	15	8	20	17	14	9	12	18
		HA		10		14		17		26		44		56		62		119
Methyl-4-carboxybenzylamine		RA	10	6	12	9	6	13	10	10	13	9	15	7	10	7	14	10
		HA		6		3		4		7		10		7		8		9
Methyl-4-cyanobenzylamine		RA	19	16	23	32	24	31	27	25	22	44	16	60	9	106	14	135
		HA		11		32		45		101		222		623		718		789

(*continued*)

Table VII (continued)

Compound	No.	S-9 Type	Control		10		25		50		µg/plate 100		250		500		1000	
			−	+	−	+	−	+	−	+	−	+	−	+	−	+	−	+
Methyl-4-nitrobenzylamine		RA	12	7	10	9	13	13	8	16	12	17	6	26	11	50	12	58
		HA		8		12		27		50		72		115		149		145
Methyl-4-fluorobenzyl amine		RA	12	12	13	14	13	12	9	17	18	17	17	21	21	36	16	72
		HA		7		8		17		23		33		39		50		77
Methyl-4-chlorobenzyl amine		RA	13	11	10	18	7	28	17	35	10	50	13	70	10	174	6	460
		HA		15		32		109		307		531		870		913		1109
Phenylbenzylamine	XIIc	RA	9	11	11	10	13	7	10	11	9	10	10	7	3	10	4	8
		HA		9		8		9		13		11		11		7		3
α-Acetoxydibenzylamine		RA	13	11	24	14	56	10	118	24	150	16	225	22	100	68	11	83

Table VIII

The Dose Responses for the Symmetrical Acyclic Nitrosamines

Compound	S-9 Type	Strain	Control		10		25		50		100		250		500		1000	
			−	+	−	+	−	+	−	+	−	+	−	+	−	+	−	+
N-nitroso-																		
Dimethylamine	RA	3	20	23	20	15	18	22	20	21	19	21	25	19	26	30	41	26
	HA	3		37		39		28		41		43		89		113		170
Diethylamine	RA	3	22	26	22	28	20	21	21	21	18	24	24	30	20	31	22	28
	HA	3		32		37		42		32		80		110		175		443
Diethanolamine	RA	5	9	13	12	11	9	7	8	13	10	16	9	10	9	15	10	21
	HA	5		17		13		8		11		11		16		16		20
Di-n-propylamine	RA	5	14	16	11	16	11	18	10	22	11	19	17	26	13	47	18	96
Bis(2-hydroxypropyl)amine	RA	5	20	8	20	9	27	7	21	10	22	12	26	15	17	16	31	24
	HA	5		9		11		14		12		22		61		113		161
Bis(2-oxopropyl)amine	RA	5	7	10	6	7	12	5	7	7	6	8	10	13	7	19	9	32
	HA	5		6		8		7		11		13		34		84		102
Diallylamine	RA	3	25	18	17	12	20	20	19	20	16	25	19	61	20	80	12	91
	HA	3		38		40		132		201		334		515		573		812
Di-n-butylamine	RA	5	13	15	18	31	16	28	19	29	7	28	18	48	12	57	8	81
	HA	5		14		NT		NT		318		587		NT		1588		1782
Di-isobutylamine	RA	5	13	15	12	19	13	38	14	46	12	74	17	74	14	39	16	136
Di-n-octylamine	RA	5	19	10	27	14	17	20	18	28	18	50	21	66	16	60	13	57
	HA	5		13		30		62		73		197		486		444		544

µg/plate

3.3.2. Unsymmetrical

The compounds that were not mutagenic with either rat or hamster S-9 mixes were N-nitrosomethyl-ethylamine, -carboxypropylamine, -neopentylamine, -*n*-octadecylamine, and N-nitrosoethanolhydroxypropylamine. The direct acting compounds were N-nitrosomethyl-methoxyamine, -2-hydroxypropylamine, and -dihydroxypropylamine, and the more potent N-nitrosoallylethanolamine, -ethanoloxopropylamine, -oxopropylallylamine, -trihydroxydipropylamine, and -allylhydroxypropylamine. Those compounds mutagenic when tested only with rat S-9 were N-nitrosomethyl-*n*-propylamine, -heptylamine, -octylamine, -nonylamine, -decylamine, and -undecylamine. Compounds essentially mutagenic only when tested with hamster liver homogenate were N-nitrosomethyl-2-oxopropylamine, N-nitroso-dihydroxypropylethanolamine, -dihydroxypropyl-2-oxopropylamine, and 2,3-dihydroxypropylallylamine. Compounds more mutagenic with hamster than with rat S-9 were N-nitrosomethyl-2-hydroxypropylamine, -*n*-butylamine, -*n*-pentylamine, -*n*-hexylamine, -*n*-dodecylamine, -*n*-tridecylamine, -*n*-tetradecylamine, N-nitrosodihydroxypropylethanolamine, and -hydroxypropyloxopropylamine. Three of the unsymmetrical N-nitroso compounds, nitrosoallylethanolamine, -ethanoloxopropylamine, and -oxopropylallylamine were of high potency and were not appreciably different in mutagenic response with either the rat or hamster S-9. See Table IX.

In general the unsymmetrical compounds were the more potent mutagens. None of the symmetrical compounds tested was direct-acting. Of the 40 acyclic nitrosamines tested, 6 were nonmutagens, 8 were mutagenic only with the hamster S-9 preparation, 8 were direct-acting and the remainder showed varying mutagenic potency with hamster or rat liver activating systems.

3.4. Nitrosamides

For the purpose of this study the nitrosamides have been divided into six groupings, nitrosoalkylcarbamate esters (many are derivatives of commonly used insecticides), nitrosoalkylnitroguanidines, nitrosoalkylureas and substituted nitrosoalkylureas, nitrosotrialkylureas, and one miscellaneous compound.

3.4.1. Nitrosoalkylcarbamates

All the compounds tested were high-potency direct-acting mutagens. See Table X.

3.4.2. Nitrosoalkylguanidine

N-nitroso-methyl- and -ethylnitroguanidine and N-nitrosocimetidine were direct-acting mutagens of high potency. The rat homogenate inactivated the mutagenicity of the N-nitrosocimetidine. See Table XI.

Table IX

The Dose Responses for the Unsymmetrical Acyclic Nitrosamines with Strain TA 1535

Compound	No.	S-9 Type	Control −	Control +	10 −	10 +	25 −	25 +	50 −	50 +	100 −	100 +	250 −	250 +	500 −	500 +	1000 −	1000 +
N-nitrosomethyl-Methoxyamine	XVII	RA	16	18	16	17	22	22	25	24	19	26	36	43	68	90	96	152
Ethylamine	Xa	RA	10	16	10	11	12	12	10	21	12	16	12	22	11	21	20	26
		HA		25		13		10		23		29		26		34		40
n-Propylamine	Xb	RA	14	16	11	16	11	18	10	22	11	19	17	26	13	47	18	96
2-Hydroxypropylamine	XLI	RA	19	20	18	21	23	25	18	20	25	22	28	21	44	40	67	56
		HA		24		95		162		171		421		617		1197		1784
Dihydroxypropylamine	XLII	RA	10	12	10	10	9	15	14	16	10	14	18	36	25	49	33	149
		HA		8		15		21		27		57		170		348		783
2-Oxopropylamine	XL	RA	17	18	17	24	16	26	13	22	14	23	19	23	15	20	16	21
		HA		27		21		29		25		37		58		75		91
3-Carboxypropylamine	XIX	RA	4	6	2	5	3	3	3	3	4	4	7	9	4	7	9	6
		HA		6		4		4		3		4		7		5		4
n-Butylamine	Xc	RA	19	9	22	9	23	11	22	15	18	9	18	32	17	35	14	55
		HA		9		34		44		78		135		183		436		684
Neopentylamine	XXIII	RA	24	26	22	15	25	21	20	21	26	21	22	21	21	27	26	23
		HA		28		10		24		15		16		24		29		51
n-Pentylamine	Xd	RA	25	14	18	11	23	16	16	13	24	27	17	48	16	58	21	95
		HA		11		53		142		258		624		1436		2103		4350
n-Hexylamine	Xe	RA	16	10	13	14	20	18	22	32	17	45	20	48	13	96	12	184
		HA		13		40		81		151		322		1188		2074		2507

(continued)

Table IX (*continued*)

Compound	No.	S-9 Type	Control −	Control +	10 −	10 +	25 −	25 +	50 −	50 +	100 −	100 +	250 −	250 +	500 −	500 +	1000 −	1000 +
n-Heptylamine	Xf	RA	15	19	14	23	13	37	15	64	12	81	13	494	15	798	10	943
n-Octylamine	Xg	RA	22	18	18	18	13	36	17	79	26	280	14	1508	6	1508	T	8
n-Nonylamine	Xh	RA	22	18	23	17	19	72	22	694	13	1103	T	496	T	T	T	T
n-Decylamine	Xi	RA	10	37	11	54	9	263	7	562	T	208	T	T	T	T	T	T
n-Undecylamine	Xj	RA	16	20	16	23	14	70	14	528	12	2929	5	447	T	35	T	15
n-Dodecylamine	Xk	RA	9	15	11	23	7	117	6	288	5	310	T	93	T	16	T	53
		HA		14		NT		NT		24		208		ND		414		1193
n-Tridecylamine		RA	19	11	19	8	17	29	16	34		117	22	900	10	1509	T	1084
		HA		12		14		25		49		176		1392		3966		4176
n-Tetradecylamine	Xl	RA	14	17	11	13	9	13	7	25	7	24	10	27	7	41	6	40
		HA		16		12		18		38		184		523		1523		559
n-Octadecylamine		RA	13	10	19	13	11	20	12	10	17	11	20	8	18	11	27	23
		HA		10		6		8		8		8		11		14		26
N-nitroso-Dihydroxypropyl ethanolamine-	XLVII	RA	9	16	12	13	12	11	16	11	19	16	10	23	17	26	30	46
		HA		17		20		34		41		115		186		565		783

Compound	L		17	18	48	41	99	74	192	118	348	235	624	406	1146	957	1276	1131
Allylethanolamine		RA																
		HA		16														
Ethanolhydroxypropyl-amine	XLIII	RA	8	14	6	14	8	18	7	10	8	11	8	13	9	14	9	18
		HA		12		13		12		10		10		15		27		26
Ethanoloxopropyl-amine	XLIV	RA	10	13	39	27	112	40	232	78	234	136	986	391	1348	913	1609	1058
		HA		12		34		48		112		212		493		754		1711
Hydroxypropyloxo-propylamine	XLIVa	RA	10	10	16	8	12	17	13	15	15	29	18	23	24	73	26	94
		HA		17		65		163		551		321		1290		1754		2653
Oxopropylallylamine	XLIX	RA	9	12	8	26	18	93	24	134	36	207	50	420	112	1044	189	1015
		HA		9		2342		3293		3696		2734		2842		2973		2219
Trihydroxydipropyl-amine	XLV	RA	14	17	14	22	18	18	15	30	43	40	44	58	94	122	176	202
		HA		24		22		61		57		113		363		537		725
Dihydroxypropyl-2-oxopropylamine	XLVI	RA	8	7	6	6	5	6	6	9	6	5	9	20	10	25	17	62
		HA		7		9		28		42		77		256		643		1282
2,3-Dihydroxypropyl-allylamine		RA	6	7	7	5	4	5	10	7	12	7	6	8	9	19	11	32
		HA		5		7		8		18		20		62		116		165
Allylhydroxypropyl-amine	XLVIII	RA	16	18	30	33	32	56	55	67	108	99	197	239	421	522	725	841
		HA		20		783		899		1624		1566		2146		2030		2451

Table X

The Dose Responses for the Nitrosoalkylguanidines with Strain TA 1535

Compound	No.	S-9 Type	Control		1		2.5		5		10		25		50		100	
			−	+	−	+	−	+	−	+	−	+	−	+	−	+	−	+
N-nitroso-																		
Methylurethane	IIIa	RA	36	20	259	26	555	17	885	13	4872	19	11M	53	13M	1842	11M	8178
Ethylurethane	IIIb	RA	33	22	45	35	41	50	99	74	86	104	185	157	418	450	754	754
Methylphenylcarbamate		RA	22	21	350	22	2741	44	15M	313	9889	3089	3009	3727	1668	8149	479	7250
Carbaryl	IIIc	RA	33	24	148	44	2088	42	4292	28	5699	711	4191	4278	3437	4655	1581	4775
Methomyl		RA	25	23	5278	136	5974	2422	3480	4727	2320	6351	624	6380	58	4292	T	1044
Aldicarb		RA	22	17	126	71	1639	1131	5452	2567	6380	6090	3828	6612	1914	4988	623	3205
Buxten		RA	20	20	319	92	3132	140	6554	653	6786	3422	3393	3814	2726	4147	2277	5133
Landrin		RA	22	15	38	14	1131	19	3074	46	4872	682	4481	3263	2871	3683	1537	3509
Carbofuran		RA	27	19	21	25	90	19	1146	52	3828	1000	5800	3035	4712	5916	1725	4640
Baygon		RA	22	13	493	50	3741	537	4669	3683	5916	6960	4321	4452	3089	4495	1784	3698
Oxazolidone		RA	12	15	93	40	177	83	433	112	722	220	1198	516	1979	1384	2679	2328
		HA		12		27		61		86		125		463		875		1642
5-Methyloxazolidone		RA	13	16	56	22	96	32	160	53	302	64	901	153	1374	648	2145	1355
		HA		17		27		36		69		71		137		275		662

Table XI
The Dose Responses for the Nitrosoalkylguanidines with Strain TA 1535 and Rat S 9

Compound	No.	Control		2.5		5		10		25		50		100		250	
		−	+	−	+	−	+	−	+	−	+	−	+	−	+	−	+
N-nitroso-																		
Methylnitroguanidine	IIa	17	35	1740	44	13M	770	16M	10M	29M	13M	28M	21M	2915	5887	NT	NT
Ethylnitroguanidine	IIb	28	27	72	50	156	115	1339	720	2963	3473	324	1760	T	T	NT	NT
Cimetidine		40	18	30	15	28	30	30	26	31	28	62	28	140	30	3103	55

3.4.3. Nitrosoalkylureas

All the compounds were direct-acting mutagens in strain TA 1535. The mutagenicity of N-nitrosophenylurea was best demonstrated in tester strain TA 1537. See Table XII.

3.4.3.1. Substituted Nitrosoalkylureas. These nitrosamines were also direct-acting mutagens and were often of high potency. The N-nitroso-3-hydroxypropylurea has comparable mutagenic activity with the other compounds at 500 to 1000 μg. These data are not shown in Table XIII.

3.4.4. Nitrosotrialkylureas

These compounds required metabolic activation. The mutagenicity of N-nitroso-ethyldimethylurea and -methyldiethylurea could only be demonstrated with hamster liver homogenate. The former was of low potency. The liquid preincubation assay using rat liver S-9 induced with phenobarbital best showed the mutagenic properties of N-nitrosotrimethylurea and -triethylurea. They were not tested with added hamster S-9. See Table XIV.

4. CONCLUSIONS/DISCUSSION

The adaptability and flexibility of the *Salmonella*/mammalian-microsome mutagenicity assay has been demonstrated by the increased reliability in the detection of the mutagenicity of nitroso compounds. The tests can be used to discern differences in mutagenicity related to chemical structure or the type of activation system.

In some cases only the highest dose tested gave a mutagenic response, but because it was the only response, a dose–response curve could not be constructed. This might indicate that if very large doses were used, a dose response could be determined. Our experiments did not exceed 1000 μg per plate, since it was conceivable that at a higher dose the activation of a trace impurity to a mutagenic form could be expressed, and this could lead to an incorrect conclusion. A test compound was not considered to be mutagenic unless a linear dose response was evident and at least two or more consecutive doses were above the level of significance established in Section 2.3 of this chapter.

The greater ability of hamster liver S-9 preparations versus rat liver S-9 to activate nitrosamines to mutagenic forms has reduced the need to use excessively high doses of the test compounds. Hamster liver preparations consistently give higher numbers of revertant colonies than rat liver homogenates.

Table XII
The Dose Responses for the Nitrosoalkylureas with Rat S-9

										μg/plate							
			Control		10		25		50		100		250		500		1000
Compound	No.	Strain	−	+	−	+	−	+	−	+	−	+	−	+	−	+	−	+
N-nitroso-																		
Methylurea	1a	5	20	27	914	1218	2451	1726	2944	3205	4089	5235	7192	4901	6612	5438	3219	3683
Ethylurea	1b	5	20	34	24	37	44	50	36	40	49	67	67	232	305	899	899	2755
n-Propylurea	1c	5	38	37	48	53	77	70	94	155	172	184	261	566	363	421	856	1146
Allylurea	1e	5	18	36	36	61	122	98	109	196	169	227	103	362	T	314	T	49
Isopropylurea	1n	5	33	39	11	26	25	38	45	40	71	85	162	192	255	315	464	566
n-Butylurea	1d	5	26	28	57	39	57	53	74	82	87	115	166	141	281	334	435	427
Isobutylurea		5	26	39	52	105	131	147	202	221	261	493	508	798	1059	1291	T	1639
Secbutylurea		5	20	22	18	13	19	17	42	30	40	30	85	41	82	89	87	118
n-Amylurea	1f	5	32	53	47	95	91	116	127	191	235	315	417	856	1030	1392	1682	1914
n-Hexylurea	1g	5	19	18	65	63	237	201	314	283	421	856	928	1262	1769	1726	1813	2552
Undecylurea	1h	5	22	41	32	49	77	55	74	68	129	116	90	213	153	276	114	316
Tridecylurea	1i	5	27	33	29	36	30	56	28	59	41	73	31	97	25	87	17	46
Phenylurea	1o	7	18	20	62	64	102	142	262	188	580	236	479	421	87	682	T	T
Cyclohexylurea	1p	5	18	33	20	23	33	43	17	32	35	51	79	115	169	171	236	334
Benzylurea	1q	5	27	33	22	43	26	51	36	48	46	53	62	102	70	130	61	125
Phenylethylurea	1r	5	20	17	192	193	493	507	769	957	943	1334	2045	2596	2828	2799	2915	2755

Table XIII
The Dose Responses for the Substituted Nitrosoalkylureas with Strain TA 1535 and Rat S-9

Compound	No.	Control		1		2.5		5		10		25		50		100	
		−	+	−	+	−	+	−	+	−	+	−	+	−	+	−	+
N-nitroso-																		
Hydroxyethylurea	Ik	13	19	29	26	73	52	164	136	348	333	696	319	2204	1319	5742	4698
Fluoroethylurea	Ij	14	20	54	36	82	98	315	436	812	1189	3190	3799	4031	4481	3553	4205
2-Hydroxypropylurea	Il	12	22	305	130	479	464	1369	1204	1436	914	2886	3103	4002	2378	5655	4277
3-Hydroxypropylurea	Im	13	13	15	12	8	16	14	16	24	22	31	39	36	56	48	62

Table XIV
The Dose Responses for Nitrosotrialkylureas

Compound	No.	S-9 Type	Strain	Control −	Control +	10 −	10 +	25 −	25 +	50 −	50 +	100 −	100 +	250 −	250 +	500 −	500 +	1000 −	1000 +
N-nitroso-																			
Ethyldimethylurea		RA	3	14	20	12	24	12	29	18	20	13	17	18	25	20	23	20	24
		HA	3		23		23		26		31		33		42		66		88
Methyldiethylurea	IVc	RA	3	17	27	14	18	13	17	17	24	10	26	19	18	15	25	14	23
	IVc	HA	3		32		27		27		24		38		142		3357		32M
Trimethylurea	IVa	RPb	3L	30	24	42	28	37	26	21	22	25	20	23	18	37	173	46	1116
Triethylurea	IVb	RPb	3L	25	24	27	31	22	27	23	17	27	174	14	153	25	96	26	150

μg/plate

The small cyclic nitrosamines were not strongly mutagenic with rat liver S-9 activation, but were strongly mutagenic when hamster liver S-9 was used. Nitroso-3-pyrroline was still much less potent than its saturated analog nitrosopyrrolidine or nitrosoazetidine. Mutagenicity was diminished by the presence of the double bond in the 3,4-position of nitroso-3-pyrroline, which contrasts sharply with the increased mutagenic effect of a double bond in the 3,4- or 2,3-position of nitrosopiperidine. The two unsaturated derivatives of nitrosopiperidine were both direct-acting. The results among this group of compounds show the great sensitivity of the mutagenic response for specific chemical structures. In general, among the cyclic nitrosamines, mutagenic activity increased with increasing molecular size in an homologous series. The exceptions were that nitrosooctamethyleneimine was a no more potent mutagen than nitrosoheptamethyleneimine, and nitrosododecamethyleneimine was a considerably less potent mutagen than its small homologs. There was no great difference in mutagenic activity among the oxygen-containing cyclic nitrosamines, nitrosooxazolidine, nitrosomorpholine, and nitrosotetrahydrooxazine. Like most nitrosamines, they were much more active with the addition of hamster liver S-9 than with rat liver S-9. That the sulfur analogs of the oxygen compounds nitrosothiazolidine and nitrosothiomorpholine were not mutagenic with either hamster or rat liver S-9 activation is a striking and unexplained effect of a small change in chemical structure on mutagenic activity. Methyl substitution at any position in this group of nitrosamines reduced or eliminated mutagenic activity. Substitution in the alpha position to the ring nitrogen is generally understood to reduce or inhibit oxidation at this site, which is essential for activation to the mutagenic agent.

Among the nitrosopiperazine derivatives, the mononitroso compounds were all weakly mutagenic or nonmutagenic with rat liver S-9 activation, but hamster liver S-9 usually resulted in potent mutagenic activity, except 4-benzoyl-3,5-dimethylnitrosopiperazine, which gave only a marginally mutagenic response with hamster S-9. An exception was the mutagenicity with rat liver S-9 of 4-acetyl-3,5-dimethylnitrosopiperazine. All of the dinitrosopiperazines were mutagenic with rat liver S-9 activation. The 2,5-dimethyl derivative was especially potent. The reasons for the differences in mutagenic potency are not clear, since there is apparently not the same relationship to steric factors that seems to be important in determining carcinogenic potency. It may be that under these conditions there are subtle effects on reactivity with the liver enzymes that would not be expressed in the animal as tumors in the liver.

Among the large number of derivatives of nitrosopiperidine examined, the general effect of most substituents was to reduce mutagenic potency. Four compounds that were more mutagenic than nitrosopiperidine were the unsaturated derivatives, nitroso-Δ^2-tetrahydropyridine, nitroso-Δ^3-tetrahydropyridine, its 3-methyl derivative, and nitroso-3,4-epoxypiperidine. With the exception of the 3-methyl derivative, all were directly acting mutagens. It is possible that the direct action of the unsaturated compounds is enzymatically mediated through formation of epoxides by oxidation within the bacteria, a process that is less likely with the 3-methyl derivative because of steric hindrance. Nitrosoguvacoline, the methyl ester of an unsaturated carboxyl acid related to nitro-Δ^3-tetrahydropyridine, is marginally activated by rat liver S-9 but significantly mutagenic with hamster liver S-9. Nitrosomethylphenidate, also an ester of a carboxylic acid derivative of nitrosopiperidine, is marginally mutagenic with hamster liver S-9 and nonmutagenic with rat liver S-9 activation. Nitroso-4-tertiarybutyl-piperidine was inactive with rat liver activation and minimally mutagenic with hamster liver S-9. Nitroso-4-cyclohexylpiperadine was not mutagenic. The analogous nitroso-4-phenylpiperidine was a mutagen with rat liver S-9, but less potent than nitrosopiperidine itself. It is unlikely that the differences in the steric configuration of the molecules account for the contrasts in mutagenic activity among the 4-substituted nitrosopiperidines. It is more likely that the specific chemical reactivity of the activated forms produced by the liver homogenate is very different. Similarly, nitroso-3,5-dimethylpiperidine, nitroso-4-piperidone, and its 3-methyl derivative are less potent mutagens than nitrosopiperidine, the metabolism of which has yet to be investigated.

The group of nitrosamines containing an aromatic ring include compounds that were part of a series that examines nitrosomethylbenzylamine and the effect of various substituents in the aromatic ring on mutagenic activity. With the exception of the 4-carboxylic acid and nitrosomethylbenzylamine itself, all were mutagenic with rat liver S-9 activation. Nitrosomethylbenzylamine was mutagenic only with hamster liver S-9 activation, while the 4-carboxylic acid was not mutagenic with either activation system. The mutagenic activity of the others was increased when hamster liver S-9 was added. The relationship of the chemistry of this group of compounds to mutagenic activity has been discussed elsewhere.[11] Nitroso-α-acetoxydibenzylamine was a directly acting mutagen, in common with the analogous α-acetoxy derivatives of the aliphatic nitrosamines. The parent compound, nitrosodibenzylamine, was not mutagenic.[12] Nitrosophenylbenzylamine and nitrosodiphenylamine, like nitrosodibenzylamine, were predictably not mutagenic because of the absence of an oxidizable alpha carbon atom, which would allow formation of a reactive alkylating agent. They were both inactive in the presence of either rat or hamster liver S-9. Like nitrosomethylbenzylamine, nitrosomethylcyclohexylamine was mutagenic only with hamster liver S-9 activation. The next higher homolog, nitrosomethyl-2-phenylethylamine, was a very potent mutagen with rat liver S-9 activation. The lower homolog of nitrosomethylbenzylamine, nitrosomethylaniline, analogous in potential reactivity to nitrosophenylbenzylamine, was nonmutagenic with either rat or hamster liver S-9 activation. Nitrosomethyl-4-fluoroaniline was also not mutagenic under any test conditions. Nitrosomethyl-4-nitroaniline was marginally mutagenic without activation, probably due to reduction of the nitro group to form the hydroxylamine.

The acyclic nitrosamines constituted the most diverse group of chemical structures examined in this study. The symmetrical compounds include nitrosodimethylamine and nitrosodiethylamine, which are the most commonly studied N-nitroso compounds. Neither of these simple compounds was mutagenic in the plate test when activated by rat liver S-9, although they were weakly mutagenic in the liquid preincubation assay. Both compounds were significantly mutagenic when hamster liver S-9 was used for activation, the diethyl compound being more mutagenic than the dimethyl compound. This might support the claim that ethylation of DNA has a greater effect than methylation on the genetics of cells. Nitrosodi-*n*-propylamine was mutagenic with rat liver S-9 activation, and nitrosodiallylamine, the unsaturated analog of nitrosodipropylamine, had approximately the same activity. Nitrosodi-*n*-butylamine, nitrosodi-iso-butylamine, and nitrosodi-*n*-octylamine showed mutagenic activity with added rat liver S-9 similar to the propyl compound. Oxidation of the alkyl groups, as in nitrosodiethanolamine, nitrosobis-(2-hydroxypropyl)amine, and nitrosobis-(2-oxopropyl)amine resulted in compounds that were not mutagenic with rat liver S-9 activation. With the exception of nitrosodiethanolamine, these compounds were significantly mutagenic with hamster liver S-9. The mutagenicity of the other symmetrical nitrosamines was greatly increased by the hamster preparation when compared with the rat liver homogenate.

The largest group of N-nitroso compounds considered were the asymmetric acyclic compounds, many of which were derivatives of nitrosomethylethylamine, which was nonmutagenic with either rat or hamster liver S-9 activation. The next higher homolog, nitrosomethyl-*n*-propylamine, was mutagenic with rat liver S-9 activation, and the activity was strikingly similar to that of nitrosodi-*n*-propylamine. The next homolog, nitrosomethyl-*n*-butylamine, was similar in potency to nitrosomethylpropylamine, as was nitrosomethyl-*n*-pentylamine. As the homologous series ascended, mutagenic potency increased with the increasing chain length to nitrosomethyltridecylamine. The next homolog, nitrosomethyltetradecylamine, was minimally mutagenic with rat liver S-9 activation but was a potent mutagen with hamster S-9. There was some indication in this series that those compounds with odd numbers of carbons in the chain were less mutagenic than those with even numbers, but this was not a consistent pattern. The hamster liver S-9 activation of the nitrosomethyl-*n*-alkylamines greatly increased mutagenic activity over the rat homogenate. Nitrosomethylneopentylamine, a polymethylated derivative of nitrosomethylethylamine, was like the latter, nonmutagenic with either rat or hamster liver S-9 activation.

Derivatives of nitrosomethyl-*n*-propylamine with substituents in the propyl group included the nonmutagenic nitrosomethyl-3-carboxypropylamine and nitrosomethyl-2-oxopropylamine, which was mutagenic only with hamster liver activation. The remaining two compounds, nitrosomethyl-2-hydroxypropylamine and nitrosomethyl-2,3-dihydroxypropylamine, were mutagenic with rat liver S-9, but much more potent when activated by the hamster preparation. These differences are difficult to explain when it is known that the 2-oxopropyl compound is reduced in the rat to the 2-hydroxypropyl compound.[13]

Among the asymmetric nitrosamines derived from nitrosodipropylamine and nitrosopropylallylamine, all were very potent mutagens with hamster liver S-9 activation and most were less mutagenic with the rat liver S-9 addition. Nitrosodihydroxypropylethanolamine and nitrosodihydroxypropylallyamine were not mutagenic with rat liver S-9 activation. The similar compound nitrosoethanolhydroxypropylamine, a homolog of nitrosodiethanolamine, was like that compound, nonmutagenic either with rat or hamster S-9 activation. The relationship of the mutagenic differences to the chemical structures of these compounds is not clear. The absence of any indication of the mechanism by which they are activated to alkylating moieties by liver homogenate enzymes, and of what the natures of these alkylating moieties may be, precludes drawing a conclusion.

Another large series of chemicals, the nitrosoalkylamides, were directly acting mutagens. The effect of incorporating the liver homogenates was to reduce or inhibit mutagenic activity by combining competitively with either the nitrosoalkylamide or the alkylating species produced. Among the nitrosoalkylcarbamates, including the N-nitroso derivatives of a number of commonly used pesticides, there were differences in mutagenic activity in the *Salmonella* bacteria, possibly related to differences in the ease of transport of the compounds into the cells. All the test compounds were extremely potent mutagens, inducing thousands of revertants per micromole of nitrosamide. Three nitrosoalkylguanidines were also very potent mutagens. The nitrosoalkylureas, which in most instances can be presumed to give rise to the same alkylating moieties as the other two groups of nitrosoalkylamides, tended to be much less potent mutagens. In particular, nitrosoethylurea was a weaker mutagen than nitrosoethylurethane or nitrosoethylnitroguanidine, and was two orders of magnitude less mutagenic than nitrosomethylurea. With few exceptions, the remaining nitrosoalkylureas were mutagens of similar potency. The exceptions were the unusually potent nitroso-2-phenylethylurea, nitroso-*n*-hexylurea, and nitrosoisobutylurea, as well as the very weakly mutagenic nitrosotridecylurea and nitrosobenzylurea. The differences in potency might be related to the differences in the reactivity of the alkylating moieties, which differ greatly in size, and these differences might also affect the ease of transport into the bacteria. In contrast with the weak mutagenic activity of nitrosoethylurea and nitroso-n-propylurea, the hydroxy- and fluoroderivatives were very potent mutagens, of comparable potency with the nitrosoalkylcarbamates. It is probable that the large increases in potency are related to the stability of the alkylating agents formed from the nitrosoalkylureas.

The four nitrosotrialkylureas were a group of stable nitrosoureas, which resist decomposition by alkali, and were not directly acting. Two activated to mutagens by rat liver S-9 were nitrosotrimethylurea and nitrosotriethylurea, the former more potent than the latter, and this may be analogous to the monomethyl- and monoethyl-nitrosoureas. The two asymmetric compounds nitrosoethyldimethylurea and nitrosomethyldiethylurea were not activated by rat liver S-9, but were mutagenic with hamster liver S-9 activation, the latter forming a very potent mutagen. This was again an indication of the greater sensitivity of the *Salmonella* bacteria to methylating compounds than to ethylating compounds.

The mode of action of the rat and hamster liver homogenates is debatable. Raineri et al.[14] suggest the possibility that the hamster enzymatic activation of nitrosamines may entail the formation of more stable alkylating intermediates, while Prival and Mitchell[15] present evidence that rat microsomal fractions contain a factor inhibitory to the mutagenicity of nitroso compounds.

In summary, the mutagenic response of the nitrosamines tested were as follows:

Classification	No. tested	No. not mutagenic	% Mutagenic
Cyclics	52	8	85
Aromatics	16	5	69
Aliphatics	40	6	85
Nitrosamides	39	0	100

Nitrosamines, as a chemical group, are considered to be carcinogens, and the results of these assays show that they may also be detected as mutagens. The correlation between the two biological effects is, therefore, a matter of great practical significance.

ACKNOWLEDGMENTS. The authors wish to thank the late Carl Valentine, Corinthia Brown, James Fornwald, Stuart Snyder, Kenneth Moats, Carol Smith, Carol Roetzel, Karen Fowler, and Robert Stauffer for their expert technical help with the mutagenesis assays; J. E. Saavedra and Stuart Snyder for their help in organizing the tables; and Nancy Forletta for typing the manuscript.

REFERENCES

1. J. McCann, E. Choi, E. Yamasaki, and B. N. Ames, Detection of carcinogens as mutagens in the *Salmonella*/microsome test: Assay of 300 chemicals, *Proc. Natl. Acad. Sci. USA 72,* 5135-5139 (1975).
2. A. W. Andrews, L. H. Thibault, and W. Lijinsky, The relationship between mutagenicity and carcinogenicity of some nitrosamines, *Mutat. Res. 51,* 319-326 (1978).
3. B. N. Ames, J. McCann, and E. Yamasaki, Methods for detecting carcinogens and mutagens with the *Salmonella* mammalian-microsome mutagenicity test, *Mutat. Res. 31,* 347-364 (1975).
4. T. Yahagi, M. Vagao, Y. Seino, T. Mateushima, T. Sugimura, and M. Okada, Mutagenicity of N-nitrosamines on *Salmonella, Mutat. Res. 48,* 121-130 (1977).
5. E. J. Green and M. A. Friedman, A modification of the Ames test which allows for calculations of true mutant frequencies, *Environ. Mutagenesis 2,* 304 (1980).
6. E. D. Thompson and P. J. Melamphy, An examination of the quantitative suspension assay for mutagenesis with strains of *Salmonella typhimurium, Environ. Mutagenesis 3,* 453-465 (1981).
7. B. N. Ames, W. E. Dursten, E. Yamasaki, and F. D. Lee, Carcinogens are mutagens: A simple test system combining liver homogenates for activation and bacteria for detection, *Proc. Natl. Acad. Sci. USA 70,* 2281-2285 (1973).
8. R. Raineri, J. A. Poiley, R. J. Pienta, and A. W. Andrews, Metabolic activation of carcinogens in the *Salmonella* mutagenicity assay by hamster and rat liver S-9 preparations, *Environ. Mutagenesis 3,* 71-84 (1981).
9. A. W. Andrews, L. H. Thibault, and W. Lijinsky, The relationship between carcinogenicity and mutagenicity of some polynuclear hydrocarbons, *Mutat. Res. 51,* 311-318 (1978).
10. A. W. Andrews, J. A. Fornwald, and W. Lijinsky, Nitrosation and mutagenicity of some amine drugs, *Toxicol. Appl. Pharmacol. 52,* 237-244 (1980).
11. G. M. Singer and A. W. Andrews, Mutagenicity and chemistry of N-nitroso-p-substituted benzylmethylamines, *J. Med. Chem. 26:* 309-312 (1983).
12. M. K. Jacobson, R. A. Barton, S. S. Yodu, S. Singh, and R. E. Lyle, Mutagenicity of the non-carcinogenic dibenzylnitrosamine and an α-acetoxy derivative, *Cancer Lett. 6,* 83-87 (1979).
13. G. M. Singer, W. Lijinsky, L. Buettner, and G. A. McClusky, Relationship of rat urinary metabolites of N-nitrosomethyl-n-alkylamine to bladder carcinogenesis, *Cancer Res. 41,* 4942-4946 (1981).
14. R. Raineri, J. A. Poiley, A. W. Andrews, R. J. Pienta, and W. Lijinsky, Greater effectiveness of hepatocyte and liver S-9 preparations from hamsters than rat preparations in activating N-nitroso compounds to metabolites mutagenic to *Salmonella, J. Natl. Cancer Inst. 67,* 1117-1122 (1981).
15. M. J. Prival and V. D. Mitchell, Influence of microsomal and cytosolic fractions from rat, mouse and hamster liver on the mutagenicity of dimethylnitrosamine in the *Salmonella* plate incorporation assay, *Cancer Res. 41,* 4361-4367 (1981).

3

Structural Basis for Mutagenic Activity of N-Nitrosamines in the *Salmonella* Histidine Reversion Assay

T. K. Rao

1. INTRODUCTION

N-nitrosamines comprise a very important group of chemicals to which human beings are constantly exposed through beverages, food, medicines, cosmetics, tobacco products, and also at the work place in industry. The first report by Magee and Barnes[1] nearly 25 years ago that nitrosodimethylamine (NDMA) caused tumors in experimental animals initiated an unbounded interest among researchers throughout the world to assess the biological hazard caused by these compounds.

Nitrosamines are not mutagenic in a number of nonmammalian test systems (with the exception of *Drosophila*), unless metabolically activated by cytochrome-linked mixed function oxidases. Thus, these compounds exhibit species and organ specificity based on metabolic capabilities. Early mutagenesis studies with nitrosopiperazines (NPZ) in *Salmonella*[2] have utilized a mammalian host-mediated assay. Since metabolism of NDMA involves an oxidative

---

T. K. Rao • Biology Division, Oak Ridge National Laboratory, Oak Ridge, Tennessee 37830. *Present address*: Environmental Health Research and Testing, Inc., Research Triangle Park, North Carolina 27709. By acceptance of this article, the publisher or recipient acknowledges the U.S. Government's right to retain a nonexclusive, royalty-free license in and to any copyright covering the article.

dealkylation, Malling[3] bubbled oxygen through an incubation medium containing *Neurospora conidia* and NDMA to convert NDMA into a mutagenic form. Similar methods were also reported by Meyer[4] for the induction of petites and canvanine-resistant mutants in yeast. Tissue homogenates contain mixed function oxidases[5] in the microsomal fraction, capable of activating promutagens/carcinogens such as NDMA into their ultimate mutagenic/carcinogenic forms. By incorporating liver homogenates into the assay, Malling[6] was capable of showing NDMA as a mutagen for the missense strains of *Salmonella typhimurium* auxotrophic for histidine. Gletten et al.[7] have presented evidence to show that a target organ such as liver possesses the activating ability, while the nontarget organs such as lung and kidney either lack or possess insignificant amounts of these activating enzymes. The effect of age and sex on inducibility of these activating enzymes will be discussed later in this volume (see Chapter 8). The readers are referred to the reviews by Neale[9] and Magee[10] for early mutagenesis work with nitroso compounds.

Nitrosamines also serve as excellent model compounds for determining the relationship between chemical structure and biological activity. The availability of an abundance of carcinogenicity data for these compounds makes them exceptionally useful in determining the empirical relationship between mutagens and carcinogens. We intend to present our studies with a number of nitrosamines that have been characterized for carcinogenicity in rats. The *Salmonella* histidine reversion assay developed by B. N. Ames[8] was utilized for the mutagenicity analysis.

2. *SALMONELLA TYPHIMURIUM*—HISTIDINE REVERSION ASSAY

B. N. Ames and co-workers have developed[8] a very sensitive and simple bacterial mutagenesis assay for detecting chemical mutagens and chemical carcinogens. The compounds are tested with several tester strains of *Salmonella typhimurium* for their sensitivity and specificity in reverting histidine auxotrophs to the wild-type (histidine independence) condition. The scope of the test has been further extended to test chemicals that require mammalian metabolic transformation by incorporating a tissue homogenate (usually rat liver S-9 with an NADPH-generating cofactor mix) into the assay.

2.1. Bacterial Strains

Histidine-requiring strains of *Salmonella typhimurium* were obtained through the courtesy of Dr. B. N. Ames. Table I lists these strains and their genotypes. In addition to their histidine requirements, these strains possess

Table 1
Genotypic and Phenotypic Characteristics of *Salmonella* Tester Strains

Strain designation	Genotype	Histidine requirement	Resistance to		
			Crystal violet	UV light	Ampicillin
—	*hisG46*	+	−	−	−
TA 1535	*hisG46;rfa;△uvrB*	+	+	+	−
TA 100	*hisG46;rfa;△uvrB;+R*	+	+	+	+
TA 1975	*hisG46;rfa*	+	+	−	−
TA 92	*hisG46; +R*	+	−	−	+
TA 1537	*hisC3076;rfa;△uvrB*	+	+	+	−
TA 1538	*hisD3052;rfa;△uvrB*	+	+	+	−
TA 98	*hisD3052;rfa;△uvrB;+R*	+	+	+	+

additional mutations resulting in the loss of the excision repair system (*uvrB*) and loss of the lipopolysaccharide barrier (*rfa*). The strains TA 98 and TA 100 carry an R factor conferring ampicillin resistance that makes them more sensitive for mutagen detection. Strains TA 1535, TA 92, TA 1975, and TA 100 are used to detect mutations caused by base substitutions while the other strains are used to detect mutations caused by frameshifts.

The bacterial cultures are maintained frozen in dimethylsulfoxide (DMSO) at −80° C.[8] Fresh overnight cultures were grown in nutrient broth (DIFCO).

2.2. Mutagenesis Assay—Procedures

The experimental procedures described by Ames *et al.*[8] were employed. In the agar overlay assay, the test material was mixed with an overnight bacterial culture (2×10^8 cells/plate) and S-9 mix, if necessary, in 2 ml of molten top agar (45° C) and poured over previously prepared Vogel–Bonner[11] minimal medium. The microsomal enzymes were obtained by homogenizing liver tissue from rats induced with either Aroclor (by injection) or phenobarbital (through drinking water) and prepared according to the procedures of Ames *et al.*[8] Revertant colonies were scored after 72 hr of incubation at 37° C. The liquid preincubation assay was essentially similar to agar overlay assay, except that the test material and the bacterial culture (and the S-9 mix if necessary) were preincubated for 1 hr at 37° C before the addition of top agar.

2.3. Source of Nitrosamines

Nitrosamines used in these studies were kindly provided by Dr. W. Lijinsky, Frederick Cancer Research Center. The compounds are of high purity (> 99%) and their synthesis has been previously reported (see Chapter 10). The test compounds were dissolved in DMSO (sterile/spectrograde, obtained from Schwarz/Mann) if insoluble in water and tested in a nontoxic dose range up to a concentration of 5 mg/plate.

3. MUTAGENESIS STUDIES IN *SALMONELLA TYPHIMURIUM*

Mutagenesis studies with nitrosamines in the *Salmonella* histidine reversion assay were analyzed to determine: (1) mechanism of mutation induction—the role of DNA repair in mutagenesis; and (2) the relationship between chemical structure and mutagenic activity. The qualitative correlation between mutagenic and carcinogenic properties was also investigated.

3.1. Mechanism of Mutation Induction by Nitrosamines

The ability of nitrosamines to alkylate macromolecules, first reported by Magee and Farber,[12] was believed to be the primary mechanism for their biological activity. Alkylation of DNA could result in errors during DNA replication causing base-pair substitutions. Mutagenesis studies with *Salmonella* support this hypothesis by exhibiting a specificity to revert missense tester strains, *hisG46*[2] or its derivatives TA 1530[13] and TA 1535.[14-18] Figure 1 illustrates the strain specificity using several tester strains of *Salmonella* and nitrosopyrrolidine (NPYR) as standard mutagen. The strains designed to detect frameshift mutations (TA 1537, TA 1538, and TA 98) were insensitive in the assay, while the missense strain TA 1535 was most sensitive for reversion. The mutagenic activity does not seem to be affected by the error-prone repair, as strains carrying on episome pKM101 (TA 100 and TA 92), which supposedly carries the genetic message for error-prone repair, are not any more sensitive than the parent strain TA 1535. However, the mutagenic lesion seems to be subject to excision repair, as strains capable of excision repair (TA 1975) are less sensitive to mutagenesis than excision repair-deficient strains, TA 1535 and TA 100.

3.2. Relationship between Chemical Structure and Mutagenic Activity

Several derivatives of cyclic nitrosamines and dialkyl nitrosamines were analyzed to determine the relationship between chemical structure and mutagenic activity. These studies were helpful in drawing conclusions about meta-

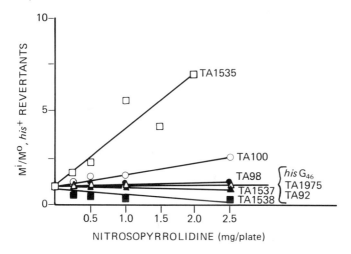

Figure 1. Strain specificity for NPYR mutagenesis. Several tester strains of *Salmonella typhimurium* were tested for NPYR-induced mutagenesis by the agar overlay technique.

bolic activation of these compounds. The mutagenic activity of these derivatives was compared with their carcinogenicity in rats (see Chapter 10). The comparisons made were purely qualitative and readers are referred to Chapter 11 for a more in-depth analysis.

3.2.1. Mutagenic Effect—Dependence on Ring Size

The relationship between mutagenic activity and molecular ring size was determined using a group of cyclic nitrosamines. The biological activity of these nitrosamines seems to depend on the ring size. The smaller ring compounds (nitrosoazetidine, -pyrrolidine, -piperidine, -morpholine, and dinitrosopiperazine) were less mutagenic than the larger ring compounds (nitroso-, hexa-, hepta-, and octamethyleneimines) (Table II). The smaller ring compounds gave similar results when tested in agar overlay or preincubation procedure. The preincubation assay was ten times more effective with the larger ring compounds. This could be due to an increased permeation of the compound or its active metabolite. The *Salmonella* assay failed to show any particular pattern with the unsaturated ring compounds. Nitrosopyrroline[18] was not a mutagen under any test conditions, while nitroso-1,2,3,6-tetrahydropyridine was a potent mutagen,[14] even in the absence of metabolic activation. The mutagenic activities in most cases paralleled the carcinogenic potencies of these compounds as shown in Table II.

Table II
Effect of Ring Size on Mutagenicity

Compound (CAS No.)	Structure	TA 1535his⁺ REV/μmol[a]		Carcinogencity[b]
		Agar overlay	Preincubation	
Nitrosoazetidine (15216-10-1)		2	5	++
Nitrosopyrrolidine (35884-45-8)		14	50	++
Nitroso-3-pyrroline (10552-94-0)		0	0	++
Nitrosopiperidine (100-75-4)		20	48	+++
Nitroso-1,2,3,6-tetrahydropyridine (55556-72-8)		105	NT	++++
Nitrosopiperazine (5632-47-3)		0	0	0
Dinitrosopiperazine (140-79-4)		35	NT	+++
Dinitrosohomopiperazine		75	NT	++++
Nitrosomorpholine		38	525	+++
Nitrosohexamethyleneimine (932-83-2)		31	360	++++
Nitrosoheptamethyleneimine (20917-49-1)		91	1080	++++
Nitrosooctamethyleneimine		120	1190	++++
Nitrosododecamethyleneimine		0	150	++

[a] Specific mutagenic activities were determined from the linear dose–response curves. [Metabolic activation: rat liver homogenates (S-9 mix) induced with phenobarbital.]
[b] Relative carcinogenicity is a subjective assessment based on dose/dose-rate/tumor incidence and time taken for tumor production (from Dr. Lijinsky; see Chapter 10 of this

3.2.2. Effect of Ring Substitution on Mutagenic Activity

Effect of ring substitution of cyclic nitrosamines by various reactive groups has been investigated in the *Salmonella* mutagenicity assay, from which several mechanistic conclusions could be derived.

3.2.2a. Alkyl Substitution. Effects of alkyl substitutions on the ring of several cyclic nitrosamines are given in Table III. Methyl substitution at the 2, 3, or 4 positions of nitrosopiperidine (NPIP) had little if any effect on mutagenicity. Substitution by two methyl groups in the 3 and 5 positions (3,5-dimethyl NPIP) significantly reduced mutagenic activity of NPIP, while substitution in

Table III
Effect of Ring Substitution by Alkyl Groups

Compound	Basic structure	TA 1535 his+ REV/μmol[a]	Carcinogenicity[b]
2,5-Dimethyl NPYR		0	0
2-Methyl NPIP		29	++
3-Methyl NPIP		48	+++
4-Methyl NPIP		85	+++
3,5-Dimethyl NPIP		12[c]	+
2,6-Dimethyl NPIP		0	0
2,2,6,6,-Tetramethyl NPIP		0	0
2,2,6,6,-Tetramethyl-4-hydroxy NPIP		0	0
2,2,6,6-Tetramethyl-4-oxy NPIP		0	0
4-Tert-butyl NPIP		0	+
N-Methyl NPZ		6	0
2-Methyl NPZ		51	++++
2,5-Dimethyl DNPZ		26	+++
2,6-Dimethyl DNPZ		53	++++
2,3,5,6-Tetramethyl DNPZ		0	0
4-Benzoyl-3,5-dimethyl DNPZ		0	+
2,6-Dimethyl NMOR		14	++++

[a] Specific mutagenic activities were determined from the linear dose–response curves. [Metabolic activation: rat liver homogenates (S-9 mix) induced with phenobarbital.]
[b] Relative carcinogenicity is a subjective assessment based on dose/dose-rat/tumor incidence and time taken for tumor production (from Dr. Lijinsky; see Chapter 10 of this volume).
[c] Metabolic activation: rat liver homogenates (S-9 mix) induced with Aroclor.

both α-positions to the N-nitroso group (2,6-dimethyl NPIP) completely eliminated the mutagenic activity. The alpha tetramethyl derivatives of NPIP and 4-hydroxy or 4-keto NPIP were inactive. Similarly, 2,5-dimethyl NPYR and 2,3,5,6-tetramethyl dinitrosopiperazine were nonmutagenic. These results suggest the involvement of positions adjacent to the N-nitroso group, probably by α-carbon hydroxylation, in the metabolic activation of these compounds. There has been evidence from *in vivo* experiments as well as *in vitro* metabolism studies that supports this theory (see Chapters 8 and 9). A close relationship was also noted between mutagenic and carcinogenic properties of these alkylated cyclic nitrosamines.

3.2.2b. Halogen Substitutions. The effect of halogen substitution on mutagenicity was investigated by examining several halogenated derivatives of NPTR and NPIP. The results are presented in Table IV. Halogen substitution at the 3 or 4 position enhanced mutagenic activity concurrent with carcinogenic activity. The exception was 3,4-dibromo NPYR, whose lack of mutagenic activity cannot be explained easily. However, it was reported to be mutagenic in another bacterial assay using *E. coli*.[18]

3.2.2c. Oxygen and Carboxy Substitution. The cyclic nitrosamines are activated by α-carbon hydroxylation. This does not rule out the possibility for oxidative metabolism to occur at positions 3 and 4. This led us to examine a series of compounds that contained hydroxy (OH) or carbonyl (CO) groups at 3 and 4 positions. The results are given in Table V. Hydroxy (OH) and keto (=o) substitutions in the 3 and 4 positions of NPIP did not affect mutagenicity.

Table IV
Effect of Halogen Substitution on the Aromatic Ring

Compound	Basic structure	TA 1535 $his+$ REV/μmol^a	Carcinogenicity^b
3,4-Dichloro NPYR		54	++++
3,4-Dibromo NPYR		0	?
3-Chloro NPIP		208	+++
4-Chloro NPIP		60	+++
3,4-Dichloro NPIP		30	++++
3,4-Dibromo NPIP		185	++++

^{a} Specific mutagenic activities were determined from the linear dose–response curves. [Metabolic activation: rat liver homogenates (S-9 mix) induced with phenobarbital.]
^{b} Relative carcinogenicity is a subjective assessment based on dose/dose-rat/tumor incidence and time taken for tumor production (from Dr. Lijinsky; see Chapter 10 of this volume).

Table V
Effect of Hydroxy and Carboxy Substitution on Cyclic Nitrosamines

Compound	Basic structure	TA 1535 his^+ REV/μmol[a]	Carcinogenicity[b]
3-Hydroxy NPYR		2	++
2-Carboxy-4-hydroxy NPYR		0	0
2-Carboxy NPYR		0	0
3-Hydroxy NPIP		22	+++
4-Hydroxy NPIP		10	+++
4-Keto NPIP		61	+++
2-Carboxy NPIP		0	0
3-Carboxy NPIP		0	0
4-Carboxy NPIP		0	0
Nitrosoguvacine		0	0
Nitrosoguvacoline		31[c]	0

[a] Specific mutagenic activities were determined from the linear dose-response curves. [Metabolic activation: rat liver homogenates (S-9 mix) induced with phenobarbital.]
[b] Relative carcinogenicity is a subjective assessment based on dose/dose-rat/tumor incidence and time taken for tumor production (from Dr. Lijinsky; see Chapter 10 of this volume).
[c] Metabolic activation: rat liver homogenates (S-9 mix) induced with Aroclor.

Similarly, hydroxy substitution of NPYR did not affect mutagenicity. All of these compounds required metabolic activation, implying that they are not the likely end products of NPIP and NPYR metabolism. A block in the α-positions to the N-nitroso group, as in 2,2,6,6-tetramethyl-4-hydroxy NPIP and 2,2,6,6-tetramethyl-4-oxy NPIP, has completely eliminated mutagenic activities (Table III), suggesting that these compounds are metabolically activated by α-carbon oxidation.

Carboxyl substitution, as in nitrosoproline (2-carboxy NPYR) or 4-hydroxy nitrosoproline and nitrosopipecolic (2-carboxy NPIP), nitrosonipecotic (3-carboxy NPIP), and nitroso-iso-nipecotic (4-carboxy NPIP) acids, completely eliminated mutagenic activities. Nitrosoguvacine (similar to nipecotic acid with an unsaturated ring structure) was also not mutagenic, while its methyl ester (nitrosoguvacoline) was mutagenic in the *Salmonella* assay. Lipophobicity could explain the lack of biological activity of these compounds.

Again the carcinogenicity results with oxygenated and carboxylated nitrosamines paralleled the *Salmonella* results, with the exception of nitrosoguvacoline, which is a mutagen but not a carcinogen.

3.2.2d. Other Derivatives. Two additional derivatives of NPIP were assayed for mutagenicity in the *Salmonella* assay. The 4-phenyl derivative of NPIP was mutagenic and moderately carcinogenic, while the other derivative with a complex substitution in the 2-position was neither carcinogenic nor mutagenic (Table VI).

3.2.3. Mutagenicity Studies with Aliphatic Nitrosamines

In contrast to the findings with cyclic nitrosamines, the aliphatic nitrosamines failed to show any consistent relationship between their mutagenic and carcinogenic properties. Results obtained[17] with approximately 25 such derivatives are given in Table VII. The most striking result from this study was that a number of carcinogenic nitrosamines for rats were not activated to a mutagenic form in the *Salmonella* assay when liver homogenates from the same species were used. These include certain relatively weak carcinogens such as nitrosodiethanolamine,[19] nitroso-bis-(-2-hydroxypropyl)-amine,[20] nitroso-bis-(-2-methoxyethyl)-amine, and some potent carcinogens, nitrosomethylethylamine[19] and nitrosomethylneopentylamine.[21] The halogenated derivatives [nitroso-bis-(2-chloroethyl)amine, nitroso-bis-(2-chloropropyl)-amine] exhibited strong mutagenic activities (active even in the absence of metabolic activation) but were weakly carcinogenic in *in vivo* studies. This discrepancy could be attributed to the chemical instability of these compounds. Results obtained with nitrosodiallylamine and nitroso-di-*n*-octylamine are more difficult to explain. These compounds were mutagenic in the *Salmonella* assay but failed to produce tumors in rats even at very high doses.[22] However, nitrosodiallylamine was reported[23] to cause respiratory tract tumors in Syrian hamsters.

A noteworthy observation from this selected group of compounds was that some asymmetrical compounds such as nitrosomethylethylamine, nitrosomethyl isopropylamine, and nitrosomethylneopentylamine were not mutagenic,

Table VI
Mutagenicity of Other Cyclic Nitrosamine Derivatives

Compound	Basic structure	TA 1535 $his+$ REV/μmol^a	Carcinogenicity^b
4-Phenyl NPIP		152^c	++
Nitrosomethylphenidate		0	0

^{a} Specific mutagenic activities were determined from the linear dose–response curves. [Metabolic activation: rat liver homogenates (S-9 mix) induced with phenobarbital.]
^{b} Relative carcinogenicity is a subjective assessment based on dose/dose-rat/tumor incidence and time taken for tumor production (from Dr. Lijinsky; see Chapter 10 of this volume).
^{c} Only when activated with Aroclor-induced rat liver S-9.

Table VII
Mutagenicity of Aliphatic Nitrosamines

Compound	TA 1535 his^+REV/100 μmol[a]		Carcinogenicity[b]
	Agar overlay	Preincubation	
Nitroso-			
Diethylamine	0	180	++++
Di-n-propylamine	15	160	+++
Di-iso-propylamine	0	0	++
Di-n-butylamine	44	210	+++
Di-iso-butylamine	60	296	++
Di-sec-butylamine	0	0	0
Di-n-octylamine	0	30[c]	0
Di-allylamine	22	250	0
Iminodiacetonitrile	0	0	0
Bis-(2-cyanoethyl)-amine	0	0	0
Bis-(2-chloroethyl)-amine	1050[d]	1750[d]	++
Bis-(2-chloropropyl)-amine	550[d]	300[d]	+
Diethanolamine	0	0	++
Bis-(2-hydroxypropyl)-amine	0	0	++
Bis-(2-methoxyethyl)-amine	0	0	+++
Bis-(2-ethoxyethyl)-amine	10	120	+++
Bis-(diethoxyethyl)-amine	0	0	0
Bis-(2-oxopropyl)-amine	0	0	+++
Bis-(trifluoroethyl)-amine	0	0	0
Methylethylamine	0	0	+++
Methyl-iso-propylamine	0	0	++
Methylneopentylamine	0	0	++++
Ethylethanolamine	0	17	+++

[a] Specific mutagenic activities were determined from the linear dose-response curves. [Metabolic activation: rat liver homogenates (S-9 mix) induced with phenobarbital.
[b] Relative carcinogenicity is a subjective assessment based on dose/dose-rate/tumor incidence and time taken for tumor production (from Dr. Lijinsky; see Chapter 10).
[c] Only when activated with Aroclor-induced rat liver S-9.
[d] Mutagenic even in the absence of metabolic activation.

even though they are quite potent in causing tumors in rats. This has led us to investigate the role of structural symmetry in mutagenic and carcinogenic properties. Table VIII illustrates results obtained with a group of asymmetrical compounds in the agar overlay and preincubation assays. With the exception of nitrosomethylethylamine and nitrosomethylneopentylamine, all the asymmetrical compounds were mutagenic, but only when tested in the preincubation assay. The potent carcinogen nitrosomethyl phenylethylamine was the most active mutagen even in the agar overlay assay. These two compounds were later reported to be mutagenic in the *Salmonella* assay when tested at acidic pH.[(24)]

Table VIII
Mutagenicity of Asymmetrical Nitrosamines

Compound	Structure	TA 1535his^+ REV/L mol[a] Agar overlay	Preincubation	Carcinogencity[b]
Nitroso-Methylethylamine	CH₃\N—NO / CH₃CH₂	0	0	+++
Methylpropylamine	CH₃\N—NO / CH₃CH₂CH₂	0	87	+++
Methylbutylamine	CH₃\N—NO / CH₃CH₂CH₂CH₂	0	93	++++
Methylneopentylamine	CH₃ CH₃\N—NO / CH₃—C—CH₂ / CH₃	0	0	++++
Methylaniline	CH₃\N—NO / C₆H₅	0	0	++
Methylphenylethylamine	CH₃\N—NO / C₆H₅—CH₂CH₂	119	1445	++++
Ethylbutylamine	CH₃CH₂\N—NO / CH₃CH₂CH₂CH₂	0	52	+++
Propylbutylamine	CH₃CH₂CH₂\N—NO / CH₃CH₂CH₂CH₂	0	230	?

[a] Specific mutagenic activities were determined from the linear dose-response curves. [Metabolic activation: rat liver homogenates (S-9 mix) induced with phenobarbital.]
[b] Relative carcinogenicity is a subjective assessment based on dose/dose-rat/tumor incidence and time taken for tumor production (from Dr. Lijinsky; see Chapter 10 of this volume).

3.3. Nitrosamines—Model Compounds for Mutagenesis Studies

With few exceptions, most nitrosamines require metabolic activation, exhibiting specificity in carcinogenesis for certain species and organs. They constitute a class of compounds that have been extensively investigated for carcinogenicity. Quantitative differences in tumorigenic potencies show a broad range of differences. Tumors are either induced[10] in a particular organ or distributed among a number of organs and tissues. For these reasons, nitrosamines constitute an important class of compounds for determining molecular mechanisms in mutagenesis, as well as for examining the relationship between mutagenesis and carcinogenesis.

Mutagenicity and carcinogenicity of N-nitrosamines was compared by using rat liver S-9 mediated mutagenesis in *Salmonella* and tumor induction in rats. The same strains of rats (Sprague–Dawley) were often used in both studies. The end points that are of interest here are very complex and completely different processes, only distantly related in that they could be caused by the same nucleophiles. The cyclic nitrosamines and their derivatives have shown a greater degree of concurrence for carcinogenic and mutagenic properties than the aliphatic nitrosamines. The correlations made are purely qualitative and any quantitative correlations made are fortuitous. The test results in the bacterial assays can be drastically changed by altering incubation conditions. For example, lowering the pH of standard incubation medium has rendered several nonmutagenic (carcinogenic) compounds to mutagenic[24] forms. Consideration of these methodological complexities argues against the use of standard test protocols, if short-term assays are to have a role as prescreens for carcinogen detection. In spite of these differences, it is noteworthy that nitrosamines are usually metabolized by liver microsomal enzymes into mutagens, even though they may produce tumors in other organs and not in the liver. The metabolic activation pathways might be the same both for mutagenesis and carcinogenesis, and the same nucleophile may be responsible for the biological activity of these compounds. The structure activity patterns are often similar for both mutagenesis and carcinogenesis, giving additional support to this hypothesis.

ACKNOWLEDGMENTS. We wish to express our deep appreciation to Dr. Lijinsky for kindly providing not only the chemicals but also his constant encouragement and support. Technical help provided by Ms. J. A. Young, Ms. D. W. Ramey, Ms. A. A. Hardigree, Ms. B. E. Allen, and Ms. J. T. Cox during various stages of this work is also deeply appreciated. Research sponsored by the Office of Health and Environmental Research, U.S. Department of Energy, under contract No. W-7405-eng-26 with the Union Carbide Corporation.

REFERENCES

1. P. N. Magee and J. M. Barnes, The production of malignant primary hepatic tumors in the rat by feeding dimethylnitrosamine, *Brit. J. Cancer 10*, 114–122 (1956).
2. E. Zeiger, M. S. Legator, and W. Lijinsky, Mutagenicity of N-nitrosopiperazines for *Salmonella typhimurium* in the host-mediated assay, *Cancer Res. 32*, 1598–1599 (1972).
3. H. V. Malling, Mutagenicity of two potent carcinogens, dimethylnitrosamine and diethylnitrosamine in *Neurospora crassa*, *Mutat. Res. 93*, 537–540 (1966).
4. V. W. Mayer, Mutagenicity of dimethylnitrosamine and diethylnitrosamine for *Saccharomyces* in an *in vitro* hydroxylation system, *Molec. Gen. Genetics 112*, 289–294 (1971).
5. E. C. Miller and J. A. Miller, The mutagenicity of chemical carcinogens: Correlation, problems and interpretations, in: *Chemical Mutagens: Principles and Methods for Their Detection* (A. Hollaender, ed.), pp. 83–119, Plenum Press, New York (1971).
6. H. V. Malling, Dimethylnitrosamine: Formation of mutagenic compounds by interaction with mouse liver microsomes, *Mutat. Res. 13*, 425–428 (1971).
7. F. Gletten, U. Weeks, and D. Brusick, *In vitro* metabolic activation of chemical mutagens. 1.

Development of an *in vitro* mutagenicity assay using liver microsomal enzymes for the activation of dimethylnitrosamines to a mutagen, *Mutat. Res. 28*, 113-122 (1975).
8. B. N. Ames, J. McCann, and E. Yamasaki, Methods for detecting carcinogens and mutagens with *Salmonella*/mammalian-microsome mutagenicity test, *Mutat. Res. 31*, 347-364 (1975).
9. S. Neale, Mutagenicity of nitrosamides and nitrosamidines in microorganisms and plants, *Mutat. Res. 32*, 229-266 (1976).
10. P. N. Magee, R. Montesano, and R. Preussmann, *N-Nitroso compounds and related carcinogens*, ACS Monograph No. 173 (C. E. Searle, ed.) pp. 491-625, American Chemical Society, Washington, D.C. (1976).
11. H. J. Vogel and D. M. Bonner, Acetylornithinase of *E. coli*: Partial purification and some properties, *J. Biol. Chem. 218*, 97-106 (1956).
12. P. N. Magee and E. Farber, Toxic liver injury and carcinogenesis. Methylation of rat-liver nucleic acids by dimethylnitrosamine *in vivo*, *Biochem. J. 83*, 114-124 (1962).
13. H. Bartsch, A. Canns, and C. Malaveille, Comparative mutagenicity of N-nitrosamines in a semi-solid and in a liquid incubation system in the presence of rat or human tissue fractions, *Mutat. Res. 37*, 149-162 (1976).
14. T. K. Rao, A. A. Hardigree, J. A. Young, W. Lijinsky, and J. L. Epler, Mutagenicity of N-nitrosopiperidines with *Salmonella typhimurium*/microsomal activation system, *Mutat. Res. 56*, 131-145 (1977).
15. T. K. Rao, J. A. Young, W. Lijinsky, and J. L. Epler, Mutagenicity of N-nitrosopiperazine derivatives in *Salmonella typhimurium*, *Mutat. Res. 57*, 127-134 (1978).
16. T. K. Rao, D. W. Ramey, W. Lijinsky, and J. L. Epler, Mutagenicity of cyclic nitrosamines in *Salmonella typhimurium*: Effect of ring size, *Mutat. Res. 67*, 21-26 (1979).
17. T. K. Rao, J. A. Young, W. Lijinsky, and J. L. Epler, Mutagenicity of aliphatic nitrosamines in *Salmonella typhimurium*, *Mutat. Res. 66*, 1-7 (1979).
18. T. K. Rao, J. T. Cox, B. E. Allen, J. L. Epler, and W. Lijinsky, Mutagenicity of N-nitrosopyrrolidine derivatives in *Salmonella* (Ames) and *Escherichia coli* K-12 (343-113) assays, *Mutat. Res. 89*, 34-43 (1981).
19. H. Druckrey, R. Preussmann, S. Ivankovic, and D. Schmähl, Organotrope carcinogene Wirkungen bei 65 verschiedenen N-Nitroso-Verbindungen an BD-Ratten, *Z. Krebsforsch. 69*, 103-201 (1967).
20. W. Lijinsky and H. W. Taylor, Comparative carcinogenicity of some derivatives of nitroso di-*n*-propylamine in rats, *Ecotoxicol. Env. Safety 2*, 421-426 (1980a).
21. W. Lijinsky, J. E. Saavedra, M. D. Reuber, and S. S. Singer, Esophageal carcinogenesis in F344 rats by Nitrosomethylethylamines substituted in the ethyl group, *J. Natl. Cancer Inst. 68*, 681-684 (1982).
22. W. Lijinsky and H. W. Taylor, Carcinogenicity tests in rats of two nitrosamines of high molecular weight, Nitrosododecamethyleneimine and Nitrosodi-*n*-octylamine, *Ecotoxicol. Env. Safety 2*, 407-411 (1978).
23. J. Althoff, C. Grandjean, and B. Gold, Diallylnitrosamine: A potent respiratory carcinogen in Syrian golden hamsters, *J. Natl. Cancer Inst. 59*, 1569-1571 (1977).
24. J. B. Guttenplan, Enhanced mutagenic activities of N-nitroso-compounds in weakly acidic media, *Carcinogenesis 1*, 439-444 (1980).

4

Effects of pH and Structure on the Mutagenic Activity of N-Nitroso Compounds

Joseph B. Guttenplan

1. INTRODUCTION

1.1. Background

N-nitroso compounds comprise a class of chemicals containing a wide variety of carcinogens and mutagens.[1-4] These compounds have been of theoretical and practical interest to researchers for a number of reasons, such as their organotropic tumor induction, the large variability in the potencies and target organs of various N-nitroso compounds, the relatively simple metabolism and breakdown of the lower molecular weight compounds, and the fact that some members of this class of compounds are environmental chemicals or can be formed *in vivo* from nitrite and secondary or tertiary amines under acidic conditions.[1-4] The contribution to human cancer caused by continued exposure to low levels of

---

Abbreviations used in this chapter: BP, benzo(a)pyrene; cfu, colony-forming units; DNB, N-nitrosodibutylamine; DEN, N-nitrosodiethylamine; DMN, N-nitrosodimethylamine; DPN, N-nitrosodipropylamine; ENNG, N-nitroso-N'-nitro-N-ethylguanidine; MBN, N-nitrosomethylbutylamine; 3MC, 3-methylcholanthrene; 7MeG, N^7-methylguanine; MEN, N-nitrosomethylethylamine; MFO, mixed function oxidase(s); MMS, methylmethanesulfonate; MNNG, N-nitroso-N'-nitroso-N-methylguanidine; NMU, N-nitrosomethylurea; NPy, N-nitrosopyrrolidine; O^6-methylguanine; PNNG, N-nitroso-N'-nitroso-N-propylguanidine; rev, revertants; S-9, 9000 × g supernatent; SDS, sodium dodecylsulfate; SSC, 0.15 M NaCl + 0.15 M Na$_3$ citrate, pH 7.0.

Joseph B. Guttenplan • Department of Biochemistry, New York University Dental Center, New York, New York 10010.

N-nitroso compounds is uncertain, but exposure of pregnant mice to levels of 10 ppb of DMN resulted in significant lung tumor formation in the male progeny.[5] A generalized scheme for the metabolism and decomposition of N-nitroso compounds is given in Figure 1.[2]

1.2. Chemistry

Decomposition of the N-nitrosamide type compounds is spontaneous and results from scission of the derivatized α-carbon. N-nitrosamines are labile only after derivatization of their α-carbon by MFO enzymes.[2,3] As the metabolites of N-nitrosamines are believed to be short-lived, the ability of the affected organs to activate N-nitrosamines is probably one factor in determining their site of action.[2-4] Other compounds, such as the N-nitrosamides and related compounds, are often active at the site of administration.[2-4] If they are able to circulate before extensive decomposition occurs, they may elicit tumors at more remote sites.[2,3]

1.3. Alkylating Properties

The lower homolog N-nitroso compounds, e.g., N-methyl and N-ethyl-N-nitroso compounds, are alkylating agents that are much more potent carcino-

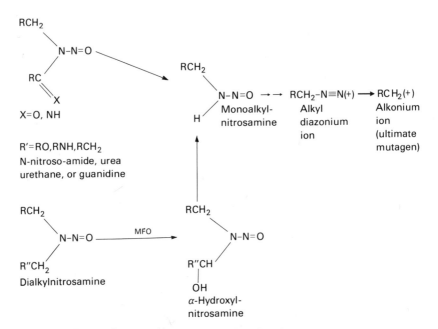

Figure 1. Decomposition and metabolism of N-nitroso compounds.

gens than many other classes of methylating or ethylating agents such as alkyl-sulfonates, dialkylsufates, alkyl bromides, or iodides.[6] This difference in potencies seems not to result from quantitative differences in total cellular alkylation but rather qualitative differences in the moieties or cellular macromolecules alkylated.[3,4,7] Current theories of carcinogenesis hold that alkylation of DNA is the primary event in tumor initiation.[8,9] For this reason, DNA alkylation patterns produced by exposure of animals to N-nitroso compounds have been compared with those obtained by administration of other simple alkylating agents.[3,4,6,7] A major difference in the alkylation patterns is the large degree of oxygen alkylation produced by the N-nitroso compounds.[3,4,6,7] This observation is believed to be highly significant since O^6-alkylated guanines have been shown to mispair in DNA and RNA polymerase systems *in vitro* and are therefore potential premutagenic and possibly precarcinogenic lesions.[10,11] The fact that N-nitroso compounds efficiently alkylate nucleotide oxygens is consistent with their proposed decomposition pathways (Figure 1). Since the oxygen sites are less nucleophilic than most of the nucleotide nitrogens, their alkylation would be expected to proceed mainly via highly reactive electrophiles such as alkyldiazonium ions or related species.[6] These species are relatively insensitive to the nucleophilicity of the acceptor and possess a high degree of S_N1 character.[6]

1.4. Organ Specificity

If O^6-alkylation of DNA guanine by N-nitroso compounds is one reason for the carcinogenic activity of N-nitroso compounds, a problem whose solution is less clear is that of their organ specificity. Current evidence suggests that DNA repair capacity of different organs is one important factor in tissue susceptibility. There are a number of studies, particularly in rats, where persistence of O^6-alkylguanine in a particular organ correlates with tumor susceptibility of the organ.[12,13] However in mice and, in particular, gerbils, O^6-alkylguanines may persist for long time periods in resistant organs.[14,15] Other factors such as cell turnover and exposure to tumor promotors have also been suggested as playing important roles in carcinogenesis.[8,16,17] The abilities of different organs to activate certain N-nitroso compounds such as N-nitrosamines is also believed to be an important factor in organ susceptibilities.[2,3] However, it cannot be excluded that some transport of activated N-nitrosamines can occur, since recent studies utilizing perfused liver to activate DMN to a mutagen for physically separated cultured cells allowed an estimate of the lifetime for activated DMN of 1–5 sec.[18] The mutagenic activity of DMN in a host-mediated bacterial assay in spleen (which possesses little metabolic activity) may also reflect the circulation of activated DMN.[19] The transported species is probably not the monomethylnitrosamine, which has a lifetime of less than 1 sec at 37°C,[1] but it may be the α-hydroxy-DMN, which, with an hydroxyl and heteroatom at the α-carbon, is somewhat analogous to a hemiacetal, which shows some limited stability.

1.5. In Vitro Mutagenesis

Partly because of the complexities involved in studying the metabolism of N-nitroso compounds and the alkylation of specific cellular macromolecules by N-nitroso compounds in experimental animals, a number of *in vitro* systems have been employed for obtaining more specific data in limited systems. Mutagenesis assays in *Salmonella* tester strains have shown high qualitative correlations with carcinogenicity data when a microsomal activation system is incorporated into the assay,[20] and it is believed by many workers that chemical mutagenesis, like carcinogenesis, is initiated by an interaction of the agent with DNA,[21,22,23] thus accounting for the correlation.

1.6. Problems with Assays: Metabolism

Although the first report of a *Salmonella*/microsome assay employed DMN,[24] studies on N-nitroso compounds using the *Salmonella* assay have been somewhat limited by the relative insensitivity of the assay under standard test conditions.[20,21] There appear to be several reasons for the limited mutagenic activity of N-nitroso compounds, in particular the N-nitrosamines, in the *Salmonella* assay. One reason is that the lower molecular weight N-nitrosamines, perhaps because of their high solubility in water, do not bind strongly to all forms of cytochrome P-450.[25] Several different K_ms have been reported for DMN demethylase, with values between 2×10^{-4} and 4×10^{-2} M, and rates of mutagenesis and DMN demethylation do not level off until about 200 mM.[25-27] For comparison, the K_m for metabolism of benzo(a)pyrene, a very hydrophobic carcinogen, is in the μM range, depending on the form of cytochrome P-450.[28] Thus, a significant and sometimes major portion of the cytochrome P-450 molecules are not able to metabolize DMN at many of the concentrations of DMN used in mutagenicity assays. This effect is particularly pronounced in microsomes isolated from Aroclor 1254-induced rats, where the proportion of the high K_m forms seems to be much greater than in microsomes isolated from untreated rats. For example, we have observed an 80% decrease in DMN demethylase activity in Aroclor-induced microsomes on going from 100 mM to 5 mM DMN, but only a twofold decrease in uninduced microsomes.[29] In addition, in mutagenesis assays this effect may be greatly exacerbated when the microsomes (or S-9) and DMN are overlayed onto agar, further diluting the DMN. This latter effect can be obviated by the use of a liquid-phase incubation mix as first described by Malling.[24]

1.7. DNA Repair

A second problem in the detection of mutagenesis by many N-nitroso-N-methyl or N-ethyl compounds is the resistance of many of the tester strains to low exposures of these compounds.[30] The reason for this resistance seems to result from a saturable, constitutive O^6-alkylguanine repair activity present in the bacteria (*vide infra*). If this activity is not saturated, then little mutagenesis is observed.

1.8. Permeation into Bacteria

A third reason for the weak mutagenic activities of many N-nitroso compounds, for which evidence will be given here, seems to result from the fact that in aqueous-liquid media the active metabolites of many N-nitroso compounds are generated extracellularly and may not possess the stability to efficiently traverse the bacterial membrane before they decompose. In a semisolid medium such as agar, diffusion is even slower, which further reduces permeation.

1.9. Optimum Conditions

From a consideration of these factors it was found that for detecting mutagenic activities of N-nitrosamines, the use of a liquid assay, high concentrations of microsomes or S-9 (to enhance metabolism), and relatively long incubation times are advantageous.[31] In addition, for DMN and DEN it is possible to saturate the constitutive repair capacity of the cells with pretreatment or cotreatment with a low concentration of NMU or a related compound.[31] For many N-nitroso compounds the use of exponentially growing cells provides 2- to 4-fold increases in sensitivity over stationary-phase cultures.[31] Further enhancement in mutagenesis by most N-nitroso compounds results when the assays are carried out in weakly acidic media.[32] This chapter deals with:

1. The effects of pH on mutagenesis induced by N-nitroso compounds.
2. The effects of structural variations on mutagenesis induced by N-nitroso compounds and some of the possible reasons for these effects.

2. METHODS

2.1. Liver Extracts

All metabolic activation was carried out using S-9 fractions from Aroclor 1254-treated (5 days before sacrifice at 500 mg/kg) Sprague–Dawley rats or occasionally Swiss–Webster mice (as indicated in the legends of the tables). For most experiments the livers were homogenized in 3 volumes of 1.15% KCl, but in one group of experiments (mutagenesis by a series of weakly mutagenic N-nitrosamines) livers were homogenized in 2 volumes of KCl.

2.2. Mutagenesis Assays

Mutagenesis assays for N-nitrosamines were carried out as previously described,[31] except that the pH of the incubation mix was made to the desired value by adjustment of the pH of the NADPH-generating system to a pH that would allow admixture with the appropriate amount of S-9 to yield a mix of the desired pH. The NADPH-generating system was first brought to a pH of 7.7 after all components were added and the solution was allowed to incubate at room temperature for 30 min, after which time the pH was 6.9. This procedure was necessary because the pH of the generating system, after dissolution of the components, was acidic, and the NADPH generation (monitored in the undiluted solution at 340 nm, using a 1-mm cell with the zero offset) was more efficient when the pH was near 7.4. The final pH to which the generating system was adjusted was dependent on the volume of S-9 in the incubation mix and was predetermined with the use of $0.1\ M\ PO_4$ solutions of various pH values that were mixed with S-9 fraction. Unless a very high ratio of S-9 to generating system was used (> 2), the pH of the entire incubation mix was nearly the same as the pH of the generating system. This procedure obviated the need for sterile titrations or the need to filter-sterilize the incubation mix, which did not readily pass through 0.45-μm filters. Only the NADPH-generating system was filtered and this readily passed through the filter. It was found that preparation of a generating system consisting of monosodium glucose-6-phosphate 18 mM; glucose-6-phosphate dehydrogenase, 18 units; NADP, 5.2 mM and $MgCl_2$, 20 mM; using a pH 7.4, 0.1 M phosphate buffer, resulted in a solution with a pH of 6.9. Mutagenesis was initiated with this generating system unless the mutagen was not soluble in water. In these cases the reaction was initiated with a concentrated solution of the mutagen in acetone. The final volume of acetone was kept below 5% of the total volume (0.5–1 ml) in order to avoid toxicity to the bacteria. Also, the generating system was combined with the S-9 fraction immediately before the assay in order to avoid MFO-catalyzed oxidation of NADPH. Although some regeneration of NADPH may occur during the incubations, additional NADPH added during the incubations results in increased mutagenesis when high concentrations of MFO enzymes are employed.[31] The mutagens were not dissolved in dimethylsulfoxide because this solvent is reported to decrease mutagenesis by N-nitrosamines when compared to aqueous solvents.[33] In all cases EDTA was present at a final concentration of 1 mM.

Mutagenesis by NMU and MNNG was measured by methods previously described.[32] Stock solutions of 10–100 times the final concentrations of MNNG were dissolved in H_2O and those of NMU in 0.01 M sodium acetate, pH

5.5. Both NMU and MNNG were stable for at least several hours under these conditions when chilled. Mutagenesis was carried out in 0.1 M phosphate buffer and was initiated by the addition of NMU or MNNG stock solution. The addition of the small volumes of the stock solution did not appreciably change the pH of the incubation mix.

All mutagenesis assays involving N-nitroso compounds were carried out using exponentially growing *Salmonella typhimurium* (TA 1535 or TA 100, generously supplied by B. Ames, Berkeley, California), which were centrifuged for 10 min at 2000 $\times$ g and resuspended in ½ volume of 0.1 M phosphate buffer. Most of the mutagenesis occurred during the liquid-phase preincubation, as mutagenesis was proportional to incubation time and in many cases no mutagenesis occurred when the incubation mix was not preincubated. Mutagenesis by MMS, BP, or 3MC was also carried out in a liquid preincubation mix, but the bacteria were used directly from the overnight culture, as the use of logarithmically growing cells did not increase mutagenesis as it did with the N-nitroso compounds. Background revertants were slightly higher at a pH of 6.5 than 7.4 and have already been subtracted from the revertant values given in the tables. Precision between duplicate assays was within 15%. Toxicity determinations were carried out on most of the mixtures containing DMN and DEN. There was no significant toxicity at low doses of these compounds but about 30% toxicity at the highest doses. The toxicity was not pH dependent.

2.3. DMN Demethylase

DMN demethylase was assayed by monitoring formaldehyde production as previously described,[31] except that the tubes were stoppered to prevent evaporation of DMN. In other studies, analysis of the head space by gas chromatography[34] showed no significant oxygen depletion during the assay. The incubation mix was the same as that used in the mutagenesis assay except that the bacteria were omitted. As the color development in the formaldehyde assay was pH dependent, standard formaldehyde solutions were added to incubation mixtures without any added DMN and these mixtures were then treated in the same manner as the full incubation mixture. These were then used as standard solutions in calculating formaldehyde production at each pH value. The intensity of the color developed was 21% greater in the solutions obtained from the pH 7.4 incubation mixtures than from the pH 6.5 incubation mixtures.

2.4. Isolation of DNA and Analysis of Alkylated Bases

Bacteria were concentrated 10-fold from pooled overnight cultures (grown in Difco Bacto-Tryptone broth) by centrifugation and resuspended in pH 7.4, 0.1 M PO_4 buffer. Ten milliliters of bacterial suspension was then added to 40 ml of the same buffer, and 0.2–1 ml of an appropriate concentration of NMU was added to initiate the reaction. NMU was recrystallized from 90% ethanol and dissolved in 0.01 M, pH 4.5 acetate buffer. Bacteria were incubated with NMU for 1 hr and an aliquot of 1 ml was taken for an assay of reversion to histidine prototrophy and determination of colony-forming units. There was no appreciable toxicity at any of the doses employed. The cells were pelleted, weighed, resuspended in SSC, and treated with 0.67% SDS at 37°C for 30 min with shaking. The suspension was brought to 1 M with Na perchlorate and incubated with shaking at 60° for 20 min. After cooling to 4°C an equal volume of 24:1 chloroform/isoamylalcohol was added, the mixture was shaken and the aqueous layer was removed and extracted with 1:1 phenol/chloroform. These two steps were repeated until the aqueous layer was clear. The DNA was then precipitated with 2 volumes of 95% ethanol, washed 2 times with 70% ethanol and dissolved in water. After addition of 1/10 volume of SSC, RNase A (Worthington Biochemical Corp., Freehold, New Jersey, 4341 μ/mg preheated at 60° for 30 min) was added to a concentration of 25 μg/ml and the solution was incubated at 37°C for 2 hr. The DNA was then precipitated and washed as described above. The DNA was then treated as described by Frei et al.,[35] first by heating to 100°C at neutral pH for ½ hr and then, after acid precipitation and separation of the supernatant, by heating at 60°C for 30 min at a pH of 1. Aliquots of the hydrolysates corresponding to 0.5–1 mg DNA (for determination of O^6-MeG and 7-MeG) or 0.1–1.0 g (for determination of guanine) were applied to a Partisil-10-SCX, 25 cm × 4.5 mm (i.d.) high-performance liquid chromatography column (Whatman Inc., Clifton, New Jersey) and eluted with pH 2.0, 0.1 M ammonium phosphate solution at a flow rate of 0.65 ml/min for 7-MeG in the neutral hydrolysate, 2 ml/min for O^6-MeG in the acid hydrolysate, and 3 ml/min for guanine in the acid hydrolysate. The eluants were monitored by their fluorescence with the excitation wavelength set at 285 nm and the emission wavelength at 365 nm.

3. EFFECTS OF pH ON MUTAGENESIS AND DNA ALKYLATION BY N-NITROSO COMPOUNDS

3.1. N-Nitrosamines

In previous reports it was shown that mutagenesis induced by low concentrations of DMN increased as the pH of the medium was reduced from 7.4 to 6.0.[31,32] At first, this effect was attributed to a higher rate of DMN metabolism at the lower pH values.[31] However, when mutagenesis and demethylation were compared at different pH values at low concentrations of DMN, differences in DMN metabolism were small relative to differences in mutagenic activities. At higher concentrations of DMN, the mutagenic potency of DMN was still greater at lower pH values despite the higher rate of DMN metabolism at pH 7.4 (Table I). The inversion in relative rates of DMN metabolism at pH 7.4 and 6.5 on going to higher concentrations of DMN is probably due to metabolism of DMN by different forms of DMN demethylase at high and low concentrations of DMN,[26,27] and these demethylases may have different pH optima. It can be calculated from the data in Table I that mutagenesis (per nmole of formaldehyde generated) increased at least severalfold on going from a pH of 7.4 to 6.5. Mutagenesis by other N-nitrosamines was also assayed at pH 7.4 and 6.5 to determine whether other N-nitroso compounds were also more efficiently activated at lower pH values. As seen in Table II, all of the N-nitrosamines tested were more potent mutagens in acid media.

Two general trends were observable. First, mutagenesis was enhanced to a greater extent at lower concentrations of mutagens, and second, with the exception of MEN and MBN, the relative increase in mutagenesis at pH 6.5 was smaller for the higher molecular weight compounds. From the data given in Table II it seems likely that most of the mutagenesis produced by MEN and MBN is attributable to the methyl portion of these molecules, as MEN and MBN are of comparable mutagenic potency to DMN and much more potent than either DEN or DBN under the conditions of the assay. (DEN was of similar potency to DMN when metabolized by high concentrations of rat liver S-9, *vide infra*.)

3.2. Direct-Acting Compounds

In order to determine whether acid pH also enhances mutagenesis by direct-acting N-nitroso compounds, mutagenesis by NMU and MNNG was also monitored at several pH values. When monitored at 60 or 90 min, as seen in Table III, mutagenesis by NMU continually increased on going from a pH of 7.4

Table I
Effect of pH on DMN Demethylase and DMN-Induced Mutagenesis[a]

pH	DMN (mM)	DMN demethylase nmol CH_2O / min × ml S-9	Mutagenesis 10^3 rev / 10^8 cfu × min × ml S-9
7.4	100	115	4.93
6.5	100	61.2	10.2
7.4	2	10.1	< 0.18
6.5	2	13.2	1.05

[a]The amount of S-9 present in the incubation mix was as follows: 62.5 μl/ml for all the DMN demethylase assays; 62.5 μl/ml in the 100 mM mutagenesis assay; and 250 μl/ml in the 2 mM mutagenesis assay. Liver S-9 from Aroclor 1254-treated mice were used. Other conditions are described in Section 2, Methods.

Table II
Effect of pH on Mutagenesis Induced by N-Nitrosamines in
Salmonella typhimurium TA 1535[a]

Compound N-nitroso	Dose (mg/ml)	Revertants/min × ml S-9	
		pH 6.5	pH 7.4
Dimethylamine	7.0	10200	4930
	0.7	1200	<200
	0.2	1050	<175
Diethylamine	10.0	85.2	18.4
	0.7	<10	<10
Dipropylamine	0.7	43.2	22.8
	0.2	36.0	14.2
Dibutylamine	0.7	4.0	2.6
	0.2	1.1	0
Pyrrolidine	2.0	60.6	36.4
	0.7	54.0	29.4
Methylethylamine	7.0	6320	1870
	0.7	2880	<140
Methylbutylamine	2.0	10350	4010
	0.2	3157	367

[a] The numbers in the table are calculated from the linear portions of the time-response curves for mutagenesis. The S-9 concentration was varied between 50 and 500 μl/ml incubation mix and the incubation times varied from 0 to 60 min. Aliquots of 50 to 500 μl were overlayed onto minimal agar at each time point. Identical conditions were used for each pair of measurements at pH 6.5 and 7.4. In some cases, as with DMN, DEN, MEN, and MBN, mutagenesis did not become appreciable until the bacteria were exposed to an apparent "threshold dose" of mutagenic metabolites, after which time the numbers of revertants increased nearly linearly with incubation time. In several cases at low doses of these mutagens no mutagenesis was observed. In these instances the upper-limit detectability under the assay conditions is indicated in the table. Liver S-9 from Aroclor 1254-induced mice was used.

to 6.0 and then leveled off. At 30 min of incubation, mutagenesis was most effective at a pH of 6.0. Similar results were obtained with MNNG (Table IV), which exhibited increased mutagenic activity at lower pH values. Under the conditions of this assay inefficiency of mutagenesis at low revertant values was apparent. This inefficiency at low revertant values has been reported for other N-nitroso compounds[30] and was observed at all pH values employed.

Mutagenesis by the higher homolog direct-acting mutagens ENNG and PNNG was also measured at pH 7.4 and 6.5. As shown in Table V, these mutagens were also more potent at a pH of 6.5.

Thus far the enhancement in mutagenesis observed at acid pH seems to be restricted to N-nitroso compounds. As shown in Table VI, the mutagenic activity of several other mutagens was reduced or unaffected by acid media. Although a different tester strain, TA 100, was used in these experiments than in the mutagenesis experiments involving N-nitroso compounds, TA 100 was also

Table III
Effect of pH and Incubation Time on Mutagenesis in TA 1535 Induced by NMU[a]

pH	Revertants/plate		
	30 min	60 min	90 min
5.5	703	1730	2253
6.0	1002	1792	2325
6.5	719	1405	1853
7.0	585	1057	1130
7.4	189	259	259

[a] The incubation mix contained 0.20 mM NMU in 0.1 M PO$_4$ buffer and 2×10^8/ml of exponentially growing cells. Aliquots of 0.1 ml were plated. Background revertants at each pH value (24–32) have already been subtracted from the above numbers.

Table IV
Effect of pH and Incubation Time on Mutagenesis in TA 1535 Induced by MNNG[a]

pH	Revertants/plate	
	30 min	60 min
6.0	7	384
6.5	19	323
7.0	0	184
7.4	4	167

[a] The incubation mix contained 1.25 μM MNNG in 0.1 M PO$_4$ buffer and 2×10^8 logarithmically growing cells. Aliquots of 0.1 ml were plated. Background revertants (34–44) were subtracted from the above values.

Table V
Effect of pH on Mutagenesis Induced by ENNG and PNNG in TA 1535[a]

Compound	Dose (μg/ml)	Rev/ml incubation mix	
		pH 6.5	pH 7.4
ENNG	1	134	51
ENNG	2	724	246
ENNG	4	3360	1125
PNNG	1	746	279
PNNG	2	1320	540
PNNG	4	1900	1014

[a] Assays were performed as described in Section 2, Methods. All tubes contained 0.5 ml of mix and were preincubated for 60 min before plating.

Table VI
Effect of pH on Mutagenesis Induced by Benzo(a)pyrene, 3-Methylcholanthrene, and Methylmethanesulfonate[a]

Mutagen	pH	Revertants at incubation time		
		0 min	30 min	60 min
MMS (0.4 µl/ml)	7.4	392	362	417
	6.5	430	373	371
BP (1.0 µg/ml)	7.4	372	991	997
	6.5	299	806	806
3MC (1.0 µg/ml)	7.4	93	606	898
	6.5	67	251	308

[a] Strain TA 100 was used with all mutagens. The incubation mix used for 3MC and BP contained per ml: 25 µl of Aroclor 1254-induced rat liver S-9, 400 µl of NADPH-generating system, 40 µl of the mutagen in acetone, 100 µl of TA 100 from an overnight culture, and 435 µl of pH 7.4 or 6.5 phosphate buffer. The MMS mix contained only bacterial culture (100 µl/ml) and MMS in buffer at the indicated concentration. Two milliliters of mix were prepared and aliquots of 0.5 ml were overlayed at the indicated times. Background revertants (130–145) have already been subtracted from the above numbers. Aliquots of all samples were taken for determination of survivors after 60 min incubation. There was no appreciable toxicity in the MMS- and 3MC-treated cells, but there was 75% toxicity in the pH 7.4, BP-treated cells and 50% toxicity in the pH 6.5 BP-treated cells.

found to be more sensitive to mutagenesis induced by N-nitroso compounds at acid pH (unpublished data).

Enhancement of mutagenic potencies of N-nitroso compounds in acidic media seems to be a general phenomenon. If DMN is a good model for other N-nitroso compounds, then as previously discussed, part of the effect may be related to a greater rate of metabolism at pH 6.5. However, this explanation is not sufficient to account for the magnitude of the effect (Table I) nor the effect at the higher substrate level (Table I) where metabolism is more rapid at a pH of 7.4. Also, the direct-acting mutagens NMU, MNNG, ENNG, and PNNG are more potent at acidic pH.

3.3. Alkylation of Isolated DNA

As DNA is believed to be the critical target in mutagenesis, experiments were conducted in which naked DNA was exposed to NMU for 1 hr at pH 6.5 and pH 7.4 and the concentrations of 7-MeG and O^6-MeG produced in the DNA were measured. As shown in Table VIIA, the pH of the medium had no significant influence on the concentrations of the methylated bases that were produced by NMU. The slight excess in the concentrations of these bases at pH 7.4 may reflect the fact that a small concentration of NMU was not decomposed at pH 6.5 because of the longer half-life of NMU at this pH.[36] These results indicate that there is no inherent difference in reactivity of guanine (at least at the O^6 and N^7 positions) with NMU decomposition products at pH 6.5 or 7.4, and the ratio of O^6-MeG to 7-MeG did not change at the two pH values.

Table VII
Effect of pH on the Formation of 7-MeG and O^6-MeG in Isolated DNA and in DNA Extracted from *Salmonella* after Treatment with NMU[a]

NMU (mM)	pH	7-MeG μmole/mole G	O^6-MeG μmole/mole G	$\dfrac{10^3 \text{ rev}}{10^8 \text{ cfu}}$
A				
0.2	7.4	330	35	—
0.2	6.5	320	32	—
1.0	7.4	1600	169	—
1.0	6.5	1400	161	—
B				
0.16	6.0	1900	62	2.8
0.16	7.4	270	<4	0.0
0.32	6.0	3500	170	7.9
0.32	7.4	340	12	1.3

[a] Calf thymus DNA, 1 mg/ml, was incubated in 2 ml of PO_4 buffer for 60 min in the presence of the indicated concentrations of NMU, added from a stock solution as described in Section 2, Methods. DNA was precipitated with 2 volumes of ethanol, washed 2× with 70% ethanol, redissolved in warm buffer, hydrolyzed, and analyzed as described in Section 2. For DNA isolated from *Salmonella typhimurium* 1535, cells were treated for 60 min in the presence of the indicated concentrations of NMU, pelleted, and worked up as described in Section 2. For comparison, revertant frequencies are also included. A, isolated DNA; B, DNA isolated from *Salmonella typhimurium*.

3.4. Alkylation of DNA in *Salmonella*

Since naked DNA was similarly affected by NMU at neutral and acidic pH values, experiments were also performed using DNA extracted from *Salmonella typhimurium* TA 1535 after treatment of the cells with NMU at two different pH values (Table VIIB). Here there was a clear difference in the concentration of O^6-MeG and 7-MeG at the two pH values, with significantly more alkylation occurring at pH 6.0. Moreover, there was a reasonable correlation between the concentration of O^6-MeG in the DNA with the revertant frequency at each pH value. The nonlinear dose-response in the formation of O^6-MeG is discussed in a later section.

3.5. Possible Explanations: NMU, MNNG

It appears then that NMU or its decomposition products are more permeable to the bacterial cells at acidic pH values. This is perhaps reasonable for NMU, which has a half-life of only several minutes at a pH of 7.4, but a half-life nearly 10 times as great at a pH of 6.5.[36] MNNG, however, has a half-life of hours at pH 6.0–7.4 (*vide infra*). Although these lifetimes would seem to be sufficient to permit permeation of the parent compounds into the cells, reactions with membrane nucleophiles could perhaps accelerate their decomposition and decrease permeation efficiency. It is known, for instance, that sulfhydryl groups accelerate the decomposition of MNNG.[3] At more acidic pH values certain potential nucleophiles such as histidine-pyrazole or lysine ε-amino groups (in a hydrophobic environment) could perhaps become protonated and lose their abilities to interact with MNNG or NMU. An alternative explanation for the enhanced mutagenesis in acidic media is that rapid equilibration between intracellular and extracellular mutagen occurs, and, because of the longer lifetime of the mutagens in acidic media, the cells in these media will be exposed to mutagen for longer times. (Intracellular pH is presumably regulated near a value of 7.4.)

At the pH values employed in this study the half-lives of NMU and MNNG are inversely related to pH[3,36] so that less decomposition (presumably to mutagenic products) occurs in acid solutions. If all other factors were equal, then lowering the pH should decrease the mutagenic potencies of NMU and MNNG. The decrease in NMU-induced mutagenesis at 30 min incubation, on going from a pH of 6.0 to 5.5, may partially reflect this effect. However, at longer incubation times, appreciable decomposition occurs in all NMU solutions and mutagenesis reflects mainly the inherent mutagenic potency at different pH values. At shorter incubation times, mutagenesis probably reflects both potency and the fraction of NMU decomposed.

The half-life of MNNG is much longer than that of NMU, but as indicated above, its decomposition can be substantially accelerated by nucleophilic catalysts. In order to estimate its half-life under the actual assay conditions, the bacteria were rapidly pelleted from the incubation mix and residual MNNG was assayed colorimetrically in the supernatant. After 60 min, residual MNNG was greatest (97%) in the pH 5.5 solution and least (80%) in the pH 7.4 solution. Despite the greater decomposition of MNNG in the pH 7.4 solution, mutagenesis was greatest in the more acid solutions, which is similar to the results observed with NMU. However, one observation that is relevant to these observations is that MNNG was less mutagenic at acidic pH values after a liquid-phase incubation if the MNNG was removed before plating (results not shown). This indicates that, in acidic media, mutagenic interactions occur during the liquid-phase incubation and in the agar. At pH 7.4 little difference in mutagenesis was observed after liquid-phase incubations if, instead of being left on the plate, MNNG was removed prior to plating. This suggests that MNNG or its short-lived decomposition products are not able to efficiently reach the bacteria in the viscous agar at pH 7.4, but the longer lifetimes at acidic pH permit greater transit in the agar. This may represent an additional explanation of the general superiority of the liquid incubation assay for detecting the mutagenic activity of N-nitroso compounds.

3.6. Possible Explanations: N-Nitrosamines

However, increased permeation of parent compounds into the cells at acidic pH cannot be the cause for increased mutagenesis by N-nitrosamines, since these compounds must first be metabolized by microsomal enzymes before they become mutagens.[1-3] As previously indicated, the α-hydroxyl-N-nitrosamines might be sufficiently long-lived to diffuse into the bacteria. A number of these compounds have recently been synthesized and are indeed more stable under acidic conditions.[37] In addition, the higher homologs were more stable than the lower homologs, which may account for the greater mutagenic potencies of the higher homolog nitrosamines in agar overlay assays[20] compared with the greater potencies of lower homologs in the liquid-phase assay (Table II). The stability of the higher homolog α-hydroxynitrosamines may also be relevant to the relatively small effects of pH on mutagenesis induced by the higher homolog N-nitrosamines (Table II).

3.7. Possible Relevance to Carcinogenesis

If the results of the bacterial assays are relevant to tumor initiation in animals, then enhanced activity of direct-acting N-nitroso compounds might be expected in acidic environments such as the urinary bladder, portions of the gastrointestinal tract, and the oral–esophageal tract. It seems unlikely that extracellular pH would have any major effects on the carcinogenic activity of N-nitrosamines in the cells that metabolize them since the intracellular pH is probably regulated, although it could conceivably vary within narrow limits. However, if activated N-nitrosamines can travel from one organ to another or one cell type to another within an organ, then the extracellular pH could noticeably affect the amount of activated compound reaching the target. It may be relevant that N-nitroso compounds are among the few classes of compounds known to elicit tumors at the above sites.[1-3] In addition N-nitroso compounds have been detected in foods,[3,4] human fecal and urine samples,[38] and they have been recovered in rat saliva at levels equal to or greater than blood levels.[39,40]

3.8. Applications to *in Vitro* Mutagenesis Assays

The increased mutagenic activities of N-nitroso compounds in acidic media may account for some of the interlaboratory differences in detecting the mutagenic activity of N-nitrosamines.[25,30,31,41,42] As described in Section 2, Methods, the NADPH-generating system is acidic despite the use of pH 7.4, 0.1 M PO$_4$ as its solvent. Since the S-9 fraction had relatively little buffering ability compared to the generating system components, the actual pH of the mutagenesis incubation mix used in many previous experiments was probably weakly acidic. There is no indication in the literature that the pH of the incubation mix is checked after addition of all the components, and even if it were, any attempt to adjust the pH would require a sterile titration. (For this reason the procedure described in the Methods section was used.) It is apparent then that differences in generating systems (components and concentrations), the use of NADPH instead of generating system, the use of different buffers, and the use of different ratios of S-9 fraction to generating system could all influence mutagenic activities of N-nitrosamines by affecting pH. In addition, studies on the effects of additives on mutagenesis could also reflect changes in pH if the additive is dissolved in a buffer at a pH different from the incubation mix.

A useful application of the pH effect on mutagenesis has been the detection at pH 6.5 of the mutagenic activities of a number of carcinogenic N-nitrosamines that were previously found to be nonmutagenic. The relative mutagenic potencies of these compounds correlated quite well the carcinogenic potencies of these compounds, and these results are discussed in a subsequent section.

4. MUTAGENESIS BY N-NITROSO-N'-NITRO-N-ALKYLGUANIDINES

4.1. How to Compare Mutagenic Potencies

As a part of our studies on mutagenesis induced by N-nitroso compounds we have been comparing mutagenic effects of different N-nitroso compounds and attempting to establish some relationships between the structures of the compounds and their potencies. An obvious question that arises is, on what basis should potencies be compared? With N-nitrosamines this presents some difficulties since these compounds, as previously indicated, bind rather weakly to some forms of cytochrome P-450 and are only partially metabolized in a liquid-preincubation assay. (In order to determine metabolism in an agar overlay assay, reactants and products would have to be extracted from the agar and analyzed. This has not yet been reported.) For instance, using the DMN demethylase activities given in Table I, it can be calculated that in a 2-mM incubation mix containing 50% S-9 fraction, only 5% of the DMN would be metabolized after 30 min. It might be possible to attain somewhat higher conversion at lower initial concentrations of DMN if a similar absolute conversion of DMN occurred, but this could only occur if the DMN concentration were well above K_m. This is possible for the low K_m DMN demethylase[26] but the contribution to metabolism by the higher K_m DMN demethylase[25] would be lost, and this contribution is the major one in S-9 fractions from Aroclor 1254-treated animals and a very important one in S-9 fractions from untreated animals[25] (see also Introduction). In fact, under many conditions (e.g., low S-9 concentration and nonacidic pH), it is not possible to observe mutagenic activity of DMN at low concentrations of DMN, and this is true of other N-nitrosamines as well. Thus any comparisons between mutagenic potencies of N-nitrosamines at a particular concentration will be dependent on the inherent mutagenic activities of the metabolites and the fraction of the parent compound metabolized. This fraction is likely to be different for each nitrosamine since each compound will probably have a different value for K_m. (The fraction metabolized may well approach 100% for hydrophobic compounds with low values for K_m, such as the aromatic hydrocarbons. [28, 43]) Thus, failure to observe mutagenic activity could reflect lack of metabolism or lack of inherent activity of metabolites. Calculating relative mutagenic activities from slopes of dose–response curves is equally ambiguous since compounds with low K_m values will bind to cytochrome P-450(s) over a relatively narrow concentration range, but those compounds with high value(s) for K_m will bind over a broad range yielding relatively small slopes.

As a first step, then, in attempting to compare mutagenic activities, the approach taken was to compare mutagenic activities of different compounds on a basis of equivalent amounts of metabolites generated, so that the mutagenic potencies reflected only inherent differences in mutagenic activities. For this purpose direct-acting mutagens were utilized since their metabolism is easily monitored. In addition the mutagens were chosen because they all decomposed via the same pathway and yielded alkylating agents that differed from each other only in chain length. Three N-nitroso-N'-nitro-N-alkylguanidines, methyl, ethyl, and propyl, were compared in three different bacterial strains, TA 1975, TA 1535, and TA 100, each containing different repair capacities.[21] TA 1975 contains a functional $uvrB$ DNA repair system, TA 1535 is similar to TA 1975 but lacks the $uvrB$ DNA repair system, and TA 100 is strain TA 1535 into which a plasmid conferring several characteristics has been incorporated. Among these characteristics are believed to be increased levels of error-prone DNA repair enzymes.[44] Exponentially growing bacteria were pelleted, resuspended in buffer, incubated with each mutagen for one hour, again pelleted, and plated in the absence of mutagen. Residual mutagen in the supernatant was monitored colorimetrically. In this manner the relative mutagenic effects of methylating, ethylating, and propylating N-nitroso compounds were compared.

4.2. MNNG

The results of these experiments are shown in Figure 2 and the slopes of the linear portions of the dose–response curves are given in Table VIII. With MNNG all three strains exhibited similar dose–response behavior, with TA 1975 showing somewhat more activity at the lower doses. All three strains

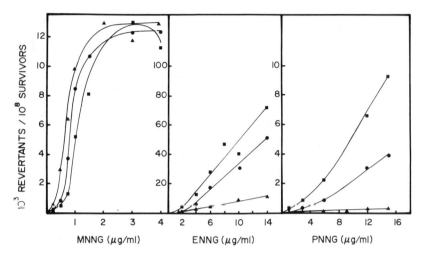

Figure 2. Concentration-dependent mutagenesis induced by MNNG, ENNG, and PNNG in *Salmonella typhimurium* TA 1975, TA 1535, and TA 100. Bacteria were exposed to the mutagen for 1 hr in pH 7.4, 0.1 M PO$_4$ buffer, separated from the residual mutagen by centrifugation, and assayed for revertants and cfu. Although there was high toxicity (50%–90%) at the high doses of ENNG and PNNG, control experiments showed that the mutation fraction induced by these compounds was independent of the number of cells in the assay. ■, TA 100; ●, TA 1535; ▲, TA 1975.

Table VIII
Mutagenic Activities of MNNG, ENNG, and PNNG in *Salmonella typhimurium* TA 1975, TA 1535, and TA 100, Expressed as Revertant Frequency/μg of Mutagen Decomposed

Mutagen	Strain	rev frequency / μg mutagen decomposed $\left[\dfrac{10^3 \text{ rev}^a}{10^8 \text{ cfu} \times \mu g}\right]$	Percent decomposition[b]	rev frequency / μg mutagen plated $\left[\dfrac{10^3 \text{ rev}}{10^8 \text{ cfu} \times \mu g}\right]$
MNNG	TA 1975	15	27	56
MNNG	TA 1535	15	27	56
MNNG	TA 100	11	27	41
ENNG	TA 1975	0.86	12.5	6.8
ENNG	TA 1535	4.2	12.5	34
ENNG	TA 100	4.4	12.5	35
PNNG	TA 1975	0.025	10.5	0.24
PNNG	TA 1535	0.38	10.5	3.6
PNNG	TA 100	0.88	10.5	8.4

[a] Taken from the linear portion of the dose–response curves in Figure 2, which are based on an incubation time of 60 min.
[b] After 60-min incubation.

exhibited disproportionately little mutagenic activity at the lower doses, an approximately linear increase at higher doses, and a leveling off at still higher doses. All strains were also tested while in stationary phase (data not shown) and were all less sensitive while in stationary phase of the growth cycle. Strains TA 100 and TA 1535 were about two times less sensitive in stationary phase, while strain 1975 exhibited a sixfold decrease in stationary phase. This latter observation suggests that the $uvrB$–DNA repair system, present in this strain but absent in the others, somehow plays a role in the repair of MNNG-induced lesions if replication does not occur too rapidly.

4.3. ENNG

Mutagenesis by ENNG was also inefficient at lower doses but did not level off at higher doses. Strain TA 100 was clearly most sensitive at lower doses but was of similar sensitivity as TA 1535 at higher doses. TA 1975 was much less sensitive than the other strains at higher doses. As seen with MNNG, this strain was many times less sensitive in stationary phase than in exponential growth phase. In preliminary experiments the other strains had similar sensitivities in exponential or stationary phase. The slopes of the linear portion of the dose–response curves for TA 100 and TA 1535 were several times higher with MNNG than with ENNG, but when these values were expressed as revertant frequencies/μg of compound decomposed, the difference in potency between MNNG and ENNG was less than a factor of two. TA 1975 was less mutable by ENNG than by MNNG, suggesting that the $uvrB$–DNA repair system functions in the repair of ENNG-induced DNA lesions. As seen with MNNG, this system appears to be more important in slower growing cells. ENNG was most toxic to TA 1535, followed by TA 100 and TA 1975. This result suggests that both $uvrB$ and error-prone DNA repair contribute to cell viability in the ENNG-treated cells.

4.4. PNNG

PNNG was the least potent of the three mutagens and exhibited greatest activity in TA 100. The slope of its dose–response curve was 0.06 that of MNNG in TA 100 but was 0.15 that of MNNG when expressed as revertant frequency/μg of compound decomposed. TA 1535 was about half as sensitive as TA 100, and TA 1975 about 1/40 as sensitive. In stationary phase TA 1975 was even three times less sensitive. The toxicity of PNNG to TA 1975 was also significantly less than to TA 100 or TA 1535. Apparently the $uvrB$–DNA repair system is also capable of coping with PNNG-induced damage to DNA. The greater sensitivity of TA 100 at all doses also suggests that error-prone DNA repair contributes to the mutagenic activity of PNNG. The dose–response curves for PNNG-induced mutagenesis in TA 100 and TA 1535 also showed biphasic behavior at the lower dose regions, although this effect was not as pronounced as with MNNG and ENNG. Mutagenesis induced by PNNG, unlike mutagenesis by MNNG, did not level off at high revertant frequencies, although revertant frequencies above those induced by MNNG were not measured due to the high toxicity of PNNG at doses that induced such high revertant frequencies. The reason for that effect seen only with MNNG is not yet clear.

4.5. Comparisons between Mutagens

From the results of these experiments it appears that methylating and ethylating N-nitroso compounds possess similar inherent mutagenic potencies above a certain "threshold" level, and propylating compounds are considerably weaker, particularly in the UV-repair competent strain, TA 1975, and in strains not containing enhanced error-prone repair activity, TA 1975 and TA 1535. Mutagenesis by PNNG was also somewhat less efficient at lower doses than higher doses, but this effect was less pronounced than that seen with MNNG and ENNG. One interpretation of this result is that there may be several pathways for mutagenesis induced by PNNG, and a saturable DNA repair process is capable of removing lesions leading to one pathway but not others. Other interpretations are also possible.

Strain TA 1975 possesses a nitroreductase activity that TA 1535 and TA 100 lack,[21] which could contribute to the relative insensitivity of this strain by reduction of the N'-nitro group in the N'-nitro-N-nitrosoguanidines, particularly when in stationary growth phase. In order to rule out this possibility, TA 1975 was compared with TA 1535 and TA 100 when all strains were in stationary phase and exposed to N-ethyl-N-nitrosourea. The relative sensitivities of these strains to this compound were the same as they were to ENNG, indicating that reduction of the nitro group is not an important factor in inhibiting mutagenic activity in TA 1975. However, the possibility that the nitroso group could be reduced in TA 1975 cannot yet be ruled out.

5. REPAIR OF O^6-METHYLGUANINE AND ITS EFFECT ON MUTAGENESIS

5.1. O^6-MeG and 7-MeG in *Salmonella* DNA

The inefficiency of low doses of N-nitroso-N-methyl compounds in generating mutations shown in Figure 2, Table VII, and in previous work[30,31] seemed of interest since it might indicate that low doses of these compounds produced repairable lesions in DNA. In order to investigate this possibility DNA was isolated from TA 1535, treated with NMU at several different doses, and analyzed for O^6-MeG and 7-MeG.

5.2. Threshold in O⁶-MeG Formation

Both O^6-MeG and 7-MeG were detected in the isolated DNA (Figure 3), but the dose dependencies of their production were quite different. The concentration of O^6-MeG generated by NMU in the bacterial DNA was biphasic, with detectable levels of O^6-MeG formed only at concentrations of NMU equal to or above 0.24 mM. Mutagenesis by NMU was similarly biphasic with significant levels seen only at NMU concentrations of 0.24 mM or greater. In order to rule out the possibility of complex decomposition behavior of NMU accounting for the biphasic behavior, levels of 7-MeG were also monitored to determine if 7-MeG was also produced in a biphasic manner. In contrast to O^6-MeG, 7-MeG levels increased near-linearly with dose with no evidence of an apparent threshold and some indication of a decrease in the efficiency of formation of 7-MeG at higher doses (Figure 3). A detectable level of 7-MeG was also present in the DNA isolated from untreated bacteria. This was presumably attributable to normal cellular methylation processes. While it might be thought that this 7-MeG could result from a small concentration of tightly bound RNA that does not separate from the DNA and is not digested by RNAase, this possibility appears unlikely since treatment of an aliquot of DNA under depurinating conditions for RNA (0.5 M NaOH at 4°C for 16 hr) failed to release any

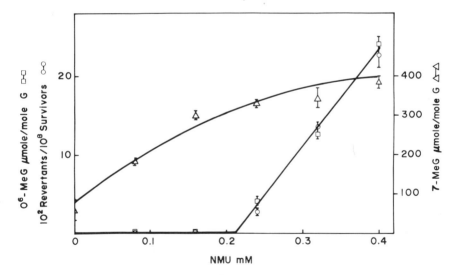

Figure 3. Formation of O^6-MeG, 7-MeG, and *his*+ revertants in *Salmonella typhimurium* TA 1535 treated with NMU. Experimental details are given in Section 2, Methods. Spontaneous revertants, $20/10^8$ survivors, have been subtracted from the revertant frequencies. The error bars for O^6-MeG, 7-MeG, represent the values obtained from two different injections of the same sample, and those for the mutagenesis represent the values obtained from duplicate plates of the same sample.

detectable guanine. Since 7-MeG is produced near-linearly with dose, whereas O^6-MeG and reversion frequency show a biphasic dependence on dose, it appears that there is a dependency of mutation on the formation of O^6-MeG.

5.3. Attempt to Observe Repair of O^6-MeG

The fact that low doses of NMU did not result in appreciable production of O^6-MeG or revertants may indicate that low concentrations of O^6-MeG in the *Salmonella* DNA are repairable. In an attempt to demonstrate repair of O^6-MeG in bacterial DNA, the bacteria were treated for 5 min with a high dose (0.4 mM) of NMU, rapidly pelleted, resuspended, and incubated in buffer. Samples were taken at selected time intervals up to 1 hr (the same incubation time used in the previous experiments) and analyzed for O^6-MeG content in DNA and revertant production. As seen in Table IX, revertant frequencies and O^6-MeG levels remained constant during a 1-hr incubation, suggesting that removal of O^6-MeG occurred during the treatment with NMU and/or during the time necessary to resuspend and lyse the bacteria.

5.4. Dependence of Mutagenesis on O^6-MeG

The results of these experiments indicate that in *Salmonella typhimurium* TA 1535, mutagenesis induced by N-methyl-N-nitroso compounds is dependent on the formation of O^6-MeG and that the bacteria can tolerate low doses of NMU before the onset of mutagenesis and the accumulation of O^6-MeG. As similar mutagenic behavior has been observed in TA 1535 and several other tester strains treated with N-methyl and N-ethyl-N-nitroso compounds, these tester strains may all be somewhat resistant to O^6-alkylation. A possible explanation for this behavior would be the existence of a saturable O^6-MeG repair activity. If the failure to observe a decrease in O^6-MeG levels in bacterial DNA, following treatment with NMU and subsequent incubation for 0–60 min in its absence, was due to rapid removal of O^6-MeG during the NMU treatment, this would suggest that the O^6-MeG repair activity in TA 1535 is constitutive since repair must have occurred during the 5-min treatment in buffer and/or the time necessary to lyse the bacteria. In contrast, removal of O^6-MeG from certain strains of adapted or unadapted *E. coli* is observable over the same time period, but the cells must be incubated in a nutrient-rich medium, indicating that removal of O^6-MeG requires synthesis of a new protein. The alternative explanations for the failure to observe repair—namely, that no repair occurs at all, or that repair is very slow—would fail to account for the biphasic production of O^6-MeG.

From the data presented in Figure 2 it can be seen that the cells were able to withstand exposure of up to 0.2 mM NMU before they began to accumulate O^6-MeG. If this threshold is due to repair then it corresponds to the removal of 26 μmol of O^6-MeG per mol guanine (obtained from the y-intercept of Figure 2).

Table IX
Effect of Posttreatment Incubation on Mutagenesis and O^6-MeG Levels in TA 1535 Pretreated with NMU[a]

Posttreatment incubation time	rev/10^8 survivors	O^6-MeG/G $\times 10^{-6}$
0	572	1.5
5	676	1.4
10	494	1.6
20	754	1.6
60	494	1.7

[a] Cells were treated with NMU (0.4 mM) for 5 min, pelleted, resuspended, and incubated in pH 7.4 PO_4 buffer. At each time point an aliquot was taken for assay of his+ revertents and a portion of this aliquot was diluted for subsequent determination of surviving cells. The remaining cells were brought to 0.02 M in sodium azide, lysed, and analyzed for O^6-MeG in DNA as described in Section 2, Methods. The average deviation between duplicate mutagenesis assays was ±15%. The average deviation between O^6-MeG/G ratios was ±20%.

This value is 4- to 20-fold less than reported by Warren and Lawley for *E. coli* challenged with high doses of NMU and subsequently incubated for 20–45 min in growth medium (when the data from both experiments are expressed in the same units), but is greater than the amount removed by *E. coli* resuspended in growth medium containing chloramphenicol or resuspended in salts.[45] Apparently constitutive levels of O^6-MeG repair activity vary considerably from strain to strain, as do maximum levels. There may be some relevance of these observations to the varying levels of O^6-MeG repair activities observed in different organs and different animal species.[3,12,15,16,46]

The biphasic nature of NMU-induced mutagenesis is useful as a probe for determining premutagenic lesions in that it allows correlations of mutagenesis with specific DNA alkylation products (such as O^6-MeG), which are also produced in a biphasic manner. In contrast, 7-MeG is produced in a near-linear relationship to dose and probably most other DNA-alkylation products exhibit similar behavior. Further studies involving dose–response curves for the generation of other DNA alkylation products would be important in establishing O^6-MeG as the major premutagenic lesion produced by NMU.

6. ALTERNATE MECHANISMS OF MUTAGENESIS BY N-NITROSO COMPOUNDS

Although mutagenesis in the *hisG46* (base-pair substitution) strains by lower homolog N-nitroso compounds is biphasic in the lower dose regions, there is a report in the literature that DEN is mutagenic in TA 98 (a predominantly frameshift—sensitive strain containing the *hisD3052* mutation) and that the dose–response in this strain is linear.[41] As DEN must be metabolically activated before it becomes mutagenic, dose–response curves with this mutagen are

Table X
Mutagenesis Induced by N-Nitroso-N-Ethylurea in
Salmonella typhimurium TA 98 and TA 100[a]

Strain	N-nitroso-N-ethylurea (mM)	10^2 rev/10^8 cfu
TA 100	0.5	1.0
TA 100	1.0	2.5
TA 100	2.0	69
TA 100	3.0	140
TA 98	1.0	0.50
TA 98	2.0	0.81
TA 98	3.0	1.2
TA 98	4.0	1.7

[a] Bacteria, 20 μl for TA 100 and 100 μl for TA 98, were transferred from overnight cultures to 0.1 M, pH 7.0, PO$_4$ buffer (0.5 ml final volume), and 20–50 μl of N-nitroso-N-ethylurea stock solution in 0.01 M, pH 5.0 acetate buffer was added to initiate mutagenesis. The mixture was incubated for 30 min at 37°C and assayed for revertants and viable cells. Spontaneous revertants were subtracted from revertant numbers before calculation of revertant frequencies. Results from control experiments showed that the number of spontaneous revertants was nearly independent of the number of cells plated, so that the same spontaneous revertant number was subtracted from each individual revertant number for a particular strain. Toxicity ranged from 5% to 62% for TA 100 and from 0 to 49% for TA 98. Precision between duplicate assays was ±15%.

difficult to interpret for reasons discussed above relating to the multiple values of K_m for nitrosamine-metabolizing proteins. However, this difficulty could be obviated by the use of a direct-acting compound such as N-nitroso-N-ethylurea. As shown in Table X, mutagenesis by this compound was approximately linear with dose. This is in contrast to mutagenesis by NEU in TA 100 or by ENNG in strains TA 100, TA 1535, and TA 1975. If the biphasic dose–response for mutagenesis by NEU in TA 100 corresponds to saturation of repair of O[6]-ethylguanine, and TA 98 contains a repair capacity similar to TA 100, then these results would indicate that NEU can also produce a premutagenic lesion, different from O[6]-ethylguanine. Although TA 100 was much more sensitive than TA 98 to high doses of NEU, this difference in sensitivity was considerably smaller at lower doses and might even be reversed at very low doses if such low mutagenic activities could be accurately measured.

7. MUTAGENESIS BY A SERIES OF "WEAKLY MUTAGENIC" N-NITROSAMINES: EFFECTS OF STRUCTURE

7.1. Conditions

By making use of weakly acidic incubation mixtures and the principles described in the Introduction, it was possible to detect the mutagenic activities of a number of N-nitrosamines whose mutagenic activities were not previously detectable. Many of these were of interest since the carcinogenic activities of these compounds in rats had already been measured and could then be compared with the mutagenic activities of these compounds, in an attempt to discern qualitative or quantitative relationships. As most of these compounds were structurally related it was also of interest to determine the effects of small structural changes on mutagenesis.

For reasons previously indicated, the approach taken was to attempt to metabolize all of the compounds to approximately the same extent so that differences in mutagenic activities resulted from differences in mutagenic potencies of the metabolites and not differences in the concentration of mutagens metabolized. In order to accomplish this, relatively high concentrations of the N-nitrosamines were employed in an attempt to saturate the various forms of cytochrome P-450. The actual concentration used, 2 mg/ml (~15–30 mM), was a compromise since it did not completely saturate the high-K_m cytochrome P-450 for some of the compounds, but it was already somewhat toxic for some of the compounds. The actual toxicity values are given in Table XI. Two of the compounds, N-nitroso-N-methylaniline and nitroso-4-*t*-butylpiperidine, had to be tested at lower doses because they were extremely toxic at 2 mg/ml. Presumably these two compounds would have higher mutagenic potencies if they could be tested at high doses. The S-9 fraction was isolated from livers of Aroclor 1254–treated rats. Rat S-9 was used because carcinogenesis studies with N-nitrosamines had been carried out on rats and any attempted correlations between mutagenesis and carcinogenesis would then involve the same animal species. Induced animals were used because S-9 fractions from these animals showed a higher activating capacity, which was necessary if the mutagenic activity of the weaker compounds was to be detected. Although induced animals were not used in the carcinogenesis studies that were reported, Aroclor 1254 induces several forms of cytochrome P-450,[47] and it is likely that some of the forms present in induced liver are also present in the target organs of the rats used in the carcinogenesis studies.

7.2. Results: Di-N-alkylnitrosamines and Related N-Nitrosamines

As seen in Table XI, most of the compounds tested were mutagenic and some very potent. The N-nitroso-N-methylalkyl-derivatives were most potent and these compounds all showed the biphasic kinetics typical of N-nitroso-N-methyl compounds, indicating that mutagenesis induced by these compounds proceeded mainly via the methyl moiety. The higher molecular weight methylalkylnitrosamines were of equal or greater potency than DMN at 2 mg/ml, although at even higher doses the differences in potencies decreased (unpublished results and Table II). These findings suggest that these higher molecular weight, more hydrophobic compounds bound more tightly to the high-K_m

Table XI

Mutagenic Activities of Some N-Nitrosamines in *Salmonella typhimurium* TA 100[a]

N-Nitroso	Revertants 10^8 cfu	Carcinogenicity
Methyl-*n*-propylamine[b]	140,000	++++
Methylneopentylamine[c]	44,000	++++
Methylethylnitrosamine	40,000	+++
Dimethylamine[c]	28,000	++++
Diethylamine	25,000	++++
Piperidine	6,640	+++
Di-*n*-propylamine	5,250	+++
Morpholine	5,050	+++
Di-*n*-butylamine	1,203	++
Pyrrolidine	1,010	++
4-*t*-butyl-piperidine[d]	660	+
Bis-(2-methoxy)-diethylamine	460	+++
Methylaniline[e]	300	++
Diisopropylamine	80	+
Ethanolisopropanolamine	78	++
Diethanolamine	49	+
Bis-(2-cyano)-diethylamine	37	0
2,6-Dimethylpiperidine	12	0
Di-*sec*-butylamine	0	0

[a] All compounds were tested at a dose of 2 mg/ml of incubation mix containing 0.5 ml of S-9 fraction (from a 2:1 homogenate of Aroclor 1254-induced rat liver), 0.4 NADPH-generating system (containing a final measured concentration of NADPH of 6 mg/ml), 0.05 ml of exponentially growing bacteria ($10 \times$ concentrated), and 0.05 ml of mutagen (in H_2O or acetone). The final pH of the mix was 6.5, and 0.2–0.5 ml was diluted if necessary and plated. Mutagenesis induced by the methylalkyl compounds increased near-linearly with time until a revertant frequency of $2.5 \times 10^4/10^8$ cfu was reached and then it leveled off. A similar effect was seen with MNNG (Figure 2). Mutagenic activities for these compounds were calculated for the times indicated and then normalized to 40 min. For most of the other compounds mutagenesis increased nearly linearly with time. Toxicity was also determined and was <50%. In control experiments, reducing the number of cells on plates without mutagen by 50% did not affect the number of spontaneous revertants, so that revertant frequencies were calculated by subtracting the same number of spontaneous revertants from each revertant plate and dividing by the number of surviving cells. Carcinogenicity values are taken from references 49, 50, 51, and 52. The potencies increase with number of pluses from 0 through ++++.
[b] Determined at 5 min incubation time.
[c] Determined at 15 min incubation time.
[d] Determined at 1 mg/ml due to extensive toxicity at 2 mg/ml.
[e] Determined at 0.25 mg/ml due to extensive toxicity at higher doses.

cytochrome P-450 than did DMN, and were metabolized to yield methyldiazonium ion at a greater rate than DMN. This effect may be quite significant since each oxidation of DMN should give rise to a methyldiazonium ion (Figure 1), which, along with ethyldiazonium ion, appears, to be more effective than the higher molecular weight alkyldiazonium ions at producing mutations (Figure 2). With the unsymmetrical compounds, only oxidation on the higher molecular weight side of the molecule could directly give rise to a methyldiazonium ion (assuming that oxidation on the methyl side of the N-nitrosamine does not generate a methylating agent in high efficiency, via scission of the resultant alkyldizonium ion or alkonium ion). Thus, despite the loss of potential mutagen-producing oxidations to relatively unproductive metabolites, the unsymmetrical nitrosamines are still more potent or as potent as DMN. The above hypothesis should be testable now that convenient methods are available for the determination metabolites of N-nitrosamines.[48]

Mutagenesis by N-nitroso-dimethyl, methylpropyl, and methylneopentyl amines all leveled off at a revertant frequency of about $2.5 \times 10^4/10^8$ survivors. A similar effect was seen in mutagenesis by the methylating compound MNNG (Figure 2) and appears to be characteristic of N-methyl-N-nitroso compounds. Only N-nitrosomethylethylamine did not show this effect and generated revertants at least up to a level of $6.9 \times 10^4/10^8$ survivors. As revertant frequencies attained with the ethylating nitroso compound, ENNG, reached higher values than those obtained with MNNG (Figure 2), it appears that the ethyl side of N-nitrosomethylethylamine also contributes significantly to its mutagenic activity under conditions of high metabolizing activity. ENNG and MNNG were of similar mutagenic potencies after their apparent thresholds were exceeded (Table VIII), and N-nitrosodiethylamine was of similar potency to DMN. This observation may indicate that metabolism of N-nitrosodiethylamine at the β-position is relatively unimportant. Metabolism at the β-position should not directly give rise to mutagenic metabolites and, therefore, the portion of the MFO-catalyzed oxidation of this compound that occurs at this position should be nonproductive. On a statistical basis this portion would be 60% since there are six β-hydrogens and four α-hydrogens.

7.3. β-Substituted N-Nitrosamines

Substitution at the β-position of N-nitrosodiethylamine led to a marked decrease in its mutagenic activity. Methoxy substitution could result in a decrease in the percentage of α-oxidation, leading to nonproductive metabolism because of the increased number of hydrogens or enhanced reactivity adjacent to the methoxyl oxygen. It might also result in the formation of a diazonium ion, which, because of its increased size (perhaps analogous to the large difference in mutagenic potencies between ENNG and PNNG) or altered electronic distribution, reacts at DNA-base sites that do not lead to mutations. In addition, other possibilities such as more efficient repair of bulky substituents cannot be ruled out.

Substitution of cyano groups at the β-positions of N-nitrosodiethylamine reduced mutagenic activity to almost zero. Since the size of the cyano group is less than that of a methoxyl group, factors other than size are probably involved in the reduced mutagenic activity of N-nitroso bis-(2-cyano)-diethylamine. The increased polarity of this compound may decrease its binding to cytochrome P-450 or it might result in a higher percentage of nonproductive metabolism, either at the β-hydrogens or the cyano group. In addition, the diazonium ion or alkonium ion that would be formed from this compound should be extremely unstable due to presence of two adjacent electropositive carbons. This ion might be too short-lived to penetrate the bacteria or might not be formed in appreciable yield due to its instability. Other factors such as site specificity and DNA repair could also be important.

7.4. α-Substituted N-Nitrosamines

N-nitrosodiisopropylamine was weakly mutagenic and much less mutagenic than its di-*n*-propyl analog. The branching in this compound may result in reduced efficiency of its activation, perhaps because of steric or statistical factors, or it may result in reduced efficiency of reaction of the resultant diazonium ion with critical sites on DNA bases. With respect to this latter possibility, secondary carbonium ions are more stable than primary carbonium ions and could show greater site selectivity. As a consequence of this behavior sites with less nucleophilic character, such as the O^6 position on guanine, might not be efficiently alkylated. A similar effect was observed when the mutagenic activity of N-nitroso-di-*sec*-butylamine was compared with that of its *n*-butyl analog. Also similar is the lack of significant mutagenic activity of nitroso-2,6-dimethylpiperidine, which is methyl-substituted at both α-positions, whereas the unsubstituted analog, N-nitrosopiperidine, is strongly mutagenic. On the other hand, the 4-*t*-butyl derivative of N-nitrosopiperidine was of intermediate activity, indicating that a large substituent at the γ-carbon does not completely prevent mutagenic activity. The weak activity of N-nitrosoethanolisopropanolamine and N-nitrosodiethanolamine apparently somehow reflects the polarity of the hydroxyl groups. N-nitrosomethylaniline was very toxic at 2 mg/ml and was therefore tested at 0.25 mg/ml. At this concentration it caused an increase over the background revertant rate but it was still somewhat toxic. Since activation via the pathway given in Figure 1 can only occur by demethylation, the intermediate monosubstituted nitrosamine would be an arylating agent and might be expected to exhibit different characteristics from the other compounds. The high toxicity of this compound could be indicative of arylation. On the other hand, many aromatic amines are mutagenic[20,21] and the mutagenesis of N-nitrosomethylaniline may reflect the aromatic amine pathway(s).

8. CORRELATION BETWEEN MUTAGENIC POTENCIES AND CARCINOGENIC ACTIVITIES OF SOME N-NITROSO COMPOUNDS

When the relative mutagenic potencies of the compounds listed in Table XI were compared with their reported carcinogenic activities, a strong positive correlation emerged, with all carcinogens exhibiting mutagenic activity and all noncarcinogens showing little or no mutagenic activity. Where semiquantitative evaluations of carcinogenic potencies of these compounds were available,[49,51,52] these correlated well with their mutagenic potencies. In general, since there was a correlation between mutagenesis and carcinogenesis, the same factors that affected carcinogenesis affected mutagenesis, e.g., α or β substitution, branching, chain length, etc.

As there are many reasons not to expect even semiqualitative correlations between carcinogenesis and mutagenesis, it is somewhat surprising that such a reasonable correlation emerged. Most of the compounds tested were chosen because they were carcinogens that had failed to elicit a mutagenic response in previous studies, and the choice was perhaps somewhat biased. Some positive and negative controls were also included for structural comparisons. It will be interesting to extend these studies to larger numbers of compounds to determine if this correlation continues to hold and to attempt to establish a pattern and a mechanistic basis for the effects of structural changes on mutagenesis and carcinogenesis.

ACKNOWLEDGMENTS. Many of the experiments described here were carried out by S. Lee, S. Milstein, and M. Miranda. The work described here was supported by Grant No. CA 19023 from the NCI.

REFERENCES

1. P. N. Magee, and J. M. Barnes, Carcinogenic nitroso compounds, *Adv. Cancer Res. 10*, 163–256 (1967).
2. H. Druckrey, R. Preussmann, S. Ivankovic, and D. Schmähl, Organtrope carcinogene wirkungen bei 65 verschiedene N-nitroso-verbindungen an BD-Ratten, *Z. Krebsforsch. 64*, 103–201 (1967).
3. P. N. Magee, R. Montesano, and R. Preussmann, N-Nitroso compounds and related carcinogens, in: *Chemical Carcinogens* (C. Searle, ed.), ACS Monograph 173, pp. 491–625, American Chemical Society, Washington, D.C. (1976).
4. E. J. Olajos and F. Coulston, Comparative toxicology of N-nitroso compounds and their carcinogenic potential to man, *Ecotoxic. Environ. Safety 2*, 317–367 (1978).
5. L. M. Anderson, L. J. Priest, and J. M. Budinger, Lung tumorigenesis in mice after chronic exposure in early life to a low dose of dimethylnitrosamine, *J. Natl. Cancer Inst. 62*, 1553–1555 (1979).
6. P. D. Lawley, Carcinogenesis by alkylating agents, in: *Chemical Carcinogens* (C. E. Searle, ed.), ACS Monograph 173, pp. 83–244, American Chemical Society, Washington, D. C. (1976).
7. B. Singer, N-Nitroso alkylating agents; formation and persistence of alkyl derivatives in mammalian nucleic acids as contributing factors in carcinogenesis, *J. Natl. Cancer Inst. 62*, 1329–1339 (1979).
8. J. H. Weisburger, Bioassays and tests for chemical carcinogens, in: *Chemical Carcinogens* (C. E. Searle, ed.), ACS Monograph 173, pp. 1–23, American Chemical Society, Washington, D. C. (1976).
9. E. C. Miller and J. A. Miller, The metabolism of chemical carcinogens to reactive electrophiles

and their possible mechanisms of action in carcinogenesis, in: *Chemical Carcinogens* (C. E. Searle, ed.), ACS Monograph 173, pp. 736–762, American Chemical Society, Washington, D. C. (1976).
10. L. L. Gerchman and D. B. Ludlam, The properties of O^6-methylguanine in templates for RNA polymerase, *Biochim. Biophys. Acta 308*, 310–316 (1973).
11. P. J. Abbott and R. Saffhill, DNA synthesis with methylated poly (dC-dG) templates: Evidence for a competitive nature to miscoding by O^6methylguanine, *Biochim. Biophys. Acta. 562*, 51–61 (1979).
12. J. W. Nicoll, P. F. Swann, and A. E. Pegg, Effects of dimethylnitrosamine on persistence of methylated guanines in rat liver and kidney DNA, *Nature (London) 254*, 261–262 (1975).
13. R. Montesano, H. Bresil, P. M. Ghyslaine, G. P. Margison, and A. E. Pegg, Effect of chronic treatment of rats with dimethylnitrosamine on the removal of O^6-methylguanine from DNA, *Cancer Res. 40*, 452–458 (1980).
14. J. Buechler and P. Kleihues, Excision of O^6-methylguanine from DNA of various mouse tissues following a single injection of N-methyl-N-nitrosourea, *Chem.-Biol. Interactions 16*, 325–333 (1977).
15. P. Kleihues, S. Bamborschke, and G. Doerjer, Persistence of alkylated DNA bases in Mongolian gerbil (Meriones unguiculatus) following a single dose of methylnitrosourea, *Carcinogenesis 1*, 111–113 (1980).
16. B. L. Van Duuren, Tumor-promoting and co-carcinogenic agents, in: *Chemical Carcinogens* (C. E. Searle, ed.), ACS Monograph 173, pp. 24–51, American Chemical Society, Washington, D. C. (1976).
17. H. C. Pitot, Biological and enzymatic events in chemical carcinogenesis, *Ann. Rev. Med. 30*, 25–29 (1979).
18. D. Jenssen, B. Beije, and C. Ramel, Mutagenicity testing on Chinese Hamster V79 cells treated in the *in vitro* liver perfusion system. Comparative investigation of different *in vitro* metabolizing systems with dimethylnitrosamine and benzo(a)pyrene, *Chem.-Biol. Interactions 27*, 27–39 (1979).
19. K. S. Bakshi and D. J. Brusick, Bioactivation of DMN in intrasanguinous host-mediated assay in male and female mice, *Environ. Mutagenesis 1*, 150 (1979).
20. J. McCann, E. Choi, E. Yamasaki, and B. N. Ames, Detection of carcinogens as mutagens in the *Salmonella*/microsome test: Assay of 300 chemicals, *Proc. Natl. Acad. Sci. USA 72*, 5135–5139 (1975).
21. B. N. Ames, J. McCann, and E. Yamasaki, Methods for detecting carcinogens and mutagens with the *Salmonella*/mammalian microsome mutagenicity test, *Mutat. Res. 31*, 347–364 (1975).
22. J. McCann and B. N. Ames, Detection of carcinogens as mutagens in the *Salmonella*/microsome test: Assay of 300 chemicals: Discussion, *Proc. Natl. Acad. Sci. USA 73*, 950–954 (1976).
23. E. C. Miller and J. A. Miller, The mutagenicity of chemical carcinogens, correlations, problems and interpretations, in: *Chemical Mutagens, Principles and Methods for Their Detection* (A. Hollaender, ed.), Vol. 1, pp. 83–119, Plenum, New York (1971).
24. H. V. Malling, Dimethylnitrosamine: Formation of mutagenic compounds by interaction with mouse liver microsomes, *Mutat. Res. 13*, 425–429 (1971).
25. P. Czygan, H. Greim, A. J. Garro, F. Hutterer, F. Schaffner, H. Popper, O. Rosenthal, and D. Y. Cooper, Microsomal metabolism of dimethylnitrosamine and the cytochrome P-450 dependency of its activation to a mutagen, *Cancer Res 33*, 2983–2986 (1973).
26. J. C. Arcos, D. L. Davies, C. E. L. Brown, and M. F. Argus, Repressible and inducible enzymic forms of dimethylnitrosamine-demethylase, *Z. Krebsforsch 86*, 171–183 (1976).
27. I. G. Sipes, M. L. Slocumb, and G. Holtzman, Stimulation of microsomal dimethylnitrosamine-N-demethylase by pretreatment of mice with acetone, *Chem. Biol. Interactions 21*, 155–166 (1978).

28. A. Wood, W. Levin, A. Y. Lu, H. Yagi, O. Hernandez, D. M. Jerina, and A. H. Conney, Metabolism of benzo(a)pyrene and benzo(a)pyrene derivatives to mutagenic products by highly purified hepatic microsomal enzymes, J. Biol. Chem. 251, 4882–4890 (1976).
29. S. Milstein, and J. B. Guttenplan, unpublished results.
30. J. B. Guttenplan, Comutagenic effects exerted by N-nitroso compounds, Mutat. Res. 66, 25–32 (1979).
31. J. B. Guttenplan, Detection of trace amounts of dimethylnitrosamine with a modified Salmonella/microsome mutagenicity assay, Mutat. Res. 64, 91–94 (1979).
32. J. B. Guttenplan, Enhanced mutagenic activities of N-nitroso compounds in weakly acidic media, Carcinogenesis 1, 439–444 (1980).
33. T. Yahagi, M. Nagao, Y. Seino, T. Matsushima, T. Sugimura, and M. Okada, Mutagenicities of N-nitrosamines on Salmonella, Mutat. Res. 48, 121–130 (1977).
34. S. Milstein and J. B. Guttenplan, Near quantitative production of molecular nitrogen from metabolism of dimethylnitrosamine, Biochem. Biophys. Res. Commun. 87, 337–342 (1979).
35. J. L. Frei, D. H. Swenson, W. Warren, and P. D. Lawley, Alkylation of deoxyribonucleic acid in vivo in various organs of C57BL mice by the carcinogens N-methyl-N-nitrosourea, N-ethyl-N-nitrosourea and ethylmethanesulfonate in relation to induction of thymic lymphoma, Biochem. J. 174, 1031–1044 (1978).
36. E. R. Garrett, S. Goto, and J. B. Stubbins, Kinetics of solvolyses of various N-alkyl-N-nitrosoureas in neutral and alkaline solutions, J. Pharm. Sci. 54, 119–123 (1965).
37. M. Mochizuki, T. Anjo, T. Sone, and M. Okada, Isolation and characterization of alpha-hydroxydialkylnitrosamines, in: Proceedings of the 38th Annual Meeting of the Japanese Cancer Association, p. 321, Japanese Cancer Association, Tokyo (1979).
38. T. Wang, T. Kakizoe, P. Oion, T. Furrer, A. J. Varghese and W. R. Bruce, Volatile nitrosamines in normal human faeces, Nature 276, 280–281 (1978).
39. J. B. Guttenplan, R. Bonomo, and J. Fliessbach, Recovery of benzo(a)pyrene (BP) metabolites and dimethylnitrosamine in rat saliva, Proc. Am. Assoc. Cancer Res. 21, 124 (1980).
40. J. Fliessbach and J. B. Guttenplan, unpublished results.
41. M. J. Prival, V. D. King, and A. T. Sheldon, Jr., The mutagenicity of dialkylnitrosamines in the Salmonella plate assay, Environ. Mutagenesis 1, 95–104 (1979).
42. H. Bartsch, A. Camus, and C. Malaveille, Comparative mutagenicity of N-nitrosamines in a semi-solid and in a liquid incubation system in the presence of rat or human tissue fractions, Mutat. Res. 37, 149–162 (1976).
43. J. Unger and J. B. Guttenplan, Kinetics of benzo(a)pyrene induced mutagenesis in a highly sensitive Salmonella/microsome assay, Mutat. Res. 77, 221–228 (1980).
44. J. McCann, N. E. Spingarn, J. Kobari, and B. Ames, The detection of carcinogens as mutagens: Bacterial tester strains with R factor plasmids, Proc. Natl. Acad. Sci. (USA) 271, 5–13 (1975).
45. W. Warren and P. D. Lawley, The removal of alkylation products from DNA of Escherichia coli cells treated with the carcinogens N-ethyl-N-nitrosourea and N-methyl-N-nitrosourea: Influence of growth conditions and DNA repair defects, Carcinogenesis 1, 67–78 (1980).
46. G. P. Margison, J. A. Swindwell, L. H. Ockey, and A. W. Craig, The effects of a single dose of dimethylnitrosamine in the Chinese hamster and the persistence of DNA alkylation products in selected tissues, Carcinogenesis 1, 91–96 (1980).
47. D. E. Ryan, P. E. Thomas, D. Kornzeniowski, and W. Levin, Separation and characterization of highly purified forms of liver microsomal cytochrome P-450 from rats treated with polychlorinated biphenyls, phenobarbitol and 3-methylcholanthrene, J. Biol. Chem. 254, 1365–1374 (1979).
48. J. B. Farrelly, A new assay for the microsomal metabolism of nitrosamines, Cancer Res. 40, 3241–3244 (1980).
49. T. K. Rao, A. A. Hardigree, J. A. Young, W. Lijinsky, and J. L. Epler, Mutagenicity of

N-nitrosopiperidines with *Salmonella typhimurium*/microsomal activation system, *Mutat. Res.* 56, 131–145 (1977).
50. W. Lijinsky, Structure–activity relations in carcinogenesis by N-nitroso compounds, Chapter 10, this book.
51. T. K. Rao, J. A. Young, W. Lijinsky, and J. L. Epler, Mutagenicity of aliphatic nitrosamines in *Salmonella typhimurium*, *Mutat. Res.* 66, 1–7 (1979).
52. W. Lijinsky, *et al.*, personal communication.

5

Induction of Bacteriophage Lambda by N-Nitroso Compounds

ROSALIE K. ELESPURU

1. MECHANISM OF INDUCTION

Induction of bacteriophage lambda is one of the consequences of the "SOS response" in *E. coli* resulting from damage to the host DNA.[1,2] This complex process appears to be mediated by the *recA* protein,[3,4] resulting in the cleavage of the lambda repressor[5] and ensuing expression of phage genes. The repressor-cleavage activity of the *recA* protein is dependent on the presence of single-stranded DNA,[6] presumably generated as a result of DNA damage. DNA damage, then, leads to the appearance of an "activated" *recA* protein,[7] which cleaves the repressor of the genes of the lytic cycle (see Reference 8 for review). Agents that damage DNA can thus be monitored by looking for the appearance of bacteriophage. Other SOS functions that have been monitored after DNA damage include mutagenesis, filamentation, and DNA repair activity.[2,9]

While mutagenesis at a specific locus occurs in a very small proportion of

---

Abbreviations for N-nitroso compounds used in this chapter: MNNG, N-methyl-N'-nitro-N-nitrosoguanidine; ENNG, N-ethyl-N'-nitro-N-nitrosoguanidine; DEN, diethylnitrosamine; DMN, dimethylnitrosamine; MEN, methylethylnitrosamine; MNU, methylnitrosourea; NPip, nitrosopiperidine; Nitroso-*product*, N-nitroso derivative of *product* indicated; NPy, nitrosopyrrolidine; A, nitroso-*Aldicarb*; M, nitroso-*Methomyl*; C, nitroso-*Carbaryl*; CF, nitroso-*Carbofuran*; L, nitroso-*Landrin*; B, nitroso-*Baygon*; Bux, nitroso-*Buxten*; MMS, methyl methanesulfonate; EMS, ethyl methanesulfonate.

ROSALIE K. ELESPURU • Biological Carcinogenesis Program, NCI-Frederick Cancer Research Facility, Frederick, Maryland 21701. *Present address*: Fermentation Program, NCI-Frederick Cancer Research Facility, Frederick, Maryland 21701. By acceptance of this article, the publisher or recipient acknowledges the right of the U.S. Government to retain a nonexclusive, royalty-free license in and to any copyright covering the article.

bacteria, induction of bacteriophage occurs in the majority of bacteria at optimum doses. It is therefore possible to measure specific enzymes synthesized as part of the induction process, as an alternative to measurement of viable phage particles. This has been accomplished in several assays (generally impractical for screening), in which a natural product of induction[10] or a product resulting from genetic design [11,12] is monitored.

2. GENETICS AND METHODOLOGY

Several phage induction assays have been developed for carcinogen screening. The Inductest of Raymond Devoret[13] involves the use of a strain of *E. coli* with mutations affecting DNA repair and permeability analogous to those of the Ames *Salmonella typhimurium* strains. The lambda lysogen used by Ho and Ho[10] expresses a mutant repressor that is sensitive to lower inducing concentrations than the wild-type. Miroslav Radman[14] has constructed a strain of *E. coli* allowing phage induction to be measured by the appearance of colored colonies on agar. Although all of these assays appear promising, few reports concerning their use in testing chemicals have appeared in the literature. In our experience, documented in the next section, bacteriophage induction assays cannot be used as previously described to detect, with reasonable sensitivity, many classes of carcinogen, including N-nitroso compounds.

We[15] have developed a biochemical assay of prophage induction, designed for screening, in which the product of the *lacZ* gene, β-galactosidase, is produced as an immediate consequence of cleavage of the lambda repressor after DNA damage. β-Galactosidase production can be monitored in a spot test on agar or in a quantitative test tube assay. Such a "biochemical" assay of phage induction does not require full phage development (or colony formation) and therefore may be less subject than other systems to the toxic effects of chemicals and solvents. The assay is complete in a few hours and is not affected by contaminating microbes or samples in rich media, in contrast to most mutagenesis assays. The genetics of the *E. coli* strain, including mutations affecting permeability, DNA repair capacity, and duration of induction, have been described.[15] A detailed methodology for performance of the assay has also been published.[16] As part of our validation of the assay, we have tested more than 100 N-nitroso compounds, the report of which is presented here.

3. FACTORS AFFECTING INDUCTION BY NITROSO COMPOUNDS

In the course of our experiments, we discovered a number of factors that were critically important in the detection of nitroso compounds, particularly those that require metabolic activation. Some of these factors have been noted

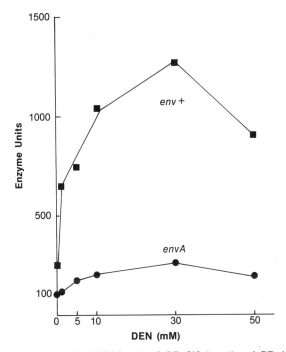

Figure 1. Prophage induction by DEN in *E. coli* BR 513 (*envA*) and BR 469 (wild-type cell envelope). Bacteria were incubated with aeration for 5 hr in the presence of chemical and Aroclor-induced male Syrian golden hamster liver S-9, prepared by R. Pennington. Activation mix (10% S-9 by volume) contained 3.5 mg/ml protein, 3 mM NADP, 10 mM glucose-6-phosphate, and 10 mM $MgCl_2$ in a total volume of 0.5 ml. Samples were then assayed for β-galactosidase by addition of buffer and substrate to the tubes.[16]

before by others and some are reported here for the first time. It is likely that other important factors, both genetic and biochemical, governing the response of bacterial systems to nitroso compounds still remain to be discovered.

3.1. Bacterial Strain

We observed that the standard strain constructed for our assay was not very sensitive to the commonly used nitroso compound and DNA-damaging agent, MNNG. In addition, the magnitude of induction seen was not as great as expected.[15] When we tested DEN, a compound requiring metabolic activation, very little induction was observed (Figure 1). Several different mutations were found to affect the system profoundly. As in the case of *S. typhimurium*,[17] a strain with a normal bacterial envelope often gave better results than a strain with a permeability mutation, *envA* in our case, as shown in Figure 1. The refractory character of the *envA* strain is more pronounced in the tube assay

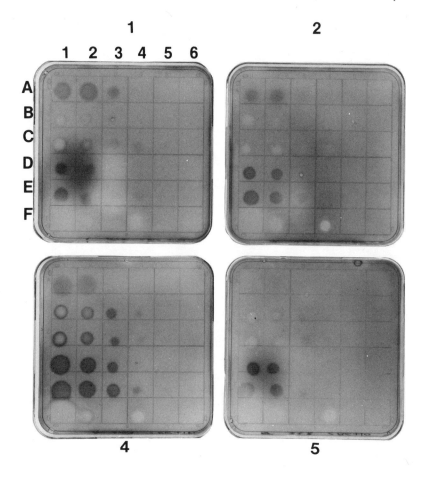

(Figure 1) than the agar plate assay (Figure 2) and may be the result of solvent effects on a defective cell membrane. The difference in behavior between *envA* and *env*$^+$ strains is greater with compounds such as DEN and DMN, both solvents, than with solid N-nitroso compounds. Except in the case of some of the long-chain derivatives, the *envA* mutation does not appear to affect the permeability of the bacteria to N-nitroso compounds, since lower concentrations of chemical are not detected in the *envA* strain (Figure 2).

A surprising enhancement in induction was seen in a strain, BR 339 (provided by Michael Yarmolinsky), tentatively identified as containing a *lexA*

3

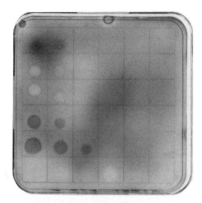

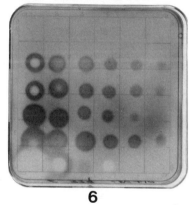

6

Figure 2. Spot test for prophage induction using mutant lambda lysogens of *E. coli*. [15] Plate 1: BR 513 (*envA*, $\Delta uvrB$); Plate 2: BR 469 ($\Delta uvrB$); Plate 3: BR 475 (wild-type); Plate 4: BR 293 ($\lambda P_R + S_7 \Delta uvrB$); Plate 5: BR 338 *(mal, $\Delta uvrB$)*; Plate 6: BR339 *(mal, lexA3, $\Delta uvrB$)*. Bacteria were poured in agar on LBE plates containing 10 µg/ml ampicillin followed by spotting of chemical solutions. After incubation at 38 °C for 4 hr, plates were overlayed with agar containing substrate BNG plus Fast Blue RR salt. Colored spots indicating the presence of β-galactosidase appeared within 10 min. See Reference 16 for detailed methods. For all chemicals, solution concentrations (and quantities spotted in 5 µl) are given. Row A, lanes 1-3: Bleomycin at 30, 12, 5 µg/ml (0.15, 0.06, .003 µg); lane 4: water 5µgl; lane 5: DMSO 5µl; lane 6: no addition. Rows B-E, lanes 1-6: Chemicals at 1.2, 0.63, 0.31, 0.15, 0.075, 0.038 mg/ml(60, 30, 15, 7.5, 3.8, and 1.9 µg). Row B: MNNG in water. Row C: MNNG in DMSO. Row D: ENNG in water. Row E: ENNG in DMSO. Row F, lanes 1-3: MMS in DMSO at 100%, 10%, 1% by volume; lanes 4-6: EMS in DMSO at 100%, 10%, 1% by volume.

mutation (Figures 2 and 3). A two- to five-fold enhancement was evident in both spot tests and quantitative assays for both nitrosamines and nitrosamides. A similar enhancement has not been observed for other classes of carcinogen tested in the assay and appears to be limited to alkylating agents.

3.2. Source and Amount of Activating Enzymes

We have found, as have others,[18,19] that the quantity and type of S-9 activation mix used in the assay is a critical factor in detection of nitroso compounds requiring metabolic activation. We were able to detect only a few

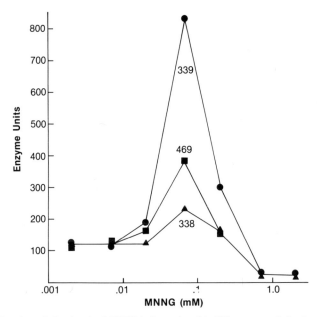

Figure 3. Prophage induction by MNNG in bacteria with different genetic backgrounds. Strain genotypes are described in Figure 2. Log phase bacteria were incubated with chemical for 5 hr, followed by the assay of enzyme activity (see Figure 1).

compounds using rat liver S-9 fraction, but were much more successful with Syrian golden hamster liver S-9 (Figure 4). In addition, the optimum detection of most nitroso compounds was dependent on the presence of five to ten times as much S-9 as is used in a standard *Salmonella* mutagenesis assay. Induction with DEN as a function of protein concentration of hamster liver S-9 fraction is shown in Figure 5 (quantitative assay). The same effect is seen in the spot test with ten different compounds (Figure 4). Very little induction is seen with nitrosamines unless the protein concentration exceeds 10 mg/plate.

Because of the substantial differences seen between results with rat and hamster liver S-9 fractions, a few experiments were performed with combinations of rat and hamster preparations. In these experiments we hoped to learn whether the differences were due to the presence of extra factors in hamster or inhibitory factors in rat preparations. Employing different quantities of S-9, we explored the influence of total protein and different absolute amounts of each S-9 preparation. From the results of Figure 4, a spot test with 10 nitrosamines, it appears that the hamster preparation contains important factors missing from rat preparations, since the addition of rat liver S-9 did not diminish the effect of hamster-mediated activation. An additive effect between hamster and rat S-9 was not seen, even though the total protein concentration was high enough to be in the optimal range (for induction by hamster S-9 alone). The same result was seen in the liquid suspension assay for DMN, DEN (Figure 6), and NPip. These observations are different from those of Prival and Mitchell with *Salmonella* (Reference 20 and Chapter 8 of this volume) employing DMN, in which rat liver microsomes appeared to inhibit activation mediated by hamster liver S-9. This dichotomy suggests the possibility that the metabolic intermediates responsible for mutagenesis and phage induction are not the same. Alternatively, a toxic factor detected in the colony assay may not affect phage induction.

3.3. Media

Although the media variable has not been extensively studied, some media were found to be better than others in facilitating induction by nitrosamines. The rich medium LBE was preferable to the less rich medium TB used in standard plaque assays, and the presence of ampicillin appeared to facilitate induction in some but not all cases. This drug has a number of different effects on the assay that have not been adequately characterized. We have evidence that ampicillin influences (positively) uptake of substrate, *lexA3*-mediated enhanced induction, and *envA* inducing capacity, while eventually causing cell lysis that terminates synthesis of β-galactosidase (Figure 7). In addition, ampicillin has a small inducing effect of its own. Greater activity of nitrosamines is seen in a soluble system than in semisolid agar, as has been observed for the *Salmonella* mutagenesis assay.[21,22] The difference may be the result of improved aeration (with shaking) in the liquid incubation system, which should facilitate both enzymatic activation and bacterial metabolism.

3.4. Kinetic Parameter

Because enzyme synthesis continues for at least 5 or 6 hr after induction, incubation time affects the amount of enzyme detected after induction by any DNA-damaging agent. However, some chemicals are affected more than others, owing to different lag periods as well as different rates of enzyme synthesis. These factors may be monitored in a kinetic assay of induction, as shown in Figure 7. Effects not apparent in assays with end points that are days, weeks, or years after chemical treatment may be noticeable in an assay, such as this one, monitored only a few hours after administration of chemical. For example, induction after MNNG treatment was associated with a longer lag period than other potent DNA-damaging agents, an effect that partly explains the difficulty in detecting this compound after a 2- or 3-hr incubation. (An irreducible lag period of one cell generation time, approximately 1 hr, already exists for the events of the SOS response; see Figure 7). With the exception of DEN, the

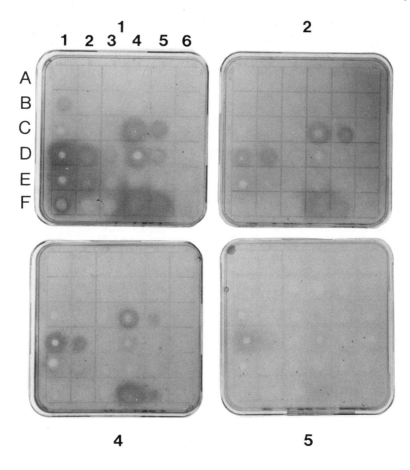

nitroso compound giving the greatest magnitude of induction, N-nitroso compounds in general required an incubation time of 4 or 5 hr for reasonable detection.

4. STRUCTURE–ACTIVITY RELATIONSHIPS

The test of more than 100 nitroso compounds for phage induction is shown in Tables I–VII, in which chemicals are organized by class. The data presented are from spot tests performed on plates in the presence (Figure 4) or absence

3

6

Figure 4. Spot test assay for prophage induction using N-nitroso compounds activated by different combinations of rat and hamster liver S-9 fractions in the presence or absence of ampicillin. Strain BR 339 (*lexA*, $\Delta uvrB$) was plated in a 3.3 ml soft agar overlay containing S-9 activation mix. Cofactor concentrations were constant, consisting of 2.5 micromoles each of NADP, glucose-6-phosphate, and $MgCl_2$ per plate. Stock liver S-9 preparations, provided by Wes Andrews, were Aroclor-induced male Syrian golden hamster, 28 mg/ml protein, and Sprague-Dawley Aroclor-induced male rat, 25 mg/ml protein, prepared in KCl. Where indicated, LBE plates contained 10 µg/ml sodium ampicillin. Plate 1 (+ amp), hamster liver S-9, 14 mg protein; Plate 2 (+ amp), hamster liver S-9, 7 mg protein; Plate 3 (+ amp), rat liver S-9, 13 mg protein; Plate 4 (+ amp), hamster liver S-9, 7 mg protein + rat liver S-9, 6.5 mg protein; Plate 5 (no amp), hamster liver S-9, 14 mg protein; Plate 6 (no amp), rat liver S-9, 13 mg protein. For all chemicals, solution concentrations (and quantities spotted in 5 µl) are given. Nitrosamine solutions are in DMSO; bleomycin solutions in water. Nitrosamines that are liquids were made up into solutions at 50%, 10%, and 1% by volume ≈ 500, 100, and 10 mg/ml (2.5, 0.5, and 0.05 mg per spot). Row A, lanes 1-3: Bleomycin at 8, 4, and 2 µg/ml (0.04, 0.02, and 0.01 µg); lane 4: water; lane 5: DMSO; lane 6: no addition. Row B, lanes 1-3: Nitrosomorpholine at 100, 10, and 1 mg/ml (0.5, 0.05, and 0.005 mg); lanes 4-6: dinitrosopiperazine 100, 10, and 1 mg/ml (0.5, 0.05, and 0.005 mg). Row C, lanes 1-3: *cis*-2, 6-dimethylnitrosomorpholine (liquid); lanes 4-6: nitrosopyrrolidine (liquid). Row D, lanes 1-3: Nitrosopiperidine (liquid); lanes 4-6: *trans*-2,6-dimethylnitrosomorpholine (liquid). Row E, lanes 1-3: Dipropylnitrosamine (liquid); lanes 4-6: nitroso-4-piperidone at 10, 1, and 0.1 mg/ml (0.05, 0.005, and 0.0005 mg). Row F, lanes 1-3: Dimethylnitrosamine (liquid); lanes 4-6: diethylnitrosamine (liquid).

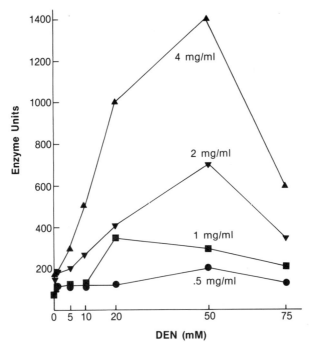

Figure 5. The induction of prophage by DEN as a function of concentration of hamster liver protein. S-9 stock at 43 mg/ml was prepared in *tris*-sucrose by Robin Pennington from Aroclor-induced, male Syrian golden hamsters. Bacterial strain BR 469 was incubated for 5 hr with chemical and activation mix and assayed for β-galactosidase, as described in Figure 1.

(Figure 2) of hamster liver S-9 activation mix, as required. Chemicals from each group were also tested in the quantitative assay, as shown in Figures 1 and 8–10, to confirm spot test results. Results from the two assays, such as order of potency, were usually consistent.

The directly acting nitrosoalkyl-guanidines, -carbamates, and -ureas were good inducers, as expected (Figure 8 and Tables I and II). Relative inducing capacity correlates very well with mutagenicity of these compounds in *E. coli*,[23,24] the order of potency and concentrations required being almost identical. For instance, the dose of nitrosoalkyl-ureas at peak response for both mutagenesis and phage induction is around 3 mM (Figure 8 and Reference 24); for nitrosoalkyl-guanidines the dose is in the range of tenths of millimolar, while for several of the nitroso derivatives of carbamate insecticides it is in the range of hundredths of millimolar (Figure 8, References 23 and 24). In general the ethyl compounds were better inducers and mutagens than the methyl compounds, as judged by the magnitude of response observed (Figures 2 and 8, Reference 24).

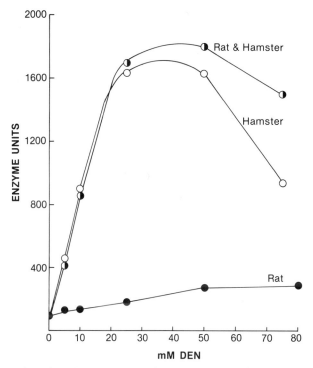

Figure 6. Induction of prophage by DEN activated by S. D. rat liver S-9 (stock at 31 mg/ml prepared in *tris*-sucrose, courtesy of Dr. Roger Shaw) or S. G. hamster liver S-9 (stock at 35 mg/ml prepared as above). S-9 present at 5% by volume (1.6 mg/ml rat, 1.8 mg/ml hamster, half of each in the combination).

This difference seemed to be due at least in part to the greater toxicity of methyl compounds, observed as toxic zones on spot tests (Figure 2). Inducing activity of the nitrosoalkyl-ureas decreased as a function of increasing chain length beyond three carbon atoms. Compounds with chain lengths of 6, 11, and 13 carbon atoms were not detected, although the aromatic derivatives possessed substantial activity (Table II).

Nitroso compounds requiring metabolic activation were, in general, much more difficult to detect as inducers than the directly acting compounds. Rat liver S-9 preparations were not adequate to activate more than a few compounds (notably DMN, DEN, and NPy) to inducing agents detectable in our spot test assay. Whether this is due to a general lack of sensitivity of the assay, compared with mutagenesis assays, or is a result of different mechanisms will be considered later. The use of hamster liver S-9 did allow the detection of most nitrosamines in a reasonably quantitative manner (Tables III–VII; and Figures 4, 9, and 10).

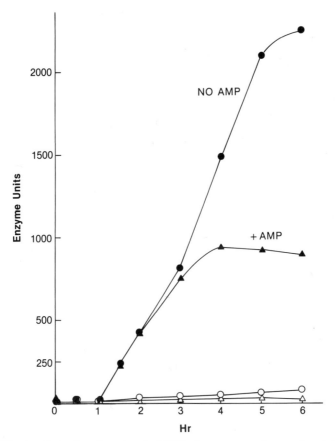

Figure 7. Kinetics of induction of prophage by DEN in the presence or absence of ampicillin (10 μg/ml). BR 469 (ΔuvrB) was incubated with 30 mM DEN as described in Figure 1. Hamster liver S-9, as described in Figure 5, was present at 11% by volume, equivalent to a protein concentration of 4–5 mg/ml. Aliquots were removed at the times shown and assayed for β-galactosidase activity. Inhibition by amp after 3 hr is the result of cell lysis.

Among the dialkyl nitrosamines, a decrease in activity similar to that seen for the nitrosoalkyl-ureas was noted as the chain length increased, in both the symmetrical and the unsymmetrical groups (Tables III and IV). Several isomeric dibutylnitrosamines gave borderline activity, detectable in some experiments but not others. Among the structural analogs tested, hydroxy compounds were generally without activity and most other derivatives (chloro-, cyano-, methoxy-) had less activity than the parent (Table III).

The cyclic nitrosamines showed a broad range of activity, depending on the class and on substituents (Tables V and VII; Figures 9 and 10). The smallest

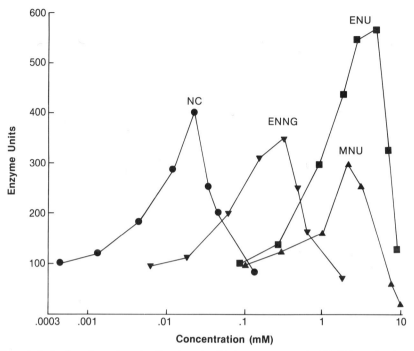

Figure 8. Induction of prophage by directly acting N-nitroso compounds. BR 339 was incubated with chemicals for 5 hr, followed by β-galactosidase assay. Compounds tested were nitrosocarbaryl, NC; methyl- and ethyl-nitrosourea, MNU and ENU; and ethylnitrosonitroguanidine, ENNG.

cyclic compound, nitrosoazetidine, was weakly inducing at high concentrations. Compounds with 5-, 6-, 7-, 8-membered rings were very good inducers. Activity seemed to increase slightly with ring size. The 9-membered ring compound, nitroso-octamethyleneimine, was detected only in the strain with the mutation *envA*, affecting membrane permeability.

Because the nitrosopiperazine analogs are rather weak carcinogens and mutagens, it is not surprising that they are difficult to detect in our assay system, but the results with the nitrosomorpholines are surprising in this regard. Nitrosomorpholine is detected easily as a mutagen in *E. coli*[25] and *S. typhimurium*,[26,27] but its activity as a phage inducer was weak or undetectable in our plate assay. *Trans*-dimethylnitrosomorpholine was a much stronger inducer than the *cis* isomer, but both were reproducibly positive in the plate assay even when the parent compound was not. The nitrosooxazolidines were the most potent inducers among the morpholine analogs (Table VI).

Nitrosopyrrolidine was notable in exhibiting a strong response in the test on agar and a weak response in the liquid incubation assay (Figures 4 and 9;

Table I
Phage Induction by Nitroso Compounds: Directly Acting Nitrosoguanidines and Nitrosocarbamates[a]

Classification		Compound	Phage induction	Minimum dose/spot (μg)
A.	1. Directly acting nitroso-guanidines	MNNG	++	5
		ENNG	+++	1
		Nitroso-*Cimetidine*	+++	50
	2. Nitroso-carbamates	Nitrosomethylurethane	++	5
		Nitrosoethylurethane	+++	5
		Nitroso-*insecticide*		
		Carbaryl	+++	1
		Methomyl	+++	0.5
		Carbofuran	+	10
		Landrin	+	~50
		Temik (=aldicarb)	+++	0.5
		Buxten	±	~50
		Methylphenylcarbamate	+++	0.5

[a]Spot test results for induction of bacteriophage lambda with N-nitroso compounds. All compounds were tested in strain BR 513 and many in BR 339 as well, which generally gave superior results (see Figure 2). Log phase bacteria were poured in 2.5 ml agar onto square petri dishes measuring 100 mm on a side. Quantities indicated of chemical solutions in DMSO were spot tested in aliquots of approximately 5 μl. Plates incubated for 5 hr at 38° C were overlaid with 2.5 ml top agar containing a substrate, BNG plus Fast Blue, for the enzyme β-galactosidase. Red spots represent positive areas of induction. Spot intensities, judged by comparison with controls on the same plate (Figures 2 and 4) were recorded as weak (+), moderate (++), or strong (+++) at the optimum inducing dose. The lowest concentration tested giving a reproducible positive response was noted as the minimum inducing dose (quantity/5 μl).

Table VI) under the conditions used. Optimum conditions for individual compounds in the latter assay have not been established, although our plate assay results indicated that nitroso compounds in general respond in the same way to variations in conditions (Figure 4). Of the few structural analogs of nitrosopyrrolidine tested, the β-halogenated derivations showed more activity than the parent, while the α-carboxy-, β-hydroxy-, and unsaturated derivatives were essentially inactive in the spot test.

The substituted piperidines showed dramatic differences in activity (Table V and Figures 9 and 10). Substitution of methyl at the 2 position reduced activity, while methyl substitution at the 4 position showed little difference from the parent compound. Substitution of all of the hydrogens alpha to the nitroso group eliminated activity. Substitution of phenyl at the 4 position increased

Table II
Phage Induction by Nitroso Compounds: Directly Acting Nitrosoalkylureas[a]

Classification	Compound	Phage induction	Minimum dose/spot (μg)
A. Directly acting			
3. Nitrosoalkylureas	Methyl	+++	10
	Ethyl	+++	10
	n-propyl	+++	10
	Iso-propyl	+++	20
	n-butyl	++	10
	Iso-butyl	–	
	Amyl	++	10
	Hexyl	±	1000
	Allyl	+++	20
	Cyclohexyl	++	50
	Undecyl	–	
	Tridecyl	–	
	Phenyl	++	< 10
	Benzyl	+++	10
	Phenylethyl	+++	10
	Fluoroethyl	+++	10

[a] See footnote a to Table I.

activity (active at one twentieth the concentration of the parent compound), but greatly decreased activity of the 4-cyclohexyl derivative was observed. Derivatives with carboxyl groups at the 3 and 4 positions showed good activity at lower doses than the parent compound, as did the 3,4-dibromo and 3,4-epoxy derivatives. Substitutions of hydroxyl and keto groups at the 3 or 4 positions substantially reduced activity. Of the partially unsaturated compounds, the Δ^2 compound showed reduced activity, while the Δ^3 compounds were very active, at least as active as the parent compound (it is possible that low levels of active oxidation products were present). Nitroso*guvacoline* did not show reproducible activity in the spot test.

5. RELATIONSHIPS BETWEEN MUTAGENICITY, CARCINOGENICITY, AND PHAGE INDUCTION

Although the results of mutagenesis, carcinogenesis, and phage induction assays generally correlate, an examination of the data in detail might yield some insight into factors limiting the response of each system. For instance, the order of potency (based on dose-response data) of the directly acting nitrosocarbam-

Table III
Phage Induction by Nitroso Compounds for Which Activation Is Required:
Symmetrical Aliphatic Nitrosamines[a]

Classification	Compound	Phage induction	Minimum dose/spot (μg)
B. Activation required			
1. Aliphatic nitrosamines			
(a) Symmetrical	Dimethyl	+	2000
	Diethyl	+++	250
	bis-(2-chloroethyl)	+	500
	bis-(2-cyanoethyl)	–	
	bis-(2-methoxyethyl)	–	
	bis-(2-ethoxyethyl)	–	
	bis-(2,2-diethoxyethyl)	–	
	Di-n-propyl	++	250
	bis-(2-hydroxypropyl)	–	
	bis-(2-oxopropyl)	–	
	bis-(2-chloropropyl)	+	< 500
	Diisopropyl	–	
	Di-n-butyl	±	500
	Di-isobutyl	±	500
	Di-sec-butyl	±	500
	Diallyl	+++	< 500
	Diethanol	–	

[a] See footnote a to Table I. S-9 activation mix in a 3.3 ml soft agar overlay contained 0.5 ml Syrian golden hamster liver S-9 (12–18 mg protein) and 2.5 μM each NADP, glucose-6-phosphate, and $MgCl_2$.

ate insecticides for mutagenicity[21] or phage induction (Table I) in *E. coli* (M ≳ A > C > CF > L > B > Bux) is rather dissimilar from the order of activity for carcinogenicity (CF > A ≳ B > L > C = M = Bux).[28] The strongest mutagens and phage inducers are the sulfur-containing *A* and *M* and the naphthyl derivative, *C*, but the latter two compounds are among the weakest carcinogens. The strongest carcinogens, *CF* and *B*, are among the weakest mutagens and inducers of bacteriophage. *Bux* is weak in both systems; *A* is strong in both systems. Studies on the uptake of strong and weak mutagens of this type revealed order of magnitude differences in affinity of the bacteria for these compounds that correlated reasonably well with the differences in mutagenicity observed.[29] The strongest mutagen was concentrated within the bacteria, while the weakest one was nearly excluded, and the compound with intermediate activity showed intermediate uptake. It seems possible, therefore, that the noncorrelation between the results in bacteria and animals with this set of

Table IV
Phage Induction by Nitroso Compounds for Which Activation Is Required:
Unsymmetrical Aliphatic Nitrosamines[a]

Classification	Compound	Phage induction	Minimum dose/spot (μg)
B. Activation required			
1. Aliphatic nitrosamines			
(b) Unsymmetrical	Methoxymethyl	+++	N.D.
	Methylethyl	+++	500
	Methylpropyl	+++	250
	Methylbutyl	+++	500
	Methylhexyl	+++	250
	Methylheptyl	+	500
	Methyloctyl	+[b]	< 500
	Methylnonyl	+[b]	< 500
	Methyldecyl	+[b]	500
	Methylundecyl	+[b]	500
	Methyldodecyl	−	
	Methyl-2-phenylethyl	+	500
	Methylcylohexyl	++	500
	Methylbenzyl	++	< 500
	Methylcarboxypropyl	±	500
	Ethanolisopropanol	−	
	Methylaniline	+	< 500

[a] See footnote to Table III. N.D., not determined.
[b] Positive results in BR 513 (envA) only.

compounds is related to differential uptake of the compounds in the two systems. In *Salmonella*,[(30)] the order of activity of nitroso carbamates for mutagenesis was the same as that for *E. coli*, with the exception of *Bux*, which was strong in *Salmonella* and weak in *E. coli*.

The relative activity of nitrosoalkyl-ureas also differs in different systems. Activity as a function of increasing chain length falls off for both carcinogenicity and phage induction, a pattern that was not seen for mutagenicity in *Salmonella*.[(30)] Nitrosoallylurea, a strong phage inducer and a mutagen,[(30)] is not detectably carcinogenic. On the other hand, nitrosohexylurea, a good carcinogen and mutagen,[(30,31)] exhibits very weak inducing activity. The nitrosoalkyl-ureas require quite high concentrations for mutagenesis and phage induction in bacteria, consistent with their inefficient uptake into the bacterial cells,[(29)] whereas relatively small quantities are required for induction of tumors.[(31)]

The simple alkylating agents with methyl or ethyl groups behave quite

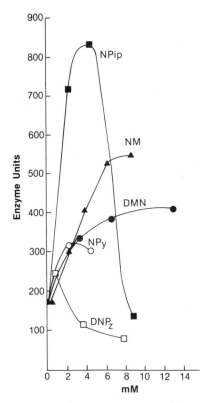

Figure 9. Induction of prophage by N-nitroso compounds requiring metabolic activation. BR 339 (*lexA*) was incubated with chemicals and hamster liver S-9 activation mix for 5 hr, followed by β-galactosidase assay. Hamster liver S-9, described in Figure 4, was present at 10% by volume, at a protein concentration of 2.9 mg/ml. Procedure and cofactor concentrations were as described in Figure 1. Compounds tested were nitrosopiperidine, NPip; nitrosomorpholine, NM; dimethylnitrosamine, DMN; nitrosopyrrolidine, NPy; and dinitrosopiperazine, DNPz.

differently in *Salmonella* and *E. coli*. In *Salmonella*, methylating compounds produce a greater response than ethylating agents.[19,30] The reverse is true for both mutagenesis[22,24] and phage induction (Figure 9) in *E. coli* and tumor production in animals.[32,33] Differences in host repair capacities and genomic targets are clearly implicated here. Methylethylnitrosamine (MEN), which possesses some properties of both DMN and DEN[33,34] and undergoes oxidation at both alkyl groups,[34] was not detected as readily as DMN or DEN as a mutagen in *Salmonella*[35-37] or *E. coli* (Elespuru, unpublished and Reference 38).

A large number of nitrosopiperidines (Table V, Figure 10) have been examined in several systems. Substitution of methyl at the 2 position usually decreases activity, and elimination of all of the hydrogens alpha to the nitroso

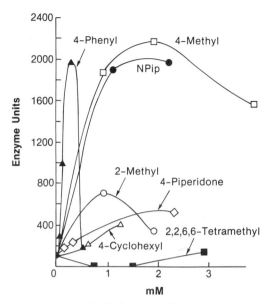

Figure 10. Structure-activity study with N-nitrosopiperidine derivatives. Compounds were incubated with BR 339 in the presence of hamster liver S-9 activation mix as described in Figure 9. Compounds tested were nitrosopiperidine, NPip, and the nitrosopiperidine derivatives indicated.

group eliminates activity consistently in all of the biological systems studied.[38–42] Substitution of cyclohexyl at the 4 position also eliminates activity. The 4-phenyl derivative exhibits increased activity for both mutagenesis in *Salmonella* and phage induction in *E. coli*, but carcinogenic activity declines. Substitution of hydroxy or keto in a number of positions eliminates phage-inducing activity, but not mutagenic or carcinogenic activity. On the other hand, carboxy derivatives have lost mutagenic and carcinogenic activity, but are quite strong phage inducers. The results of Rao indicated that more than one enzyme system, with different substrate preferences, may be involved in the activation of nitrosopiperidines.[38]

The same appears to be true for some morpholine derivatives. The *trans* isomer of dimethylnitrosomorpholine is more active than *cis* for esophageal tumor production in rats,[43] for mutagenesis in *Salmonella* (via rat liver S-9, Reference 36), and phage induction in *E. coli* (via hamster liver S-9, Table VI). However, the *cis* isomer is more potent than the *trans* for tumor production in hamsters.[42] Liver cell tumors are not induced in either animal, although liver enzymes from both are capable of activating the compounds to DNA-damaging agents.

Clearly, several different factors are important in determining biological

Table V
Phage Induction by Nitroso Compounds for Which Activation Is Required:
Heterocyclic Nitrosamines—Nitrosopiperidines[a]

Classification	Compound	Phage induction	Minimum dose/spot (μg)
B. Activation required			
2. Heterocyclic nitrosamines			
(a) Nitroso-piperidines	Nitrosopiperidine	+++	250
	2-methyl	+	2500
	2-carboxy	±	500
	3-hydroxy	−	
	3-chloro	+	< 500
	3-carboxy	+++	50
	4-methyl	++	500
	4-phenyl	+++	10
	4-cyclohexyl	−	
	4-hydroxy	−	
	4-keto	−	
	4-carboxy	+++	50
	3,4-dibromo	+++	50
	3,4-epoxy	+++	< 500
	2,6-dimethyl	−	
	2,2,6,6-tetramethyl	−	
	1,2,3,6-tetrahydropyridine (Δ^3)	+++	< 500
	1,2,3,6-tetrahydropyridine-3-methyl (Δ^3)	+++	< 500
	1,2,3,4-tetrahydropyridine (Δ^2)	++	
	Nitroso-*guvacoline*	±	500

[a]See footnote to Table III.

activity. A structural change that abolishes activity in all test systems may affect chemical reactivity, as is the case when hydrogens alpha to the nitroso group are blocked. Structural changes that affect a set of activities in one system, such as mutagenesis and phage induction in *E. coli*, may affect uptake of the chemical. Chemicals, such as nitrosomethoxymethylamine, that work only in systems utilizing *in vitro* metabolic activation, but not in animals, may be inactivated *in vivo* before reaching a necessary target. Structural changes that lead to loss of activity for mutagenesis but not phage induction may indicate the existence of different mechanistic pathways leading to different end points. Documentation

Table VI
Phage Induction by Nitroso Compounds for Which Activation Is Required:
Heterocyclic Nitrosamines—Nitrosopyrrolidines and Nitrosomorpholines[a]

Classification	Compound	Phage induction	Minimum dose/spot (μg)
B. Activation required			
2. Heterocyclic nitrosamines			
(b) Nitroso-pyrrolidines	Nitrosopyrrolidine	+++	100
	3,4-dichloro	+++	50
	3,4-dibromo	+++	50
	2-carboxy (nitroso*proline*)	±	
	2-carboxy-4-hydroxy (hydroxy*proline*)	−	
	3-pyrroline	−	
	3-pyrrolidinol	−	
(c) Nitroso-morpholines	Nitrosomorpholine	±	1000
	2-methyl	+	2500
	2,6-dimethyl (*cis*)	+	2000
	(*trans*)	+++	500
	Thiomorpholine	±	1000
	Tetrahydro-1,3-oxazine	+	50
	Nitroso-oxazolidine	++	500
	-5 methyl	+++	50
	-2 methyl	−	
	Nitroso-*phenmetrazine*	−	

[a]See footnote to Table III.

of these possibilities would be useful in assessing results in different systems, particularly those that generate negative correlations.

6. MECHANISM OF ACTION OF NITROSO COMPOUNDS AS INDUCERS OF BACTERIOPHAGE LAMBDA

The generation of mutations by MNNG or MNU in a *recA* or *lexA* strain[45-48] indicated that mutagenesis was not entirely dependent on the induction of the SOS pathway, in contrast with UV and most other carcinogens. Additional evidence on the chemical activity and biological effects of alkylating agents reinforced the concept of two pathways,[49-53] one "direct" or SOS-independent, resulting from mispairing of bases alkylated at crucial hydrogen-bonding positions,[52] and the other SOS-dependent, resulting from errors in

Table VII
Phage Induction by Nitroso Compounds for Which Activation Is Required:
Heterocyclic Nitrosamines—Nitrosopiperazines and Other Heterocyclic Nitrosamines[a]

Classification	Compound	Phage induction	Minimum dose/spot (μg)
B. Activation required			
2. Heterocyclic nitrosamines			
(d) Nitroso-piperazines	1-Nitrosopiperazine	−	
	Dinitrosopiperazine	−	
	2-methyl	−	
	2,6-dimethyl	−	
	4-benzoyl-3,5-dimethyl	+	< 500
	Homopiperazine	−	
(e) Other heterocyclic nitrosamines	Azetidine	+	2500
	Hexamethyleneimine	+++	500
	Heptamethyleneimine	+++	250
	Octamethyleneimine	+++[b]	50
	Dodecamethyleneimine	−	

[a] See footnote to Table III.
[b] Positive only in BR 513 (envA).

DNA repair.[53] A correlation was found between the ability of certain alkylating agents to alkylate the O^6 position of guanine and the capacity for SOS-independent mutagenesis.[49,51–53] Transition-type mutations (GC → AT), expected from mispairing of guanine with thymine, predominated after MNNG treatment.[54] Furthermore, the removal of O^6-alkylguanine has been associated with the "adaptive response,"[55] a process that reduces the lethal and mutagenic effects of alkylating agents in E. coli.[48,56–60]

While so-called "direct" mutagenesis appears to depend on O^6 alkylation, less is known about the lesion responsible for induction of the SOS response. That it may be different from O^6 alkylation is suggested by the activity of compounds such as MMS and EMS, the former a reasonably good phage inducer[61] and lexA-dependent mutagen[53] yielding little O^6 alkylation,[49,62,63] the latter a poor inducer[45–61] and a lexA-independent mutagen[49] yielding relatively high levels of O^6-ethylguanine.[49,62,63] Indeed, the addition of a small alkyl group would not seem to qualify as a "bulky lesion," thought necessary to inhibit DNA synthesis and thereby yield an inducing signal.[51] On the other hand, the calculations of Strauss indicate that MMS yields considerable quantities of O^6-methylguanine,[64] contradicting the assessment of others.[49,62,63] In addition, Schendel has shown[48] that the "adaptive response" lowers both *lexA*-dependent and *lexA*-independent mutagenesis in the case of both EMS and MMS. This rather surprising observation indicates that O^6-alkylguanine is important for both types of mutagenesis or that some other lesion is also being removed. Indeed, recent genetic[65] and biochemical[59,60] evidence has shown that more than one enzyme is induced, affecting more than one lesion, during the "adaptive response." Besides a guanine O^6-methyltransferase, a 3-methyladenine glycosylase is induced, the latter enzyme playing a role (in conjunction with *polA*) in the alleviation of toxicity but not mutagenicity.[59,60] This glycosylase might also play a role in generating the SOS response after alkylating agents, a possibility that could be tested genetically.[59,60]

The involvement of *polA*[59,60,66] suggests the presence of gaps in the DNA, a substrate that one might reasonably expect to generate the SOS response. Indeed, the induction of lambda prophage[67] and *lex*-dependent mutagenesis[53] is more efficient in *polA* strains. The observations in the literature[68-71] noting the presence of single-strand breaks in the DNA after treatment with various nitroso compounds are consistent with this hypothesis. Alkylations leading to strand breaks have attracted little attention, but this type of damage, generally considered not to be important for *lex*-independent (base mispairing) mutagenesis,[49,62,63,75] may be relevant for the induction of the SOS response.[53]

The induction of prophage in a *lexA* lysogen is a surprising result worth noting. We expected to see little or no induction in this strain since it is noninducible with UV[1,2,73]; we observed instead an enhancement of response with alkylating agents (Figures 2 and 3). Although UV sensitivity and greatly reduced capacity for induction by UV are consistent with the *lexA* phenotype, our mutant requires further characterization before its genetic assignment is assured. The capacity for induction appears to be altered in the mutant, while the optimum dose for induction is the same as that for wild-type (Figure 3). From this we conclude that the mutation probably does not affect cell permeability or repair of lesions, either of which would result in a shift of the dose–response curve to lower doses. It is possible that the mutant is deficient in a pathway affecting toxicity and thereby the synthesis of enzyme. Alternatively, DNA damage in mutant bacteria may result in more efficient cleavage of the lambda repressor.

Although several hundred N-nitroso compounds have been tested and found to be mutagens, it was not known whether the majority of compounds, particularly those requiring metabolic activation, could induce the SOS pathway. Most of the nitroso compounds tested did induce bacteriophage lambda, including many of the cyclic compounds that have an alkylation capacity that is much smaller than that of the simple open chain compounds. The conditions under which phage induction could be detected differed notably from those required for mutagenesis in several respects, including genetic background and conditions for metabolic activation. For instance, mutagenesis by most alkylating agents is diminished in *lexA* strains,[48,56] whereas phage induction in our *lexA* mutant is enhanced (Figure 3); the *uvr* system has no effect on phage induction by ethylating agents in our system (Figure 2), whereas removal of O^6-ethylguanine is diminished[73] and mutagenesis is enhanced in *uvrA* mutants.[46,74] Rat liver S-9 stocks with proven activity in *S. typhimurium* mutagenesis assays activate few if any nitroso compounds to compounds detectable as phage inducers. The enzyme activity of hamster liver S-9, a superior source of activating capacity for both mutagenesis and phage induction, can be inhibited by rat liver S-9 mix for mutagenesis[20] but not for phage induction (Figures 4 and 6). The relative order of potency for phage induction was similar to that for mutagenesis with some groups of compounds and different for other groups. It should prove interesting to determine whether different chemical-DNA interaction products are responsible for these two end points of DNA damage, as the circumstantial evidence would suggest.

7. REFERENCES

1. M. Radman, SOS repair hypothesis: Phenomenology of an inducible DNA repair which is accompanied by mutagenesis, in: *Molecular Mechanisms for Repair of DNA* (P. Hanawalt and R. B. Setlow, eds.), Part A, pp. 355–367, Plenum Press, New York (1975).
2. E. M. Witkin, Ultraviolet mutagenesis and inducible DNA repair in *Escherichia coli*, *Bact. Rev.* 40, 869–907 (1976).
3. K. Brooks and A. J. Clark, Behavior of λ bacteriophage in a recombination deficient strain of *Escherichia coli*, *J. Virol.* 1, 283–293 (1967).
4. K. McEntee, Protein X is the product of the *recA* gene of *Escherichia coli*, *Proc. Natl. Acad. Sci. USA* 74, 5275–5279 (1977).
5. J. W. Roberts, C. W. Roberts, and N. L. Craig, *Escherichia coli recA* gene product inactivates phage λ repressor, *Proc. Natl. Acad. Sci. USA* 75, 4714–4718 (1978).
6. N. L. Craig and J. W. Roberts, *E. coli recA* protein-directed cleavage of phage λ repressor requires polynucleotide, *Nature* 283, 26–30 (1980).
7. A. Bailone, A. Levine, and R. Devoret, Inactivation of prophage λ repressor *in vivo*, *J. Mol. Biol.* 131, 553–572 (1979).
8. R. Devoret, Inducible error-prone repair and induction of prophage lambda in *Escherichia coli*, *Prog. Nucleic Acid Res. Mol. Biol.* 26, 251–263 (1981).
9. C. J. Kenyon and G. C. Walker, DNA-damaging agents stimulate gene expression at specific loci in *Escherichia coli*, *Proc. Nat. Acad. Sci. USA* 77, 2819–2823 (1980).
10. Y. L. Ho and S. K. Ho, The induction of a mutant prophage in *Escherichia coli*: A rapid screening test for carcinogens, *Virology* 99, 257–264 (1979).
11. C. L. Smith and M. Oishi, The molecular mechanism of virus induction. I. A procedure for the biochemical assay of prophage induction, *Mol. Gen. Genet.* 148, 131–138 (1976).
12. A. Levine, P. L. Moreau, S. G. Sedgwick, B. Sedgwick, R. Devoret, S. Adhya, M. Gottesman, and A. Das, Expression of a bacterial gene turned on by a potent carcinogen, *Mutat. Res.* 50, 29–35 (1978).
13. P. L. Moreau, A. Bailone, and R. Devoret, Prophage λ induction in *Escherichia coli* K12 *envA*

uvrB: A highly sensitive test for potential carcinogens, Proc. Natl. Acad. Sci. USA 73, 3700–3704 (1976).
14. Z. Toman, C. Dambly, and M. Radman, Induction of a stable, heritable epigenetic change by mutagenic carcinogens: A new test system, in: *Molecular and Cellular Aspects of Carcinogen Screening Tests* (R. Montesano, H. Bartsch, and L. Tomatis, eds.), IARC Scientific Publications No. 27, pp. 243–255, International Agency for Research on Cancer, Lyon France (1980).
15. R. K. Elespuru and M. B. Yarmolinsky, A colorimetric assay of lysogenic induction designed for screening potential carcinogenic and carcinostatic agents, *Environ. Mutagenesis 1*, 65–78 (1979).
16. R. K. Elespuru, A biochemical phage induction assay for carcinogens, in: *Short-Term Tests for Chemical Carcinogens* (H. Stich and R. H. C. San, eds.), Topics in Environmental Physiology and Medicine, pp. 1–11, Springer-Verlag, New York (1981).
17. J. McCann, E. Choi, E. Yamasaki, and B. N. Ames, Detection of carcinogens as mutagens in the Salmonella/microsome test: Assay of 300 chemicals, *Proc. Natl. Acad. Sci. USA 72*, 5135–5139 (1975).
18. H. Bartsch, C. Malaveille, and R. Montesano, *In vitro* metabolism and microsome mediated mutagenicity of dialkylnitrosamines in rat, hamster and mouse tissues, *Cancer Res. 35*, 644–651 (1975).
19. M. J. Prival, V. D. King, and A. T. Sheldon, Jr., The mutagenicity of dialkyl nitrosoamines in the Salmonella plate assay, *Environ. Mutagenesis 1*, 95–104 (1979).
20. M. J. Prival and V. D. Mitchell, Influence of microsomal and cytosolic fractions from rat, mouse, and hamster liver on the mutagenicity of dimethylnitrosamine in the *Salmonella* plate incorporation assay, *Cancer Res. 41*, 4361–4367 (1981).
21. T. Yahagi, M. Nagao, Y. Seino, T. Matsushima, T. Sugimura, and M. Okada, Mutagenicities of N-nitrosamines on *Salmonella, Mutat. Res. 48*, 121–130 (1977).
22. H. Bartsch, A. Camus, and C. Malaveille, Comparative mutagenicity of N-nitrosamines in a semi-solid and in a liquid incubation system in the presence of rat or human tissue fractions, *Mutat. Res. 37*, 149–162 (1976).
23. W. Lijinsky and R. K. Elespuru, Mutagenicity and carcinogenicity of N-nitroso derivatives of carbamate insecticides, in: *Environmental N-Nitroso Compounds, Analysis and Formation* (E. A. Walker, P. Bogovski, and L. Griciute, eds.), IARC Scientific Publications No. 14, pp. 425–428, International Agency for Research on Cancer, Lyon, France (1976).
24. T. A. Hince and S. Neale, A comparison of the mutagenic action of the methyl and ethyl derivatives of nitrosamides and nitrosamidines on *Escherichia coli, Mutat. Res. 24*, 383–387 (1974).
25. R. K. Elespuru and W. Lijinsky, Mutagenicity of cyclic nitrosamines in *Escherichia coli* following activation with rat liver microsomes, *Cancer Res. 36*, 4099–4101 (1976).
26. R. F. Gomez, M. Johnston, and A. J. Sinskey, Activation of nitrosomorpholine and nitrosopyrrolidine to bacterial mutagens, *Mutat. Res. 24*, 5–7 (1974).
27. E. Zeiger and A. T. Sheldon, The mutagenicity of heterocyclic N-nitrosamines for *Salmonella typhimurium, Mutat. Res. 57*, 1–10 (1978).
28. W. Lijinsky and D. Schmähl, Carcinogenesis by nitroso derivatives of methylcarbamate insecticides and other nitrosamides in rats and mice, in: *Environmental Aspects of N-Nitroso Compounds* (E. A. Walker, M. Castegnaro, R. E. Lyle, and L. Griciute, eds.), IARC Scientific Publications No. 19, 495–501, International Agency for Research on Cancer, Lyon, France (1978).
29. R. K. Elespuru, Mutagenicity of nitrosocarbaryl and other methylating nitrosamides as related to uptake in Haemophilus influenzae, *Environ. Mutagenesis 1*, 249–257 (1979).
30. W. Lijinsky and A. W. Andrews, The mutagenicity of nitrosamides in *Salmonella typhimurium, Mutat. Res. 68*, 1–8 (1979).

31. W. Lijinsky and C. Winter, Skin tumors induced by painting nitrosoalkylureas on mouse skin, *Cancer Res. Clin. Oncology 102*, 13–20 (1981).
32. H. Druckrey, R. Preussmann, S. Ivankovic, and D. Schmähl, Organotrope carcinogene Wirkungen bei 65 verschiedenen N-nitroso-verbindungen an Bd-Ratten, *Z. Krebsforsch. 69*, 103–201 (1967).
33. W. Lijinsky and M. D. Reuber, Comparative carcinogenesis by some aliphatic nitrosamines in Fischer rats, *Cancer Lett. 14*, 297–302 (1981).
34. J. G. Farrelly, A new assay for the microsomal metabolism of nitrosamines, *Cancer Res. 40*, 3241–3244 (1980).
35. T. K. Rao, J. A. Young, W. Lijinsky, and J. L. Epler, Mutagenicity of aliphatic nitrosamines in *Salmonella typhimurium*, *Mutat. Res. 66*, 1–7 (1979).
36. A. W. Andrews and W. Lijinsky, The mutagenicity of 45 nitrosamines in *Salmonella typhimurium*, *Teratogenesis, Mutagenesis and Carcinogenesis 1*, 295–303 (1980).
37. S. Y. Lee and J. B. Guttenplan, A correlation between mutagenic and carcinogenic potencies in a diverse group of N-nitrosamines: Determination of mutagenic activities of weakly mutagenic N-nitrosamines, *Carcinogenesis 2*, 1339–1344 (1981).
38. T. K. Rao, B. E. Allen, W. Winton, W. Lijinsky, and J. L. Epler, N-nitrosamine induced mutagenesis in *Escherichia coli* K-12 (343/113): I. Mutagenic properties of certain aliphatic nitrosamines, *Mutat. Res. 89*, 209–215 (1981).
39. W. Lijinsky and H. W. Taylor, Carcinogenicity of methylated nitrosopiperidines, *Int. J. Cancer 16*, 318–322 (1975).
40. T. K. Rao, A. A. Hardigree, J. A. Young, W. Lijinsky, and J. L. Epler, Mutagenicity of N-nitrosopiperidines with *Salmonella typhimurium*/microsomal activation system, *Mutat. Res. 56*, 131–145 (1978).
41. C. E. Nix, B. Brewen, R. Wilkerson, W. Lijinsky, and J. L. Epler, Effects of N-nitrosopiperidine substitutions on mutagenicity in *Drosophila melanogaster*, *Mutat. Res. 67*, 27–38 (1979).
42. F. W. Larimer, D. Ramey, W. Lijinsky, and J. L. Epler, Mutagenicity of methylated N-nitrosopiperidines in *Saccharomyces cerevisiae*, *Mutat. Res. 57*, 155–161 (1978).
43. W. Lijinsky and M. Reuber, Comparison of carcinogenesis by two isomers of nitroso-2,6-dimethylmorpholine, *Carcinogenesis 1*, 501–503 (1980).
44. M. S. Rao, D. G. Scarpelli, and W. Lijinsky, Carcinogenesis in Syrian hamsters by N-nitroso-2,6-dimethylmorpholine, its *cis* and *trans* isomers, and the effect of deuterium labeling, *Carcinogenesis 2*, 731–735 (1981).
45. S. Kondo, H. Ichikawa, K. Iwo, and T. Kato, Base-change mutagenesis and prophage induction in strains of *Escherichia coli* with different DNA repair capacities, *Genetics 66*, 187–217 (1970).
46. Y. Ishii and S. Kondo, Comparative analysis of deletion and base change mutabilities of *Escherichia coli* B strains differing in DNA repair capacity (wild-type *uvrA, polA, recA*) by various mutagens, *Mutat. Res. 27*, 27–44 (1975).
47. T. A. Hince and S. Neale, Physiological modification of alkylating-agent induced mutagenesis. I. Effect of growth rate and repair capacity on nitrosomethylurea-induced mutation of *Escherichia coli*, *Mutat. Res. 46*, 1–10 (1977).
48. P. F. Schendel, M. Defais, P. Jeggo, L. Samson, and J. Cairns, Pathways of mutagenesis and repair in *Escherichia coli* exposed to low levels of simple alkylating agents, *J. Bacteriol. 135*, 466–475 (1978).
49. P. D. Lawley, Some chemical aspects of dose-response relationships in alkylation mutagenesis, *Mutat. Res. 23*, 283 (1974).
50. J. W. Drake and R. H. Baltz, The biochemistry of mutagenesis, *Ann. Rev. Biochem. 45*, 11–37 (1976).
51. M. Radman, G. Villani, S. Boiteux, M. Defais, P. Callet-Fauquet, and P. Spadari, On the mechanism and control of mutagenesis due to carcinogenic mutagens, in: *Origins of Human*

Cancer (J. P. Watson and H. Hiatt, eds.), pp. 903–922, Cold Spring Harbor Laboratory, Cold Spring Harbor, New York (1977).
52. A. Loveless, Possible relevance of O^6 alkylation of deoxyguanosine to the mutagenicity and carcinogenicity of nitrosamines and nitrosamides, Nature (London) 223, 206–207 (1969).
53. B. A. Bridges, R. P. Mottershead, M. H. L. Green, and W. J. H. Gray, Mutagenicity of dichlorvos and methyl methanesulphonate for Escherichia coli WP2 and some derivatives deficient in DNA repair, Mutat. Res. 19, 295–303 (1973).
54. C. Coulondre and J. H. Miller, Genetic studies of the lac repressor. IV. Mutagenic specificity in the lacI gene of Escherichia coli, J. Mol. Biol. 117, 577–606 (1977).
55. P. F. Schendel and P. E. Robbins, Repair of O^6-methylguanine in adapted Escherichia coli, Proc. Nat. Acad. Sci. USA 75, 6017–6020 (1978).
56. P. Jeggo, M. Defais, L. Samson, and P. Schendel, An adaptive response of E. coli to low levels of alkylating agents: Comparison with previously characterized DNA repair pathways, Mol. Gen. Genet. 157, 1–9 (1977).
57. P. Jeggo, Isolation and characterization of Escherichia coli K12 mutants unable to induce the adaptive response to simple alkylating agents, J. Bacteriol. 139, 783–791 (1979).
58. J. Cairns, P. Robbins, B. Sedgwick, and P. Talmud, Inducible repair of alkylated DNA, Prog. Nucleic Acid Res. Mol. Biol. 26, 237–243 (1981).
59. P. Karran, T. Hjelmgren, and T. Lindahl, Induction of a DNA glycosylase for N-methylated purines is part of the adaptive response to alkylating agents, Nature 296, 770–773 (1982).
60. I. G. Evensen and E. Seebert, Adaptation to alkylation resistance involves the induction of a DNA glycosylase, Nature 296, 773–775 (1982).
61. P. Moreau and R. Devoret, Potential carcinogens tested by induction and mutagenesis of prophage λ in Escherichia coli K12, in: Origins of Human Cancer (H. H. Hiatt, J. D. Watson, and J. A. Winsten, eds.), Vol. B, pp. 1451–1472, Cold Spring Harbor Laboratory, Cold Spring Harbor, New York (1978).
62. B. Singer, The chemical effects of nucleic acid alkylation and their relation to mutagenesis and carcinogenesis, Prog. Nucleic Acid Res. Mol. Biol. 15, 219–284 (1975).
63. A. E. Pegg, Formation and metabolism of alkylated nucleosides: Possible role in carcinogenesis by nitroso compounds and alkylating agents, in: Advances in Cancer Research, Vol. 25 (G. Klein and S. Weinhouse, eds.), pp. 195–269, Academic Press, London (1977).
64. B. Strauss, D. Scudiero, and E. Henderson, The nature of the alkylation lesion in mammalian cells, in: Molecular Mechanisms for Repair of DNA (P. C. Hanawalt and R. B. Setlow, eds.), Part A, pp. 13–24, Plenum Press, New York (1974).
65. P. Jeggo, Isolation and characterization of Escherichia coli K12 mutants unable to induce the adaptive response to simple alkylating agents, J. Bacteriol. 139, 783–791 (1979).
66. P. Jeggo, M. Defais, L. Samson, and P. Schendel, The adaptive response of E. coli to low levels of alkylating agent: The role of polA in killing adaptation, Mol. Gen. Genet. 162, 299–305.
67. M. Blanco and L. Pomes, Prophage induction in Escherichia coli K12 cells deficient in DNA polymerase I, Mol. Gen. Genet. 154, 287–292 (1977).
68. A. O. Olson and K. M. Baird, Single-strand breaks in Escherichia coli DNA caused by treatment with nitrosoguanidine, Biochim. Biophys. Acta 179, 513–514 (1969).
69. R. F. Kimball, M. Liu, and J. K. Setlow, Effects of posttreatment on single-strand breaks in DNA of Haemophilus influenzae exposed to nitrosoguanidine and methyl methanesulphonate, Mutat. Res. 13, 289–295 (1971).
70. B. W. Stewart and E. Farber, Strand breakage in rat liver DNA and its repair following administration of cyclic nitrosamines, Cancer Res. 33, 3209–3215 (1973).
71. H. S. Rosenkranz, S. Rosenkranz, and R. M. Schmidt, Effects of nitrosomethylurea and nitrosomethylurethan on the physical chemical properties of DNA, Biochim. Biophys. Acta 195, 262–265 (1969).

72. R. F. Kimball, J. K. Setlow, and M. Liu, The mutagenic and lethal effects of monofunctional methylating agents in strains of *Haemophilus influenzae* defective in repair processes, *Mutat. Res. 12*, 21–28 (1971).
73. J. J. Donch and J. Greenberg, The effect of *lex* on UV sensitivity, filament formation and λ induction in *lon* mutants of *Escherichia coli, Mol. Gen. Genet. 128*, 277 (1974).
74. W. Warren and P. D. Lawley, The removal of alkylation products from the DNA of *Escherichia coli* cells treated with the carcinogens N-ethyl-N-nitrosourea and N-methyl-N-nitrosourea: Influence of growth conditions and DNA repair defects, *Carcinogenesis 1*, 67–78 (1980).
75. R. C. Garner, C. Pickering, and C. N. Martin, Mutagenicity of methyl-, ethyl-, propyl- and butylnitrosourea towards *Escherichia coli* WP2 strains with varying DNA repair capabilities, *Chem. Biol. Interact. 26*, 197–205 (1979).

6

The Relationship between the Carcinogenicity and Mutagenicity of Nitrosamines in a Hepatocyte-Mediated Mutagenicity Assay

CAROL A. JONES AND ELIEZER HUBERMAN

Carcinogenesis is a complex process in which tumors develop from cells that have undergone a hereditary change from the normal phenotype to malignancy. A fraction of chemicals in the environment are capable of efficiently inducing such a change. While the exact mechanism of the malignant transformation is as yet not understood, it is widely believed that it is initiated as a result of interactions between the carcinogen and the cellular DNA, resulting in mutation(s) or mutationlike changes in the genes controlling cell growth and differentiation.[1-4]

In recent years, short-term *in vitro* assays have been devised for determination of the mutagenic activity of chemical agents as a means of identifying suspect carcinogenic chemicals.[5,6] These systems have demonstrated a qualitative relationship between mutagenicity and carcinogenicity of a wide variety of chemicals,[6-8] but in order to obtain the best possible risk assessment, they must also exhibit a quantitative relationship between carcinogenicity and mutagenicity. In the case of chemicals that require metabolism for their biological activity, the same metabolites should be responsible for both events. Chemical carcino-

---

CAROL A. JONES AND ELIEZER HUBERMAN • Argonne National Laboratory, Division of Biological and Medical Research, Argonne, Illinois 60439.   The submitted manuscript has been authored by a contractor of the U. S. Government under contract No. W-31-109-ENG-38. Accordingly, the U. S. Government retains a nonexclusive, royalty-free license to publish or reproduce the published form of this contribution, or allow others to do so, for U. S. Government purposes.

gens also exhibit marked organ and species specificity.[9] Unfortunately, most *in vitro* systems do not exhibit a quantitative organ or cell-specific response that is representative of the carcinogenic potential of a chemical. Therefore, we have undertaken to develop *in vitro* mutagenesis systems in which this organ/species specificity can be considered. Our approach is based upon the cell-mediated mutagenesis system described originally by Huberman and Sachs,[12] which utilizes intact mammalian cells both as the genetic targets and to metabolize the chemicals. Most chemical carcinogens require metabolic activation to electrophilic species that are capable of interacting with DNA.[11,12] In many *in vitro* systems this metabolizing capacity is supplied by liver S-9 fractions that can differ from intact cells in their scope of metabolic competence and can generate artifactual metabolites.[13-15] Therefore, it is better to use intact cells as the metabolizing component. Furthermore, since the critical factors determining the susceptibility of the target cell type to a chemical carcinogen may involve alterations in the organization of the genetic material, membrane permeability, or DNA repair systems,[16] the choice of a mammalian cell type over bacteria might be advantageous.

In the original cell-mediated system, embryonic fibroblasts were employed to activate the carcinogens, and mutable Chinese hamster V79 cells, which are incapable of metabolizing most carcinogens, were used as the target cells for the mutagenic intermediates. Using this system, a quantitative correlation between the mutagenic and carcinogenic activity of polycyclic hydrocarbons was demonstrated.[10-17] The capacity of fibroblasts to metabolically activate chemical carcinogens, however, is limited almost entirely to the polycyclic hydrocarbons. In order to determine the mutagenic potential of other chemical classes, it is necessary to use activating cells with a broader spectrum of metabolizing enzymes. For this purpose hepatocytes are the best choice,[18] and either primary rodent hepatocytes or liver-derived epithelial lines have been incorporated into a number of short-term screening systems.[19,20] The scope of the cell-mediated mutagenesis system was considerably expanded by the use of primary rat hepatocytes in place of the fibroblasts.

Nitrosamines, many of which are potent carcinogens in experimental animals, are distributed widely in the human environment[24-28] and hence may represent a serious carcinogenic hazard to humans. However, this particular chemical class has not been consistently detected in the various short-term assays, and consequently the correlation between the mutagenicity of the nitrosamines and their *in vivo* carcinogenicity is very poor.[7,29]

We therefore undertook to develop an hepatocyte-mediated mutagenesis system in which the mutagenicity of a series of nitrosamines would be related to their carcinogenic activity *in vivo*. To accomplish this, we prepared primary hepatocytes from adult male Fischer rats to metabolize the nitrosamines.[30] In the cell-mediated system, hepatocytes were cocultivated with the mutable V79

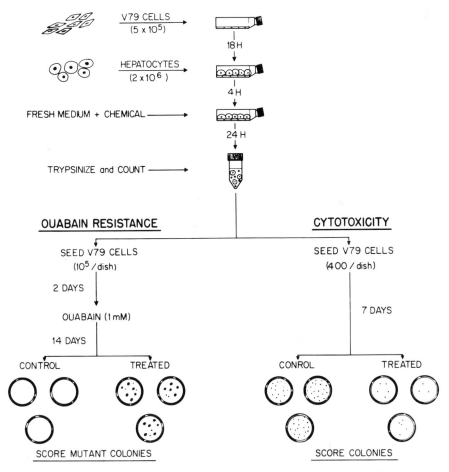

Figure 1. Scheme for the induction of ouabain-resistant mutants in the hepatocyte-mediated assay. (From Jones and Huberman.[22])

cells in the presence of nitrosamines,[22] and mutations were characterized by resistance to ouabain[31] and 6-thioguanine.[32] The protocol is outlined in Figure 1. Using this assay we were able to show that carcinogenic nitrosamines were mutagenic in a dose-dependent manner (Figure 2). The potent liver carcinogens nitrosodimethylamine and nitrosodiethylamine could be detected with doses lower than 1 μM (Figure 2), whereas the noncarcinogenic nitrosamines such as 1-nitrosopiperazine and nitrosodiallylamine were nonmutagenic at doses as high as 10 mM. To analyze the relationship between carcinogenicity and mutagenicity of these chemicals we calculated indices of potency for these two events.

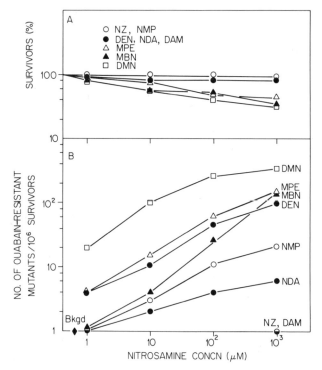

Figure 2. Induction of ouabain-resistant mutants by carcinogenic nitrosamines. Chinese hamster V79 cells were seeded at 5×10^5 cells/25 cm^2 T-flask. After 18 hr, 2×10^6 primary rat hepatocytes were seeded on the V79 cells in 4 ml Leibovitz's L-15 medium containing 2 mM glutamine, 10% fetal calf serum, 100 units penicillin/ml, and 100 μg streptomycin/ml. After 3 hr, the medium was changed and the nitrosamines were added in 4 ml fresh L-15 medium. The cells were dispersed 18 hr later with 0.05% trypsin and 0.02% EDTA. In A, the V79 cells were reseeded at 400 cells/60 mm Petri dish in Dulbecco's modified Eagle's medium containing 2 mM glutamine, 10% fetal calf serum, 100 units penicillin/ml, and 100 μg fungizone/ml to determine the percentage of cell survival. In B, V79 cells were seeded at 10^5 cells/60 mm Petri dish to determine the frequency of ouabain-resistant colonies. The average cloning efficiency of the untreated controls was $57 \pm 8\%$, and this value was considered as 100% for the determination of the percentage of surviving cells after treatment. DMN, nitrosodimethylamine; DEN, nitrosodiethylamine; DAM, nitrosodiallylamine; NMP, nitrosomethyl-n-propylamine; MBN, nitrosomethyl-n-butylamine; MPE, nitrosomethyl-2-phenylethylamine; NDA, nitrosomethyldodecylamine; NZ, 1-nitrosopiperazine. (From Jones and Huberman.[37])

The mutagenic potency unit (D_{10c}) was defined as the concentration (nM) that yielded a mutant frequency ten times higher than the spontaneous mutant frequency.[33] The D_{10c} values were calculated from the linear portion of the dose–response curves (Figure 2). The index of mutagenicity was described as $1/D_{10c}$ where D_{10c} = mean value calculated for each compound from at least three experiments.

The carcinogenicity index was obtained from experiments in which groups of male and female Fischer rats were treated with nitrosamines for 50 weeks unless the animals died earlier. These rats are genetically very similar to those used in the mutagenesis studies. A detailed account of these studies, which were performed by W. Lijinsky and co-workers, is presented in Chapter 10. In expressing the carcinogenic potency, it was necessary to consider both the total dose of compound administered and the time for tumor development as measured by the time until death from tumors for 50% of the animals. Analysis of the data from extensive long-term studies with nitrosamines led Druckrey and others to deduce the following relationship: dose $\times$ time^n = const, where n has an average value of 3 for nitrosamines.[34-36] For the purposes of the present study, the following index of carcinogenic potency was used:

$$\left(\ln 1 + \frac{100}{DT^3} \right)$$

where D = total dose (mol) of compound administered per rat and T = time (weeks) for death of 50% of the animals from tumors.[34] The logarithmic transformation was used to scale the data linearly. The index included 1 and 100 so that a numerical value could be assigned for noncarcinogenic chemicals, but without attaching excessive importance to them.

Using these indices, a linear relationship could be established between the degree of carcinogenicity of nitrosamines and the degree of their mutagenicity at two genetic markers, with a P value of 0.0001 (Figures 3 and 4).[37] At the 6-thioguanine marker, the correlation coefficient r was 0.85, whereas at the ouabain marker, where the degree of correlation was less strong, r = 0.66.

The better correlation between carcinogenicity and mutagenicity observed with the 6-thioguanine marker is presumably due to its ability to detect a wider variety of mutagens relative to the ouabain marker.[26,27] Ouabain resistance, which presumably results from very specific mutations, probably base substitutions, cannot detect chemicals that cause mainly frameshifts or deletions.[38] Since the nitrosamines selected for this study constitute a diverse chemical group and may induce different types of genetic damage resulting in mutations, the 6-thioguanine marker would offer a broader means for detecting this damage and hence provide a more significant correlation. Furthermore, the mutant frequency at the 6-thioguanine marker is less affected by the cytotoxicity of the test chemical compared with the ouabain marker. Nitrosamines have varying degrees of cytotoxicity, which do not correspond to their mutagenic activity (Figure 2). Therefore, the mutant frequencies at the 6-thioguanine marker, where cytotoxicity is less influential, probably represent a more accurate measure of the relative mutagenic activities of this group of nitrosamines.

It is interesting that the relationship between degree of carcinogenicity and degree of mutagenicity holds for several nitrosamines, namely, 1-nitrosopi-

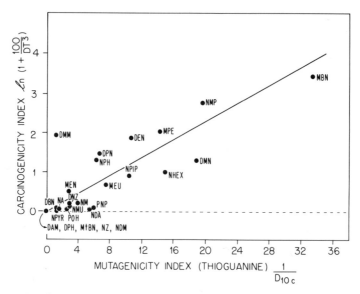

Figure 3. Relationship between carcinogenicity and mutagenicity for nitrosamines with 6-thioguanine resistance as the genetic marker. In Index of Mutragenicity. D = dose (nM) yielding a mutant frequency ten times higher than the spontaneous mutant frequency. In Index of Carcinogenicity, D = dose (mol) at which 50% of rats die from tumors in time T (weeks). DPN, nitrosodi-N propylamine; POH, nitrosobis-(2-hydroxypropyl)amine; DBN, nitrosodi-N-butylamine; DPH, nitrosodiphenylamine; MEN, nitrosomethylethylamine; MtBN, nitrosomethyl-tert-butylamine; NMU, nitrosomethylundecylamine; NA, 1-nitrosoazetidine; NPYR, 1-nitrosopyrrolidine; NPIP, 1-nitrosopiperidine; PNP, 4-phenylnitrosopiperidine; NHEX, 1-nitrosohexamethyleneimine; NPH, 1-nitrosoheptamethyleneimine; NDM, 1-nitrosododecamethyleneimine; DNZ, dinitrosopiperazine; NM, 4-nitrosomorpholine; DMM, 2,6-dimethylnitroso morpholine; MEU, nitroso-1-methyl-3,3-diethylurea. (Other abbreviations, see Figure 2.) (From Jones and Huberman.[37])

peridine, nitrosomethyl-n-butylamine, and nitrosomethyl-2-phenylethylamine, which do not induce liver tumors in Fischer rats. The critical factors that determine the susceptibility of an organ to a given carcinogen remain largely unknown. Metabolism of the carcinogen to mutagenic intermediates may be only one of the determining factors involved.[9] The mammalian cell-mediated mutagenesis assay, which can use cells from various organs to metabolically activate chemical carcinogens, could thus be used as one of the means of investigating the organ or cell specificity of chemical carcinogens.[39,41]

Studies reviewed in this paper demonstrate that the use of intact hepatocytes to activate nitrosamines, coupled with appropriate mammalian target cells to indicate the genetic damage, represents a useful system for both detecting and establishing the mutagenicity of a series of nitrosamines in a manner corresponding to the carcinogenic potency produced by the same chemicals *in vivo*.

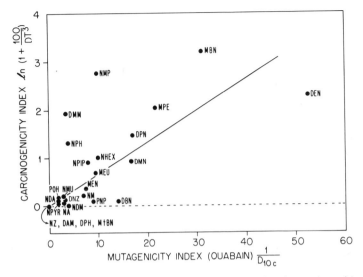

Figure 4. Relationship between carcinogenicity and mutagenicity for nitrosamines with ouabain resistance as the genetic marker. Other information is the same as for Figure 3. (From Jones and Huberman.[37])

Therefore, we suggest that mammalian cell-mediated mutagenesis assays constitute a valuable approach for the detection and evaluation of potentially carcinogenic chemicals in the environment.

ACKNOWLEDGMENTS. This chapter was prepared under the support of the U.S. Department of Energy under contract No. W-31-109-ENG-38.

REFERENCES

1. C. Heidelberger, Chemical carcinogenesis, *Ann. Rev. Biochem. 44*, 79–121 (1975).
2. E. Huberman, Mutagenesis and cell transformation of mammalian cells in culture by chemical carcinogens, *J. Environ. Pathol. Toxicol. 2*, 29–42 (1978).
3. T. Kuroki and C. Heidelberger, The binding of polycyclic aromatic hydrocarbons to the DNA, RNA and proteins of transformable cells in culture, *Cancer Res. 31*, 2168–2176 (1971).
4. H. M. Temin, On the origin of the genes for neoplasia, G. M. A. Clowes Memorial Lecture, *Cancer Res. 34*, 2835–2841 (1974).
5. H. Bartsch, C. Malaveille, A. M. Canuis, G. Brun, and A. Hautefeuille, Validity of bacterial short-term tests for the detection of chemical carcinogens, in: *Short-term Mutagenicity Test Systems for Detecting Carcinogens* (K. Norpoth and R. C. Garner, eds.), Springer-Verlag, New York (1979).
6. H. H. Hiatt, J. D. Watson, and J. A. Winston (eds.), *Origins of Human Cancer*, Cold Spring Harbor Laboratory Publications, Cold Spring Harbor, New York (1977).

7. J. McCann, E. Choi, E. Yamasaki, and B. N. Ames, Detection of carcinogens as mutagens in the *Salmonella*/microsome test: Assay of 300 chemicals, *Proc. Natl. Acad. Sci. USA* 72, 5135–5139 (1975).
8. M. Meselson and K. Russell, Comparisons of carcinogenic and mutagenic potency, in: *Origins of Human Cancer* (M. M. Hiah and J. D. Watson, eds.), Book C, pp. 1473–1482, Cold Spring Harbor Laboratory Publications, Cold Spring Harbor, New York (1977).
9. P. N. Magee, Organ specificity of chemical carcinogenesis, in: *Carcinogenesis* (G. P. Magison ed.), Vol. 1, pp. 213–221, Pergamon Press, Oxford (1979).
10. E. Huberman and L. Sachs, Cell-mediated mutagenesis of mammalian cells with chemical carcinogens, *Int. J. Cancer 13*, 326–333 (1974).
11. E. C. Miller and J. A. Miller, Mechanism of chemical carcinogenesis: Nature of proximate carcinogens and interactions with macromolecules, *Pharmacol. Rev. 18*, 805–835 (1966).
12. P. Brookes and P. D. Lawley, Evidence for the binding of polynuclear aromatic hydrocarbons to the nucleic acids of mouse skin: Relation between carcinogenic power of hydrocarbons and their binding to deoxyribonucleic acid, *Nature (London) 202*, 781–784 (1964).
13. R. F. Newbold, C. B. Wigley, M. H. Thompson, and P. Brookes, Cell-mediated mutagenesis in cultured Chinese hamster cells by carcinogenic hydrocarbons: Nature and extent of the associated hydrocarbon–DNA reaction, *Mutat. Res. 43*, 101–116 (1977).
14. J. D. Selkirk, Benzo(a)pyrene carcinogenesis—A biochemical selective mechanism, *J. Toxicol. Environ. Health 2*, 1245–1258 (1978).
15. C. A. M. Bigger, J. E. Tomaszewski, and A. Diple, Differences between products of binding of 7,12-dimethylbenz(a)anthracene to DNA in mouse skin and in a rat liver microsomal system, *Biochem. Biophys. Res. Commun. 80*, 229–235 (1978).
16. M. V. Malling and L. R. Valvoic, Gene mutation in mammals, in: *Progress in Genetic Toxicology* (D. Scott, B. A. Bridges, and F. H. Sobels, eds.), pp. 155–164, Elsevier/North-Holland Biomedical Press, New York (1977).
17. E. Huberman and L. Sachs, Mutability of different genetic loci in mammalian cells by metabolically activated carcinogenic polycyclic hydrocarbons, *Proc. Natl. Acad. Sci. USA 73*, 188–192 (1976).
18. J. W. Grisham, Use of hepatic cell cultures to detect and evaluate the mechanisms of action of toxic chemicals, *Intl. Rev. Exptl. Pathol. 20*, 123–210 (1979).
19. R. H. San and G. M. Williams, Rat hepatocyte primary cell culture-mediated mutagenesis of adult rat liver epithelial cells by procarcinogens, *Proc. Soc. Exptl. Biol. Med. 156*, 534–538 (1977).
20. J. A. Poiley, R. Raineri, A. W. Andrews, D. M. Cavanaugh, and R. J. Pienta, Metabolic activation by hamster and rat hepatocytes in the *Salmonella* mutagenicity assay, *J. Natl. Cancer Inst. 65*, 1293–1295 (1980).
21. R. Langenbach, H. J. Freed, and E. Huberman, Liver cell-mediated mutagenesis of mammalian cells by liver carcinogens, *Proc. Natl. Acad. Sci. USA 75*, 2864–2867 (1978).
22. C. A. Jones and E. Huberman, A sensitive hepatocyte-mediated assay for the metabolism of nitrosamines to mutagens for mammalian cells, *Cancer Res. 40*, 406–411 (1980).
23. G. Michalopoulos, S. C. Strom, A. D. Kligerman, G. P. Irons, and D. L. Novicki, Mutagenesis induced by procarcinogens at the hypoxanthine–guanine phosphorsybl transferase locus of human fibroblasts cocultured with rat hepatocytes, *Cancer Res. 41*, 1873–1878 (1981).
24. P. N. Magee, R. Montesano, and R. Preussmann, N-Nitroso compounds and related carcinogens, *Am. Chem. Soc. Monogr. 173*, 491–625 (1976).
25. W. Lijinsky, Significance of *in vitro* formation of N-nitroso compounds, *Oncology 37*, 223–226 (1980).
26. I. S. Krull, G. Edwards, M. H. Wolf, T. Y. Fan, and D. M. Fine, N-Nitrosamines in consumer products and in the workplace, *Am. Chem. Soc. Symp. Ser. 101*, 175–194 (1979).

27. C. C. Harris, H. Autrup, G. D. Stoner, E. M. McDowell, B. F. Trump, and P. Schafer, Metabolism of acyclic and cyclic N-nitrosamines in cultured human bronchi, *J. Natl. Cancer Inst.* 59, 1401–1406 (1971).
28. L. Mingxin, L. Ping, and L. Baorong, Recent progress in research on esophageal cancer in China, *Adv. Cancer Res.* 33, 173–249 (1980).
29. T. K. Rao, J. A. Young, W. Lijinsky, and J. L. Epler, Mutagenicity of aliphatic nitrosamines in *Salmonella typhimurium*, *Mutat. Res.* 66, 1–7 (1979).
30. J. R. Fry and J. W. Bridges, The metabolism of xenobiotics in cell suspensions and cell cultures, in: *Progress in Drug Metabolism* (J. W. Bridges and L. F. Chasseaud, eds.), Vol. 2, pp. 71–118, Wiley and Sons, London (1977).
31. R. M. Baker, D. M. Brunette, R. Mankovitz, L. H. Thompson, G. F. Whitmore, L. Siminovitch, and J. E. Till, Ouabain-resistant mutants of mouse and hamster cells in culture, *Cell 1*, 9–21 (1974).
32. A. Abbondanolo, S. Bonatti, G. Corti, R. Fiorio, N. Loprieno, and A. Mazzaccaro, Induction of 6-thioguanine-resistant mutants in V79 Chinese hamster cells by mouse-liver microsome-activated dimethylnitrosamine, *Mutat. Res.* 46, 365–373 (1977).
33. M. O. Bradley, B. Bhuyan, M. C. Francis, R. Langenbach, A. Peterson, and E. Huberman, Mutagenesis by chemical agents in V79 Chinese hamster cells: A review and analysis of the literature, *Mutat. Res.* 87, 81–142 (1981).
34. H. Druckrey, R. Preussmann, S. Ivankovic, and D. Schmahl, Organotrope carcinogene Wirkung bei 65 verschiedenen N-nitroso-Verbindungen an BD-Ratten, *Z. Krebsforsch.* 69, 103–201 (1967).
35. H. Druckrey, Quantitative aspects of chemical carcinogenesis, in: *Potential Carcinogenic Hazards from Drugs (Evaluation of Risks)* (R. Truhaut, ed.), UICC Monogr. Vol. 7, pp. 60–78, Springer-Verlag, New York (1967).
36. H. A. Guess and D. G. Hoel, The effect of dose on cancer latency period, *J. Environ. Pathol. Toxicol.* 1, 279–286 (1977).
37. C. A. Jones and E. Huberman, The relationship between the carcinogenicity and mutagenicity of nitrosamines in a hepatocyte-mediated mutagenicity assay, *Carcinogenesis 2*, 1075–1077 (1981).
38. L. Siminovitch, On the nature of heritable variation in cultured somatic cells, *Cell 7*, 1–11 (1976).
39. C. A. Jones, R. M. Santella, E. Huberman, J. K. Selkirk, and D. Grunberger, Cell specificity in the metabolism, DNA-adduct formation and mutagenicity of benzo(a)pyrene in rat hepatocytes and fibroblasts, *Carcinogenesis*, in press (1983).
40. R. Langenbach, S. Nesnow, A. Tompa, R. Gingell, and C. Kuszynski, Lung and liver cell-mediated mutagenesis systems: Specificities in the activation of chemical carcinogens, *Carcinogenesis 2*, 851–858 (1981).
41. M. N. Gould, Mammary gland cell-mediated mutagenesis of mammalian cells by organ-specific carcinogens, *Cancer Res.* 40, 1836–1841 (1980).

7

Mutagenic Activity of Nitrosamines in Mammalian Cells

Study with the CHO/HGPRT and Human Leukocyte SCE Assays

TI HO, JUAN R. SAN SEBASTIAN, AND ABRAHAM W. HSIE

1. INTRODUCTION

The potential impact of environmental chemicals on the induction of cancer in humans has become a worldwide problem. It has been estimated that more than 80% of human cancers are caused by exposure to chemical agents.[1,2] One class of chemicals that could pose a significant human health hazard is the N-nitroso compounds. These compounds are widespread in our environment,[3] occurring in food preservatives,[4] milk,[5] tobacco smoke,[6,7] meat-curing mixtures,[8] and various other foodstuffs,[9] industrial products and by-products,[10-12] and some plants.[5,13] The finding that feeding secondary amines and nitrite caused tumors in laboratory animals through nitrosamine synthesis in the stomach[14] has led to a considerable concern for nitrosamines as a major class of environmental carcinogens. Although the mode of action of nitrosamines is largely unclear, it is thought that nitrosamines act through biotransformed active metabolites. These metabolites may alkylate cellular macromolecules, particularly DNA, thereby possibly altering the genome of the target cells, which could

---

TI HO, JUAN R. SAN SEBASTIAN, AND ABRAHAM W. HSIE • Biology Division, Oak Ridge National Laboratory, Oak Ridge, Tennessee 37830. Dr. San Sebastian's present address is Pharmakon Research International Inc., Waverly, Pennsylvania 18471. By acceptance of this article, the publisher or recipient acknowledges the U. S. Government's right to retain a nonexclusive, royalty-free license in and to any copyright covering the article.

eventually result in transformation into a neoplastic state.[15] The carcinogenicity of nitroso compounds as related to their chemistry, mutagenicity, and other biological activities has been extensively studied[4,19-29] and reviewed[3,16-18]

A variety of tests have been developed in recent years to determine the mutagenic activity and, by extrapolation, the possible carcinogenic potential of environmental chemicals. Most N-nitroso compounds are mutagenic in various systems;[17] however, the mutagenicity does not always correlate with their carcinogenicity.[25] Many investigators have tested the mutagenic effect of a series of nitrosamines using the *Salmonella* histidine reversion assay,[29-36] the *Saccharomyces* gene conversion test,[37,38] and the *Drosophila* X-linked recessive lethal system.[39,40] All three systems showed good correlation between mutagenicity and carcinogenicity of nitrosamines.

The first part of this article deals with our study of the ability of eight nitrosamines to induce gene mutation at the hypoxanthine-guanine phosphoribosyl transferase (hgprt) locus in Chinese hamster ovary (CHO) cells, the CHO/HGPRT assay.[41] The second portion presents the induction of sister-chromatid exchanges (SCE) by nitrosopiperidine (NPIP) and nitrosopiperazines (NPZ) and their derivatives in a human leukocyte culture system.[42] We also present the correlation between the mutagenicity and carcinogenicity of these nitrosamines.

2. MUTAGENIC ACTIVITY OF EIGHT NITROSAMINES IN THE CHO/HGPRT ASSAY

2.1. The CHO/HGPRT Assay

CHO-K_1-BH_4 cells,[43] a subclone of the near-diploid CHO cell line, were used in these studies. These cells are well-characterized genetically and have been used extensively for studying mutagen-induced cytotoxicity and gene mutation.[44-48] They routinely exhibit a cloning efficiency higher than 80% in a well-defined medium. The population doubling time of these cells is 12-14 hr. Their stable and easily recognizable karyotype has made them one of the best choices for studies on the effect of mutagens at the chromosome level.[47-49]

The experimental procedures for cell culture, treatment with chemicals, and measurement of cytotoxicity and gene mutation have been described previously.[41,43,50] Briefly, to measure gene mutation, we quantified the frequency of mutants resistant to a purine analog, 6-thioguanine (TG). It should be noted that the CHO/HGPRT assay has been defined in terms of medium, pH, TG concentration, optimal cell density for selection (and, hence, recovery of the presumptive mutants), and expression time for the mutant phenotype.[44,50] Mutagenicity of promutagens is determined by coupling a metabolic activation system (S-9) derived from Aroclor 1254-preinduced male Sprague–Dawley rat livers to the CHO/HGPRT assay.[45,51,52]

The CHO/HGPRT assay permits determination of gene mutation and cytotoxicity induced by various promutagens and direct-acting mutagens. The quantitative nature of the system provides a basis for the study of structure–mutagenicity relationships within a given class of compounds. An analysis of mutagenicity of 68 chemicals, including alkanesulfonates, alkylsulfonates, nitrosoguanidines, and nitrosoureas,[53,54] heterocyclic mustards (ICR compounds),[55,56] platinum (II)chloroamines,[57] quinolines, nitrosamines, haloethanes, polyaromatic hydrocarbons, and miscellaneous compounds, has been summarized.[58] Mutagenic activity of a chemical can be expressed as mutation frequency on a molar basis. Using this expression for potency, useful information has been obtained comparing mutagenicity of structurally related direct-acting mutagens.[58] Studies on the interrelationships of molecular lesions and mutation induction reveal that both the nature and the quantity of mutagen binding to DNA affect chemical mutagenicity.[57,59]

The CHO/HGPRT assay can be expanded to measure mutagen-induced cytogenetic effects.[49] We term this expanded system the Multiplex CHO System,[45,47,48] in which cytotoxicity, gene mutation, chromosome aberration, and SCE can be determined in the same treated culture. Thus, this system appears to be useful as a means to gain insight on the interrelationship of these biological end points.

2.2. Mutagenic Activity of Nitrosodimethylamine as Studied in the CHO/HGPRT Assay

Because CHO cells are incapable of transforming nitrosodimethylamine (NDMA) into biologically active metabolites, NDMA (33–1333 μg/ml) is neither mutagenic nor cytotoxic to the cells when tested as a direct-acting agent. However, when CHO cells were injected into the peritoneal cavity of athymic BALB/c mice, followed by subcutaneous injection of NDMA, mutation induction increased with increasing quantity of NDMA administered.[51] Later experiments showed that athymic (nude) mice, their phenotypically hairy littermates, or BALB/c mice of both sexes are suitable as hosts.[60] In addition, using a 2-hr treatment time, we observed a linear dose–response relationship up to 250 mg of NDMA/kg body weight. At 100 and 500 mg of NDMA/kg, mutation induction increased with time up to at least 6 hr.[60]

The rat liver S-9 mix used in the *Salmonella typhimurium* histidine reversion assay[61] is capable of activating the promutagen NDMA in the CHO/HGPRT assay. We first observed that both cytotoxicity and mutagenicity of NDMA (5 mg/ml) increases with an increasing amount (1.5–3.0 mg/ml) of S-9 preparation.[51] Further study showed that addition of $CaCl_2$ (10 mM) to

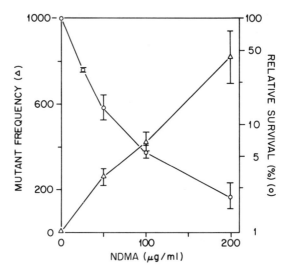

Figure 1. Mutagenicity and cytotoxicity of NDMA as a function of treatment concentration. Mutation frequency ($\triangle$-$\triangle$) is expressed as mutants/10^6 clonable cells; the spontaneous frequency is 0-10×10^{-6} mutants/cell. Cytotoxicity (O-O) is expressed in percent as the cloning efficiency of the treated culture relative to the untreated one, which exhibited an absolute efficiency of 70–95%. The vertical lines represent the standard error of the means. This graph is redrawn from data presented in Reference 52.

the S-9 mix increased the sensitivity for determining the mutagenicity of NDMA by 250-fold; this enhancement was dependent on the presence of sodium phosphate.[52] Under conditions of maximum enhancement (10 mM calcium chloride, 10 mM magnesium chloride, and 50 mM sodium phosphate in the S-9 mix), both cellular lethality and mutation induction increase with increasing the amount of S-9 protein (0.3–3.0 mg/ml). Time course studies showed that both cytotoxic and mutagenic effects were time-dependent over a range of 1–5 hr. The concentration-dependent increase of NDMA mutagenicity under optimal conditions is shown in Figure 1. The mutagenicity of NDMA can be described as 110 mutants/10^6 cells per mM NDMA per mg/ml of S-9 protein per hour.[52]

Recently, we found that the calcium phosphate effect is not observed with S-9 mix buffered with N-2-hydroxyethylpiperazine-N'-ethanesulfonic acid (HEPES). Commercially obtained calcium phosphate, tricalcium phosphate, and alumina Cγ gels could substitute for CaCl$_2$ in the S-9 mix, but diethylaminoethyl (DEAE) cellulose could not. Alumina Cγ gel can exert its effect in the absence of both CaCl$_2$ and phosphate.[62] The mode of action whereby these gels exert their biological activity is as yet unknown. It is possible that microsomes are absorbed by the gels and then settle in close proximity to the cells. This would allow the demethylated product of NDMA to enter the cells rapidly.[62]

2.3. Mutagenicity of Eight Nitrosamines in the CHO/HGPRT Assay and Its Correlation with Carcinogenicity

Having quantified the mutagenic activity of DMN, we studied the mutagenicity of eight representative nitrosamines whose carcinogenic activities are known (References 19–29, W. Lijinsky, personal communication) (Table I).

We express the mutagenicity of each nitrosamine tested in the presence and absence of S-9 in three ways: first, mutation frequency, which is the number of mutant colonies that arise from 1×10^6 cells corrected for the cloning efficiency of the cells; second, mutagenic response, which assesses a positive or negative activity of a test chemical, with a positive response demonstrated by a concentration-dependent increase of mutation frequency over the spontaneous level, and three or more concentrations tested yielding an observed frequency of 50×10^{-6} or higher mutants/clonable cell[41]; and third, mutagenic potency, which is mutation frequency per unit concentration (μM) tested (Table I).

Six of the eight nitrosamines studied are carcinogenic in test animals and all but 4-methyl-1-nitrosopiperazine, a weak carcinogen, showed mutagenic activity. All of the mutagenic nitrosamines require S-9 activation to exert their mutagenic activity. The two noncarcinogenic nitrosamines (nitroso-3-methyl-2-phenylmorpholine and nitroso-2,5-dimethyl-pyrrolidine) are not mutagenic. With this limited number of nitrosamines tested, there exists 88% (7/8) correlation between mutagenicity in the CHO/HGPRT assay and the tumorigenicity in the animal. A clear structure–activity relationship appears to exist for the two pyrrolidines; the carcinogen nitrosopyrrolidine is mutagenic whereas the noncarcinogen nitroso-2,5-dimethyl-pyrrolidine is not mutagenic.

It should be noted that determination of the mutagenicity of the weak carcinogen 4-methyl-1-nitrosopiperazine was not exhaustive; the highest test concentration (4 mM) still resulted in 80% survival. Whether this nitrosamine would exhibit a mutagenic response at test concentrations higher than 4 mM remains to be determined. This concern extends to studies with nitroso-3-methyl-2-phenylmorpholine and nitroso-2,5-dimethylpyrrolidine since the highest test concentration permits 70% survival. As for other mutation assays, a negative test result in the CHO/HGPRT assay is subject to reservations such as the point discussed, and a positive result is more easily established. For example, ethyl methanesulfonate and ICR-191[43,55] exhibit high mutation frequency at concentrations that do not show appreciable toxicity.

Of the five mutagenic nitrosamines (nitrosodimethylamine, nitrosodiethylamine, nitrosomorpholine, nitrosopyrrolidine, and dinitrosohomopiperazine) tested, a 100-fold difference in the mutagenic potency (nitrosodimethylamine vs. nitrosopyrrolidine) can be detected by the CHO/HGPRT assay. Bearing in mind the difference in the determination of the biological end points between mutagenicity and carcinogenicity, the correlation of the potency determination

Table I
The Mutagenicity and Cytotoxicity of Eight Nitrosamines in the CHO/HGPRT Assay

Compound	Concentrations (μM)	S-9	Survival (%)	Mutagenicity[a]			Carcinogenicity[b] (organ)
				Mutation frequency (MF) (mutants/10^6 cells)	Mutagenic response	Mutagenic potency (MF/μM)	
Nitrosomethylamine	400– 4,590	+	25– 0.5	120–1,780	+	0.39	+++ (liver)
	400– 4,590	–	97–89	SMF			
Nitrosoethylamine	290– 7,340	+	85– 5	33–938	+	0.13	++++ (liver, esophagus)
	290– 7,340	–	100–81	SMF			
Nitrosomorpholine	500– 5,000	+	70–57	18–96	+	0.02	++ (liver)
	500– 5,000	–	68–52	SMF			
Nitroso-3-methyl-2-phenylmorpholine	10– 1,000	+	97–67	SMF	–	NM	–
	10– 1,000	–	97–92	SMF			
Nitrosopyrrolidine	5,000–25,000	+	57–43	21–75	+	0.003	++ (liver)
	50,000	–	72	SMF			
Nitroso-2,5-dimethylpyrrolidine	5,000–25,000	+	90–74	SMF	–	NM	–
	50,000	–	72	SMF			
4-Methyl-1-nitrosopiperazine	50– 4,000	+	100–78	SMF	–	NM	+ (nasal cavity)
		–	100–99	SMF			
Dinitroso-homopiperazine	5– 2,000	+	97–69	5–68	+	0.03	+++ (esophagus)
	5– 2,000	–	100–80	SMF			

[a] Mutagenicity is expressed in three ways: (1) Mutation frequency: the number of mutants observed per 10^6 cells at the specified mutagen concentration; the spontaneous mutation frequency (SMF) is 0–15 mutants/10^6 cells. (2) Mutagenic response: a positive (+) or negative (–) response of a given chemical according to criteria set by EPA Gene-Tox Report,[41] in which a positive response requires a concentration-related increase over three or more concentrations, yielding a mutation frequency of $\geq 50 \times 10^{-6}$ mutants/clonable cell. (3) Mutagenic potency: mutation frequency per micromole of mutagen tested.
[b] Carcinogenicity is expressed as very strong (++++), strong (+++), moderate (++), and weak (+) activity in organs specified (References 19–29 and W. Lijinsky, personal communication).

of these five compounds appears to be good because potent mutagens (nitrosodimethylamine and nitrosodiethylamine) are strong carcinogens, and moderate mutagens (nitrosomorpholine, nitrosopyrrolidine, and dinitrosohomopiperazine) exhibit moderate carcinogenicity.

3. MUTAGENIC ACTIVITY OF 15 NITROSAMINES IN THE HUMAN LEUKOCYTE SCE ASSAY

3.1. Sister-Chromatid Exchanges (SCE) in the Human Leukocyte Culture System

It was reported nearly a quarter century ago that a reciprocal labeling of sister chromatids of metaphase chromosome occurs in cells that have undergone a full cell cycle of [^{3}H]thymidine incorporation followed by a replicative cycle in the nonlabeled medium.[63] The finding that 5-bromo-2'-deoxyuridine (BrdU) substitution into DNA quenches the fluorescence of bound dyes such as Hoechst 33258[64,65] has facilitated the study of SCE by differential staining of sister chromatids. The BrdU–Hoechst staining technique has allowed the effects of many mutagens and carcinogens on SCE *in vitro* and *in vivo* to be studied.[66-70] SCE appears to be a sensitive indicator of DNA damage[66-70] and it has been suggested that it correlates with the mutagenic potential.[71,72] The correlation between carcinogenicity, gene mutation, and SCE appears to be good,[66-70] with a few exceptions.[70,73,74] Although the molecular basis of SCE is largely unknown, SCE represents interchanges of homologous chromatids, presumably as a result of "errors" during DNA replication.

We used the BrdU–Hoechst technique to study the nitrosamine-induced SCE in human leukocytes in culture.[42] Briefly, heparinized human blood was obtained and processed by the buffy coat technique.[75] The leukocytes were inoculated into 5 ml of medium (RPMI 1640, GIBCO) in a T25 flask containing 27% human AB$^+$ heterologous serum, 3% phytohaemagglutinin (PHA, Wellcome), and 2×10^{-5} M BrdU (Sigma). Since different blood donors were used, concurrent untreated control samples were included in all experiments. The cultures were incubated at 37°C in a 5% CO$_2$ incubator. At 30–32 hr after initiation the indicated amount of test chemicals in dimethyl sulfoxide (DMSO) (30 μl), with or without S-9 mix, was added to the culture. Addition of 2,6-dimethyl- and 2,3,5,6-tetramethyl-DNPZ at higher concentrations (10^{-3}–10^{-2} M) to the culture medium formed a visible precipitate. S-9 mix was prepared as described previously.[61] Five hours after the addition of the test chemical, the cells were washed twice with Hank's balanced salt solution and replaced in fresh medium without the test chemical. Incubation was continued for a total of 70 hr and with treatment with 2×10^{-7} M Colcemid® (GIBCO) during the final 2.5 hr.

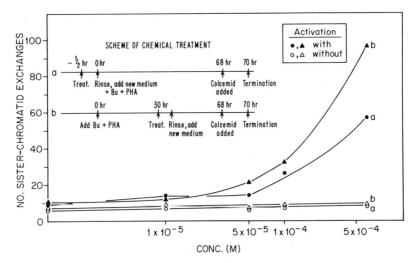

Figure 2. The effects of adding cyclophosphamide at different times relative to PHA initiation on the SCE induction in human leukocyte cultures. Cyclophosphamide (10^{-5}–5×10^{-4} M) was added ½ hr before (a) or 30 hr after (as indicated by the arrow labeled "Treat") PHA initiation. Activation refers to the presence (●, ▲) or absence (○, △) of rat liver S-9. Results are expressed as the number of SCEs/cell vs. concentrations of cyclophosphamide.

A positive control with cyclophosphamide and metabolic activation was performed. A 30-min treatment time was used with this chemical to minimize the cytotoxicity to the culture. The effect of adding cyclophosphamide at two different time intervals is shown in Figure 2. We found that cyclophosphamide is more effective in inducing SCE if it is added 30 hr after PHA initiation (curve b), rather than prior to PHA initiation (curve a). Thus, treatment during S-phase is significantly more effective in inducing SCEs. Accordingly, we chose to add the test chemical to the culture after PHA stimulation.

Cells were collected and the air-dried cytological preparations were made as previously described.[76] After 1–2 days the slides were stained by a modified fluorescence-plus-Giemsa technique.[64,67] One to two drops of Hoechst 33258 stain (50 μg/ml, America Hoechst Co.) were added to the surface of the slides. Cover glasses were placed over the stain for 20 min, and the slides were rinsed with distilled water and air dried. The slides were then mounted in citrate–phosphate buffer (0.0065 M citric acid and 0.087 M Na_2HPO_4) (pH 7.0) and illuminated with two 15-W Ge black light tubes[77-79] at a distance of 5 cm for 30 min. Immediately after the black light exposure, the slides were placed in a Coplin jar containing 2X SSC (0.3 M NaCl + 0.03 M sodium citrate at pH 7.0) and kept in a 62°C water bath for 90 min. The slides were stained with 3% Giesma in 0.1 M phosphate buffer (pH 6.8) for 12 min. After drying, the slides were mounted in Euparal (Heaton Mills) and analyzed by light microscopy.

Twenty intact and well-spread second metaphase cells were scored for SCE. Results are expressed as the mean numbers of SCE/cell ± the standard error of the mean among the 25 cells scored.

3.2. Induction of SCE by 15 Cyclic Nitrosamines in Human Leukocytes *in Vitro* and Its Correlation with Carcinogenicity

Representatives of two groups of cyclic nitroso compounds were tested for their effectiveness in inducing SCE in cultured human leukocytes. A comparison of SCE frequencies of 15 untreated controls with and without S-9 indicates that 12 of them had an increased frequency in the presence of S-9, although the magnitude of the difference is too small to be statistically significant.

The effects of nitrosopiperidines (NPIP) with and without metabolic activation are shown in Table II. There is no increase of SCE by NPIP or its derivatives without metabolic activation. With activation both NPIP and 3-methyl-NPIP produced a significant twofold dose-dependent increase in SCE. A greater than twofold dose-dependent response was produced by 3,4-dichloro-NPIP. Only a marginal effect of 4-chloro-NPIP on SCE was observed. At the lowest concentrations (10^{-6}–10^{-5} M) used, 2,6-dimethyl-NPIP was too cytotoxic to score SCEs. Of the oxygenated derivatives only nitroso-3-piperidinol increased the SCE by 50%. Nitroso-4-piperidone did not cause an increase of SCE. Nitrosonipecotic acid was the only carboxylated derivative to show a significant increase in SCE with a dose dependence. Nitrosopipecolic acid and nitrosoguvacoline showed a marginal effect, which was not dose-dependent. At the highest concentrations (1×10^{-2} M), nitrosoguvacoline was cytotoxic even without metabolic activation.

The SCE data obtained with nitrosopiperazine (NPZ) and four of its derivatives are presented in Table III. Like NPIP and its derivatives, none of the nitrosopiperazines were active in the absence of metabolic activation. The parent compound NPZ was only slightly effective in inducing SCE and its effect was not dose-dependent. Dinitrosopiperazine was the only compound in this group to produce a 50% increase at 10^{-2} M with a dose dependence. 2-methyl-NPZ did not increase the number of SCE and was cytotoxic at the highest concentration (10^{-2} M), with or without metabolic activation. 2,6-Dimethyl-NPZ and 2,3,5,6-tetramethyl-NPZ had little, if any, effect on SCE induction.

The correlation between SCE induction and tumorigenicity by these nitrosamines is summarized in Table IV. Also included in this table is the mutagenicity data from *Salmonella*,[33-36] *Saccharomyces*,[37,38] and *Drosophila*.[39,40]

All six carcinogenic nitrosopiperidines except nitroso-4-piperidone were found to be effective in inducing SCE. However, results with noncarcinogenic nitrosopiperidines are less consistent; in fact, all three carboxyl derivatives of nitrosopiperidines, which are noncarcinogenic, caused a marginal to positive response for SCE induction at the highest concentrations (10^{-3}–10^{-2} M) tested.

Table II
Effect of Nitrosopiperidines (NPIP) on SCE in Human Leukocytes in Culture

| Compounds | S-9 activation | \
Concentration (M) | | | | |
|---|---|---|---|---|---|---|
| | | 0 | 10^{-5} | 10^{-4} | 10^{-3} | 10^{-2} |
| 1. Nitrosopiperidine (NPIP) | − | 7.6 ± 0.61[a] | — | 8.6 ± 0.66 | 8.8 ± 0.66 | 7.6 ± 0.61 |
| | + | 8.0 ± 0.63 | — | 8.7 ± 0.66 | 11.4 ± 0.75 | 17.2 ± 0.93 |
| 2. Methylated compounds | | | | | | |
| 3-Methyl NPIP | − | 5.7 ± 0.53 | — | 6.4 ± 0.57 | 8.0 ± 0.63 | 7.8 ± 0.62 |
| | + | 7.7 ± 0.62 | | 7.9 ± 0.6 | 10.7 ± 0.73 | 12.8 ± 0.80 |
| 2,6-Dimethyl NPIP | − | 7.1 | t[b] | t | t | — |
| | + | 7.6 | t | t | t | — |
| 3. Halogenated compounds | | | | | | |
| 4-Chloro NPIP | − | 6.4 ± 0.56 | 6.2 ± 0.56 | 7.8 ± 0.62 | 8.0 ± 0.65 | t |
| | + | 6.9 ± 0.56 | 8.9 ± 0.67 | 9.0 ± 0.67 | 10.4 ± 0.72 | t |
| 3,4-Dichloro NPIP | − | 6.1 ± 0.55 | 5.4 ± 0.52 | 5.7 ± 0.53 | 6.3 ± 0.61 | 6.8 ± 0.58 |
| | + | 6.5 ± 0.57 | 7.7 ± 0.62 | 10.6 ± 0.73 | 16.6 ± 0.91 | 19.1 ± 0.98 |
| 4. Oxygenated compounds | | | | | | |
| Nitroso-3-piperidinol | − | 8.7 ± 0.66 | — | 8.3 ± 0.64 | 8.5 ± 0.65 | 8.0 ± 0.63 |
| | + | 9.8 ± 0.70 | | 9.3 ± 0.68 | 10.8 ± 0.73 | 12.4 ± 0.79 |
| Nitroso-4-piperidone | − | 6.1 ± 0.55 | 7.6 ± 0.62 | 7.8 ± 0.62 | 7.9 ± 0.59 | 7.0 ± 0.59 |
| | + | 8.0 ± 0.63 | 8.4 ± 0.65 | 7.4 ± 0.61 | 5.8 ± 0.54 | 7.8 ± 0.62 |
| 5. Carboxylated derivatives | | | | | | |
| Nitrosopipecolic acid (2-carboxyl-NPIP) | − | 6.5 ± 0.57 | 6.4 ± 0.56 | 7.1 ± 0.60 | 6.7 ± 0.58 | 6.2 ± 0.60 |
| | + | 8.4 ± 0.65 | 7.4 ± 0.61 | 7.9 ± 0.63 | 8.4 ± 0.65 | 10.5 ± 0.72 |
| Nitrosonipecotic acid (3-carboxyl-NPIP) | − | 7.1 ± 0.59 | 7.1 ± 0.59 | 5.8 ± 0.54 | 5.8 ± 0.54 | 6.5 ± 0.58 |
| | + | 8.8 ± 0.66 | 9.9 ± 0.70 | 9.4 ± 0.68 | 11.0 ± 0.74 | 13.7 ± 0.83 |
| Nitrosoguvacoline (3-carboxylmethyl-Δ³- | − | 5.4 ± 0.52 | 5.9 ± 0.54 | 6.0 ± 0.55 | 6.3 ± 0.56 | t |
| | + | 7.3 ± 0.60 | 8.9 ± 0.67 | 7.6 ± 0.62 | 10.7 ± 0.73 | t |

[a] Data are expressed as the number of SCEs per cell ± the standard error of the mean among cells.
[b] t: Toxic to the cells at the indicated test concentration.

Table III
Effect of Nitrosopiperazines (NPZ) on SCE in Human Leukocytes in Culture

Compounds	S-9 activation	0	10^{-5}	10^{-4}	10^{-3}	10^{-2}
1. Nitrosopiperazine						
(NPZ)	−	6.5 ± 0.57	6.0 ± 0.55	7.0 ± 0.59	5.3 ± 0.51	6.7 ± 0.58
	+	6.4 ± 0.56	8.1 ± 0.63	9.5 ± 0.69	8.7 ± 0.66	9.7 ± 0.70
2. Dinitrosopiperazine						
(DNPZ)	−	7.5 ± 0.61	6.3 ± 0.56	7.5 ± 0.61	7.9 ± 0.63	6.7 ± 0.58
	+	8.1 ± 0.63	7.2 ± 0.60	8.7 ± 0.66	9.8 ± 0.70	11.9 ± 0.77
3. Methylated compounds						
2-Methyl DNPZ	−	7.4 ± 0.61	7.9 ± 0.63	7.5 ± 0.61	7.3 ± 0.60	t[b]
	+	7.6 ± 0.61	8.1 ± 0.64	8.5 ± 0.65	8.2 ± 0.64	t
2,6-Dimethyl DNPZ	−	6.6 ± 0.57	6.3 ± 0.56	6.3 ± 0.56	6.0 ± 0.55[c]	6.0 ± 0.55[c]
	+	6.5 ± 0.57	6.5 ± 0.57	6.7 ± 0.58	8.0 ± 0.63[c]	8.4 ± 0.65[c]
2,3,5,6-Tetramethyl DNP	−	6.2 ± 0.56	7.3 ± 0.60	6.1 ± 0.55	6.6 ± 0.57[c]	6.3 ± 0.56[c]
	+	6.2 ± 0.55	8.0 ± 0.63	6.9 ± 0.59	9.0 ± 0.67	8.4 ± 0.65[c]

[a] Data are expressed as the number of SCEs per cell ± the standard error of the mean among cells.
[b] t: Toxic to the cells.
[c] Chemical precipitation in culture medium.

Table IV
Correlation between Mutagenicity and Carcinogenicity of NPIP and NPZ

Chemicals	Mutagenicity[a] (+S-9)			SCE in leukocytes (+S-9)	Carcinogenicity[b]
	Salmonella	Saccharomyces	Drosophila[a]		
Nitrosopiperidine (NPIP)					
1. Nitrosopiperidine (NPIP)	+	+	+	+	+
2. Methylated compounds					
3-Methyl NPIP	+	+	+	+	+
2,6-Dimethyl NPIP	−	−	−	t[c]	−
3. Halogenated compounds					
4-Chloro NPIP	+	NT[d]	+	+	+
3,4-Dichloro NPIP	+	NT	+	+	+
4. Oxygenated compounds					
Nitroso-3-piperidinol	+	NT	NT	+	+
Nitroso-4-piperidone	+	NT	−	−	+

5. Carboxyl derivatives				
Nitrosopipecolic acid	−	NT	−	−
Nitrosonipecotic acid	−	NT	NT	−
Nitrosoguvacoline	+	NT	−	−
Nitrosopiperazines (NPZ)				
1. 1-Nitrosopiperazine (NPZ)	−	−	−	−
2. Dinitroso piperazine (DNPZ)	+	+	+	+
3. Methylated compounds				
2-Methyl DNPZ	+	+	+	+
2,6-Dimethyl DNPZ	+	+	− ^{e}	+
2,3,5,6-Tetramethyl DNPZ	−	−	− ^{e}	−

^{a} Data on *Salmonella*, *Saccharomyces*, and *Drosophila* assays are from References 33–36, 37 and 38, and 39 and 40, respectively.
^{b} Data on carcinogenicity are from References 19–29 and personal communication with W. Lijinsky.
^{c} t: Toxic to the cells at $\geq 1 \times 10^{-6}$ M tested.
^{d} NT: Not tested.
^{e} Chemical precipitation in the culture medium.

Three carcinogenic nitrosopiperazines (dinitrosopiperazine, 2-methyl-dinitrosopiperazine, and 2,6-dimethyl-dinitrosopiperazine) showed negative to marginal responses; problems with toxicity and solubility apparently complicated these SCE studies. Two noncarcinogenic nitrosopiperazines (nitrosopiperazine and dinitroso-2,3,5,6-tetramethyl-piperazine) did not induce SCE; the study with the latter compound was complicated by the formation of a precipitate upon addition of the test compound to the culture medium. In contrast to these results, a full concordance between carcinogenicity and mutagenicity in *Salmonella, Saccharomyces,* and *Drosophila* was found for these five nitrosopiperazines.

4. SUMMARY AND CONCLUDING REMARKS

The mutagenic and cytotoxic activities of eight nitrosamines [nitrosodimethylamine, nitrosodiethylamine, nitrosomorpholine, nitroso-3-methyl-2-phenylmorpholine (nitrosophenmerazine), nitrosopyrrolidine, nitroso-2,5-dimethylpyrrolidine, 4-methyl-1-nitrosopiperazine, and dinitrosohomopiperazine] have been quantified in a mammalian cell gene mutational system, the Chinese hamster ovary cell/hypoxanthine–guanine phosphoribosyl transferase (CHO/HGPRT) assay coupled with a rat liver metabolic activation (S-9) system. Six of the eight nitrosamines are carcinogenic and all but a weak carcinogen, 4-methyl-1-nitrosopiperazine, showed mutagenic activity. The two noncarcinogenic nitrosamines (nitroso-3-methyl-2-phe-nylmorpholine and nitroso-2,5-dimethyl-pyrrolidine) are not mutagenic. The potent mutagens nitrosodimethylamine and nitrosodiethylamine are strong carcinogens, and the moderate mutagens nitrosomorpholine, nitrosopyrrolidine, and dinitroso-homopiperazine exhibit moderate carcinogenicity. High correlation appears to exist between CHO/HGPRT mutagenicity and animal carcinogenicity both qualitatively and quantitatively.

Fifteen nitrosamines [nitrosopiperidine (NPIP), 2-methyl-, 2,6-dimethyl-, 4-chloro-, and 3,4-dichloro-NPIP, nitroso-3-piperidinol, nitroso-4-piperidone, nitrosopipecolic acid, nitrosonipecotic acid, nitrosoguvacoline, N-nitrosopiperazine (NPZ), dinitrosopiperazine (DNPZ), 2-methyl-, 2,6-dimethyl-, and 2,3,5,6-tetramethyl-DNPZ] have been screened (10^{-5}–10^{-2} M) for their ability to induce sister-chromatid exchanges (SCE) in a S-9-coupled human leukocyte culture system. Three carcinogens, NPIP, 3-methyl-, and 3,4-dichloro-NPIP, exhibited a dose-dependent twofold increase of SCE induction in the presence of S-9, but there is no induction without S-9. However, three carcinogens (DNPZ, 2-methyl-, and 2,6-dimethyl DNPZ) showed negative to marginal response; toxicity and solubility complicate these studies. Two noncarcinogens (NPZ and 2,3,5,6-tetramethyl NPZ) did not induce SCE. Correlation between animal carcinogenicity and SCE induction appears to be good with these nitrosamines.

ACKNOWLEDGMENTS. We thank W. Lijinsky for supplying cyclic nitrosamines; W. Winton and J. L. Epler for S-9 preparation; and R. J. Preston, R. L. Schenley, and L. C. Waters for reviewing the manuscript. Research sponsored by the Office of Health and Environment Research, U. S. Department of Energy, under contract No. W-7405-eng-26 with the Union Carbide Corporation.

5. REFERENCES

1. E. Boyland, The correlation of experimental carcinogenesis and cancer in man, *Prog. Expt. Tumor Res. 11*, 222-234 (1969).
2. P. Bogovski, The importance of the analysis of N-nitroso compound in international cancer research, in: *N-Nitroso Compounds Analysis and Formation* (R. Preussman, E. A. Walker, and W. Davis, eds.), IARC Scientific Publications No. 3, pp. 1-5, International Agency for Research on Cancer, Lyon, France (1972).
3. W. Lijinsky and S. S. Epstein, Nitrosamines as environmental carcinogens, *Nature (London) 255*, 21-23 (1978).
4. H. Druckrey, R. Preussmann, S. Ivankovic, and D. Schmähl, Organotrope carcinogene wirkungen bei 65 versabiedenon N-nitroso verbindungen and BD ratten, *Z. Krebsforsch. 69*, 103-120 (1967).
5. L. Hedler and P. Marquardt, Occurrence of diethylnitrosamine in some samples of food, *Food Cosmet. Toxicol. 6*, 341-348 (1968).
6. H. Druckrey and R. Preussman, Zur Entstehung carcinogener Nitrosamine und beispiel des Tabakrauches, *Naturwissenschaften 49*, 498-499 (1962).
7. G. B. Neurath, Nitrosamine formation from precursors in tobacco smoke, in: *N-Nitroso Compounds Analysis and Formation* (P. Bogovski, R. Preussman, E. A. Walker, and W. Davis, eds.), IARC Scientific Publications No. 3, pp. 134-136, International Agency for Research on Cancer, Lyon, France (1972).
8. N. P. Sen, W. F. Mile, B. Donaldson, T. Panalaks, and J. R. Iyengar, Formation of nitrosamines in a meat curing mixture, *Nature (London) 245*, 104-105 (1973).
9. I. A. Wolff and A. E. Wasserman, Nitrates, nitrites and nitrosamine, food stuff and tobacco contains nitrosamines, *Science 177*, 15-19 (1972).
10. L. Fishbein, in *Studies in Environmental Sciences 4: Potential Industrial Carcinogens and Mutagens*, pp. 331-351, Elsevier Scientific Publishing Co., Amsterdam (1979).
11. P. N. Magee, Possible hazard from nitrosamines in industry, *Ann. Occup. Hyg. 15*, 19-22 (1972).
12. S. Z. Cohn, G. Zweig, M. Law, D. Wright, and W. R. Bontoyan, Analytical determination of N-nitroso compounds in pesticides by the United States Environmental Protection Agency—A preliminary study, in: *Environmental Aspects of N-Nitroso Compounds* (E. A. Walker, M. Gastegnaro, L. Griciute, R. E. Lyle, and W. David, eds.), IARC Scientific Publications No. 19, pp. 333-356, International Agency for Research on Cancer, Lyon, France (1978).
13. M. Nagao, T. Sugimura, and T. Matsushima, Environmental mutagens and carcinogens, *Ann. Rev. Genet. 12*, 117-159 (1978).
14. G. Eisenbrand, O. Ungerer, and R. Preussmann, Rapid formation of carcinogenic N-nitrosamine by interaction of nitrite with fungicides derived from dithiocarbamic acid *in vitro*

under simulated gastric conditions and *in vivo* in the rat's stomach, *Food Cosmet. Toxicol.* 12, 229-232 (1974).
15. P. N. Magee and E. Farber, Toxic liver injury and carcinogenesis: Methylation of rat liver nucleic acid by dimethylnitrosamine *in vitro, Biochem. J.* 83, 114-124 (1962).
16. P. N. Magee and J. M. Barnes, Carcinogenic nitroso compounds, *Adv. Cancer Res.* 10, 163-246 (1967).
17. R. Montesano and H. Bartsch, Mutagenic and carcinogenic N-nitroso-compound: Possible environmental hazards, *Mutat. Res.* 32, 179-228 (1978).
18. J.-P. Anselme, The organic chemistry of N-nitrosamines: A brief review, in: *N-nitrosamines* (J.-P. Anselme, ed.), ACS Symposium Series 101, pp. 1-12, American Chemical Society, Washington, D. C. (1979).
19. H. Garcia and W. Lijinsky, Tumorigenicity of five cyclic nitrosamines in MRC rat, *Z. Krebsforsch.* 77, 257-261 (1972).
20. H. Garcia and W. Lijinsky, Studies of the tumorigenic effect in feeding of nitrosamino acids and of low doses of amines and nitrite to rats, *Z. Krebsforsch.* 79, 141-144 (1972).
21. W. Lijinsky and H. W. Taylor, Carcinogenicity of methylated nitrosopiperidine, *Int. J. Cancer* 16, 318-322 (1975a).
22. W. Lijinsky and H. W. Taylor, Carcinogenicity of methylated dinitropiperazine in rats, *Cancer Res.* 35, 1270-1273 (1975b).
23. W. Lijinsky and H. W. Taylor, Tumorigenesis by oxygenated nitrosopiperidines in rats, *J. Natl. Cancer Inst.* 55, 705-707 (1975c).
24. W. Lijinsky and H. W. Taylor, Carcinogenicity of N-nitroso 3,4-dichloro and N-nitroso 3,4-dibromopiperidine in rats, *Cancer Res.* 35, 3209-3211 (1975d).
25. W. Lijinsky and R. K. Elespuru, Mutagenicity and carcinogenicity of N-nitroso derivatives of carbamate insecticides, in: *Environmental N-Nitroso Compounds Analysis and Formation* (E. A. Walker, P. Bogovski, L. Griciate, and W. David, eds.), IARC Scientific Publications No. 14, pp. 425-428, International Agency for Research on Cancer, Lyon, France (1976).
26. W. Lijinsky and H. W. Taylor, Carcinogenicity tests of N-nitroso derivatives of two drugs, phenmetrazine and methylphenidate, *Cancer Lett.* 1, 359-363 (1976).
27. W. Lijinsky and H. W. Taylor, Carcinogenesis tests of nitroso-N-methylpiperazine, 2,3,5,6-tetramethyl dinitrosopiperazine, nitrosoisonipecotic acid and nitrosomethyoxymethylamine in rat, *Z. Krebsforsch.* 89, 31-36 (1977).
28. A. L. Love, W. Lijinsky, L. F. Kufer, and H. Garcia, Chronic oral administration of 1-nitrosopiperazine at high doses to MRC rats, *Z. Krebsforsch.* 89, 69-73 (1977).
29. A. W. Andrew, L. H. Thibault, and W. Lijinsky, The relationship between mutagenicity and carcinogenicity of some nitrosamines, *Mutat. Res.* 51, 319-326 (1978).
30. W. Lijinsky and A. W. Andrews, The mutagenicity of nitrosamines in *Salmonella typhimurium, Mutat. Res.* 68, 1-8 (1979).
31. E. Zeiger, M. S. Legator, and W. Lijinsky, Mutagenicity of N-nitrosopiperazine for *Salmonella typhimurium* in the host mediated assay, *Cancer Res.* 32, 1598-1599 (1972).
32. E. Zeiger and A. Sheldon, The mutagenicity of nitropiperidines for *Salmonella typhimurium, Mutat. Res.* 57, 85-89 (1978).
33. T. K. Rao, A. A. Hardigree, J. A. Young, W. Lijinsky, and J. L. Epler, Mutagenicity of N-nitrosopiperidines with *Salmonella typhimurium*/microsomal activation system, *Mutat. Res.* 5, 131-145 (1977).
34. T. K. Rao, J. A. Young, W. Lijinsky, and J. L. Epler, Mutagenicity of N-nitrosopiperazine derivatives in *Salmonella typhimurium, Mutat. Res.* 57, 127-134 (1978).
35. T. K. Rao, D. W. Ramey, W. Lijinsky, and J. L. Epler, Mutagenicity of cyclic nitrosamines in *Salmonella typhimurium, Mutat. Res.* 67, 21-26 (1979).
36. T. K. Rao, J. A. Young, D. W. Ramey, W. Lijinsky, and J. L. Epler, Mutagenicity of alphatic nitrosamines in *Salmonella typhimurium, Mutat. Res.* 66, 1-7 (1979).

37. F. W. Larimer, D. W. Ramey, W. Lijinsky, and J. L. Epler, Mutagenicity of methylated N-nitrosopiperidines in *Saccharomyces cerevisiae, Mutat. Res. 57*, 155-161 (1978).
38. F. W. Larimer, A. A. Hardigree, W. Lijinsky, and J. L. Epler, Mutagenicity of N-nitrosopiperazine derivatives in *Saccharomyces cerevisiae, Mutat. Res. 77*, 143-148 (1980).
39. C. E. Nix, B. Brewen, R. Wilkerson, W. Lijinsky, and J. L. Epler, Effects of N-nitrosopiperidine substitutions on mutagenicity in *Drosophila melanogaster, Mutat. Res. 67*, 27-38 (1979).
40. C. E. Nix, B. Brewen, W. Lijinsky, and J. L. Epler, Effects of methylation and ring size on mutagenicity of cyclic nitrosamines in *Drosophila melanogaster, Mutat. Res. 73*, 93-100 (1980).
41. A. W. Hsie, D. A. Casiano, D. B. Couch, D. F. Krahn, J. P. O'Neill, and B. L. Whitfield, The use of Chinese hamster ovary cells to quantify specific locus mutation and to determine mutagenicity of chemicals, *Mutat. Res. 86*, 193-214 (1981).
42. J. L. Epler, F. W. Larimer, T. M. Rao, C. E. Nix, and T. Ho, Energy-related pollutants in the environment: Use of short-term test for the mutagenicity in the isolation and identification of biohazards, *Environ. Health Persp. 27*, 11-20 (1978).
43. A. W. Hsie, P. A. Brimer, T. J. Mitchell, and D. G. Gosslee, The dose-response relationship for ethyl methanesulfonate-induced mutations at the hypoxanthine-guanine phosphoribosyl transferase locus in Chinese hamster ovary cells, *Somat. Cell Genet. 1*, 247-261 (1975).
44. A. W. Hsie, J. P. O'Neill, D. B. Couch, J. R. San Sebastian, P. A. Brimer, R. Machanoff, J. C. Riddle, A. P. Li, J. C. Fuscoe, N. L. Forbes, and M. H. Hsie, Quantitative analyses of radiation and chemical-induced cellular lethality and mutagenesis in Chinese hamster ovary cells, *Radiat. Res. 76*, 471-492 (1978).
45. A. W. Hsie, J. P. O'Neill, J. R. San Sebastian, and P. A. Brimer, The CHO/HGPRT mutation assay: Progress with quantitative mutagenesis and mutagen screening, in: *Banbury Report 2: Mammalian Cell Mutagenesis* (A. W. Hsie, J. P. O'Neill, and V. K. McElheny, eds.), pp. 407-420, Cold Spring Harbor Laboratory, Cold Spring Harbor, New York (1979).
46. A. W. Hsie, Structure-mutagenicity analysis with the CHO/HGPRT system, *Food Comet. Toxicol. 19*, 617-621 (1981).
47. A. W. Hsie, R. L. Schenley, K. R. Tindall, R. Machanoff, P. A. Brimer, S. W. Perdue, J. R. San Sebastian, and E.-L. Tan, in: *Banbury Report 13: Indicators of Genotoxic Exposure* (B. A. Bridges, B. E. Butterworth, and I. B. Weinstein, eds.), pp. 487-501, Cold Spring Harbor Laboratory, Cold Spring Harbor, New York.
48. A. W. Hsie, and R. L. Schenley, Utilization of Chinese hamster cells *in vitro* and *in vivo* in genetic toxicology: A multiphasic approach, *Environ. Mutagen.* (in press).
49. J. R. San Sebastian, J. P. O'Neill, and A. W. Hsie, Induction of chromosome aberrations, sister chromatid exchanges, and specific gene mutations in Chinese hamster ovary cells by 5-bromodeoxyuridine, *Cytogenet. Cell Genet. 28*, 47-54 (1980).
50. J. P. O'Neill, P. A. Brimer, R. Machanoff, G. P. Hirsch, and A. W. Hsie, A quantitative assay of mutation induction at the hypoxanthine-guanine phosphoribosyl transferase locus in Chinese hamster ovary cells (CHO/HGPRT system): Development and definition of the system, *Mutat. Res. 45*, 91-101 (1977).
51. J. P. O'Neill, D. B. Couch, R. Machanoff, J. R. San Sebastian, P. A. Brimer, and A. W. Hsie, A quantitative assay of mutation induction at the hypoxanthine-guanine phosphoribosyl transferase in Chinese hamster ovary cells (CHO/HGPRT system): Utilization with a variety of mutagenic agents, *Mutat. Res. 45*, 103-109 (1977).
52. J. P. O'Neill, R. Machanoff, J. R. San Sebastian, and A. W. Hsie, Cytotoxicity and mutagenicity of dimethylnitrosamine in mammalian cells (CHO/HGPRT system): Enhancement by calcium phosphate, *Environ. Mutag. 4*, 7-18 (1982).
53. D. B. Couch, N. L. Forbes, and A. W. Hsie, Comparative mutagenicity of alkylsulfate and alkanesulfonate derivatives in Chinese hamster ovary cells, *Mutat. Res. 57*, 217-224 (1978).
54. D. B. Couch and A. W. Hsie, Mutagenicity and carcinogenicity of congeners of two classes of nitroso compounds in Chinese hamster ovary cells, *Mutat. Res. 57*, 209-216 (1978).

55. J. P. O'Neill, J. C. Fuscoe, and A. W. Hsie, Mutagenicity of heterocyclic nitrogen mustards (ICR compounds) in cultured mammalian cells, *Cancer Res. 38*, 506–509 (1978).
56. J. C. Fuscoe, J. P. O'Neill, R. M. Peck, and A. W. Hsie, Mutagenicity and cytotoxicity of nineteen heterocyclic nitrogen mustards, *Cancer Res. 39*, 4875–4881 (1979).
57. N. P. Johnson, J. D. Hoeschele, J. D. Rahn, J. P. O'Neill, and A. W. Hsie, Mutagenicity, cytotoxicity and DNA binding of platinum(II)-chloroammines in Chinese hamster ovary cells, *Cancer Res. 40*, 1463–1468 (1980).
58. A. W. Hsie, Quantitative mutagenesis and mutagen screening with Chinese hamster ovary cells, in: *The Predictive Value of Short-Term Screening Tests in Carcinogenicity Evaluation* (G. M. Williams, R. Kroes, H. W. Waaijers, and K. W. van de Poll, eds.), pp. 89–102, Elsevier/North-Holland Biomedical Press, Amsterdam (1980).
59. H.-W. Thielman, C. H. Schroder, J. P. O'Neill, and A. W. Hsie, Relationship between DNA alkylation and specific locus mutation induction by N-methyl- and N-ethyl-N-nitrosourea in cultured Chinese hamster ovary cells (CHO/HGPRT system), *Chem.-Biol. Interact. 26*, 233–243 (1979).
60. A. W. Hsie, R. Machanoff, D. B. Couch, and J. M. Holland, Mutagenicity of dimethylnitrosamine and ethyl methanesulfonate as determined by the host-mediated CHO/HGPRT assay, *Mutat. Res. 51*, 77–84 (1978).
61. B. N. Ames, J. McCann, and E. Yamasaki, Methods for detecting carcinogens and mutagens with the *Salmonella*/mammalian microsome mutagenicity test, *Mutat. Res. 31*, 347–364 (1975).
62. E.-L. Tan and A. W. Hsie, Effect of phosphate and alumina Cγ gels on the mutagenicity and cytotoxicity of dimethylnitrosamine as studied in the CHO/HGPRT system, *Mutat. Res. 84*, 147–156 (1981).
63. J. H. Taylor, Sister chromatid exchanges in tritium labeled chromosomes, *Genetics 43*, 515–529 (1958).
64. S. A. Latt, Localization of sister chromatid exchanges in human chromosomes, *Science 185*, 74–76 (1974).
65. S. A. Latt, Sister chromatid exchanges, indices of human chromosome damage and repair, detection by fluorescence and induction by mitomycin C, *Proc. Natl. Acad. Sci. USA 71*, 3162–3166 (1974).
66. P. Perry and H. J. Evans, Cytological detection of mutagen–carcinogen exposure by sister chromatid exchanges, *Nature 258*, 121–125 (1975).
67. P. Perry and S. Wolff, New giemsea method for the differential staining of sister chromatids, *Nature 251*, 156–158 (1974).
68. H. Kato and H. Shimada, Sister chromatid exchange induced by mitomycin C, A new method of detecting DNA damaged at the chromosomal level, *Mutat. Res. 28*, 459–641 (1975).
69. H. Kato, Induction of sister chromatid exchanges by UV light and its inhibition by caffeine, *Expt. Cell Res. 82*, 383–390 (1973).
70. S. A. Latt, J. Allen, S. E. Bloom, A. Carrano, E. Falke, D. Krahn, E. Schneider, R. Shreck, R. Tice, B. Whitefield, and S. Wolff, Sister-chromatid exchanges: A report of the Gene-Tox Program, *Mutat. Res. 87*, 17–62 (1981).
71. A. V. Carrano, L. H. Thompson, P. A. Lindle, and J. L. Minkler, Sister-chromatid exchange as an indicator of mutagenesis, *Nature 271*, 551–553 (1978).
72. A. V. Carrano, L. H. Thompson, D. G. Stetka, J. L. Minkler, J. A. Mazrimas, and S. Fong, DNA crosslinking, sister-chromatid exchange and specific-locus mutations, *Mutat. Res. 63*, 175–188 (1979).
73. S. A. Latt, R. P. Schreck, K. S. Loveday, C. P. Dougherty, and C. F. Shuler, Sister chromatid exchange, *Adv. Human Genet. 10*, 267–331 (1980).
74. S. Wolff, Sister chromatid exchanges, *Ann. Rev. Genet. 11*, 183–201 (1977).
75. M. A. Bender and D. M. Prescott, DNA synthesis and mitosis in culture of human peripheral leukocytes, *Exptl. Cell Res. 27*, 221–229 (1962).

76. P. S. Moorhead, P. C. Nowell, W. J. Mellman, D. M. Battips, and D. A. Hungerford, *Exptl. Cell Res. 20*, 613–616 (1960).
77. K. Gato, S. Malda, Y. Kano, and T. Sugiyama, Factors involved in differential Giemsa-staining of sister chromatids, *Chromosoma (Berlin) 66*, 351–359 (1978).
78. R. Tice, M. A. Bender, J. L. Ivett, and R. Drew, Cytogenetic effects of inhaled ozone, *Mutat. Res. 58*, 293–304 (1978).
79. R. Tice, E. L. Schneider, and J. M. Mary, The utilization of bromodeoxyuridine incorporation into DNA for analysis of cellular kinetics, *Expt. Cell Res. 102*, 232–236 (1976).

8

Dimethylnitrosamine Demethylase and the Mutagenicity of Dimethylnitrosamine

Effects of Rodent Liver Fractions and Dimethylsulfoxide

MICHAEL J. PRIVAL AND VALERIE D. MITCHELL

1. INTRODUCTION

Although the Ames *Salmonella* plate incorporation assay[1] is the most widely used method for screening chemicals for potential carcinogenicity, it is well known that many carcinogens are not detected in this test. Some of these "false negative" compounds, such as highly chlorinated organic carcinogens, are also difficult to detect in other mutagenicity assays. Dimethylnitrosamine (DMN), however, is negative in the standard *Salmonella* plate incorporation assay,[2,3] although it is positive in a variety of other types of mutagenicity tests. DMN is carcinogenic in at least 12 different animal species.[4]

We have found that the mutagenic activity of DMN can be detected in the plate incorporation assay if the S-9 is derived from mouse or hamster liver rather than from the more widely used rat liver. In attempting to explore the basis for the different abilities of these various types of S-9 to activate DMN to a mutagen, we investigated the properties of the DMN demethylase contained in

MICHAEL J. PRIVAL AND VALERIE D. MITCHELL • Genetic Toxicology Branch, Food and Drug Administration, Washington, D. C. 20204.

the S-9 fractions. This enzyme is generally considered to be the sole enzyme in the activation pathway of DMN, as shown in Figure 1.[5,6] Kinetic studies have indicated that freshly prepared rat and mouse liver S-9 fractions contain at least two forms of DMN demethylase that have substantially different K_m values. We set out to determine whether the frozen preparations from rat, mouse, and hamster livers used in our assays have the same kinetic properties as the fresh rat and mouse liver preparations that others have studied. We then investigated whether the differing abilities of the S-9 fractions from different rodent species to activate DMN to a mutagen in the plate incorporation assay could be explained by differences in DMN demethylase activities.

Yahagi et al.[2] found that DMN is mutagenic even with rat liver S-9 if the "preincubation" modification of the plate incorporation assay is used and the DMN is dissolved in water. However, if dimethylsulfoxide (DMSO) is used as the solvent, no mutagenic activity is observed.[2] DMSO is routinely used as the solvent in Salmonella assays with other chemicals without apparent detrimental effects. Thus, in addition to being a "false negative" in the standard plate incorporation assay when rat liver S-9 is used, DMN is also anomalous in that DMSO interferes with its mutagenic activity. We therefore studied the effects of DMSO on the activity of DMN demethylase.

2. MATERIALS AND METHODS

2.1. Chemicals and Media

The sources of the chemicals used and the composition of the media used for plate incorporation assays have been previously described.[7] For suspension assays, we used a suspension growth medium (SGM) consisting of Vogel–Bonner medium E,[8] 0.5% glucose, 0.5 mM L-histidine hydrochloride, 2.5 μM biotin, and 1 g of histidine assay medium (Difco Laboratories, Detroit, Michi-

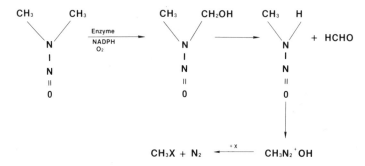

Figure 1. Presumed pathway of activation of DMN to a carcinogen or mutagen.

gan) per liter. Histidine assay medium is a complex mixture of nutrients containing amino acids but lacking histidine. SGM-H medium is the same as SGM medium but without histidine. Agar media were prepared from SGM and SGM-H by adding 1.5% agar.

2.2. Mutagenesis Assays

All assays were performed using *Salmonella typhimurium* strain TA 1530, which contains a base-pair substitution mutation resulting in a requirement for histidine.[1] Plate incorporation assays were performed as described by Ames *et al.*[1] with the following exceptions: (1) overnight cell cultures were washed once and resuspended in phosphate-buffered saline (PBS) at a density of approximately 2×10^9 cells per milliliter before testing; (2) the total volumes of the top agar overlays were adjusted to 2.7 ml in all assays; (3) the glucose 6-phosphate level was 10 μmoles per plate rather than 2.5 μmoles; and (4) 1.4 units per plate of glucose 6-phosphate dehydrogenase was added to each overlay. Colonies were counted with a New Brunswick Biotran II automatic colony counter (New Brunswick Scientific Co., Edison, New Jersey) except where hand counting is indicated in the text.

Liquid suspension mutagenesis assays were performed by preparing a 2.7-ml reaction mixture containing approximately 2×10^9 TA 1530 cells pregrown to log phase in SGM, all cofactors used in DMN demethylase assays (but no semicarbazide), SGM, 3.7 mM DMN, and S-9 containing 4.5 mg protein. This reaction mixture was shaken in a 16-mm-diam culture tube at 250 rpm in a New Brunswick water bath at 37°C. After 40 min, the cells were washed and resuspended in 0.8 ml PBS. The number of *his*+ revertants was determined by plating 0.2 ml of the resuspended cell suspension on SGM-H agar plates in triplicate. To determine the total number of cells present, an aliquot was diluted 3×10^6-fold in PBS, and 0.2-ml aliquots were plated in triplicate on SGM agar. After overnight incubation of plates at 37°C, all colonies were counted by hand, using a Quebec colony counter.

2.3. Preparation of S-9 Fractions and Microsomes

Methods for inducing animals with phenobarbital and with Aroclor 1254 have been previously described.[7] The S-9 fractions were prepared in homogenizing buffer[7] and frozen by the method described by Ames *et al.*[1] In all experiments except the one described in Figure 3, the protein concentration of each S-9 was adjusted to 30 mg/ml with homogenizing buffer before each mutagenicity assay. Unless otherwise specified, 150 μl of S-9 (4.5 mg protein) was used in each mutagenicity assay.

Microsomes and cytosol were prepared from thawed S-9 (at 30 mg protein per ml) by centrifugation for 1 hr at 105,000 $\times g$. The cytosolic fraction was the supernatant from this centrifugation. The microsomes were washed once and resuspended in a volume of homogenizing buffer equal to the volume of the original S-9.

The procedure for treatment of cytosol with trypsin has been described elsewhere.[7]

2.4. Enzyme Assays

The assays for DMN demethylase were performed as described previously.[7] The reaction volumes of the assay mixtures were 2.7 ml in all cases, which is the same volume as the top agar mixture used in plate incorporation assays for mutagenicity. All cofactors were also present at the same concentrations used in the mutagenicity assays. The enzyme assay measures the release of formaldehyde (see Figure 1).

3. RESULTS

3.1. Kinetics of DMN Demethylase in Hamster Liver S-9

DMN demethylase in rat and mouse liver S-9 fractions or microsomes has at least two apparent K_m values.[9,10] The low K_m form of the enzyme is referred to as DMN demethylase I and the high K_m form as DMN demethylase II.[10] Figure 2 is a Hofstee[11] plot for DMN demethylase in Aroclor 1254-induced hamster liver S-9; this plot shows that the enzyme in hamster liver also has at least two distinct apparent K_m values for DMN, which were determined to be 0.46 ± 0.10 (S.E.) and 74 ± 5 mM. These apparent K_m values were calculated by using a computer program based on the program "HYPER" described by Cleland.[12] Apparent K_m values determined for these and other S-9 preparations are given in Table I.

3.2. Correlation of Mutagenic and Enzyme Activities

Figure 3 shows the mutagenic activity of DMN over a wide range of doses in the plate incorporation assay, using S-9 fractions from Aroclor 1254-induced rat, mouse, and hamster livers. Mutagenic activity occurred with mouse liver at DMN doses of 25 μmoles per plate (9.3 mM) or higher. When hamster liver S-9 was used, DMN doses even below 1 μmole per plate (0.37 mM) were mutagenic. However, with Aroclor-induced rat liver S-9, which is the standard S-9 generally used for screening chemicals, little or no activity occurred at DMN doses up to 100 μmoles per plate (37 mM).

If the pathway shown in Figure 1 is the actual pathway of activation of DMN to a mutagen and if the first (enzymatic) step is rate limiting, then there

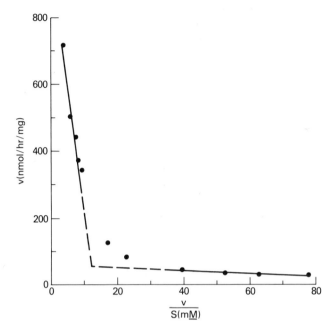

Figure 2. Hofstee plot for DMN demethylase from Aroclor 1254-induced hamster liver S-9. v = Rate of formation of formaldehyde; S = concentration of DMN.

Table I
Apparent K_m Values of DMN Demethylase I and DMN Demethylase II in Rat, Mouse, and Hamster Liver S-9 Fractions[a]

Species	Uninduced animals, DMN demethylase I	Aroclor 1254-induced animals	
		DMN demethylase I	DMN demethylase II
Rat	ND[b]	0.13 (0.17)	158 (16)
Mouse	0.71 (0.11)[c]	1.3 (0.15)	50 (3)
Hamster	0.43 (0.04) 0.35 (0.08)[c] 0.23 (0.03)	0.46 (0.10)	74 (5)

[a] All apparent K_m values are expressed as mM DMN. Numbers in parentheses are standard errors. The K_m values for DMN demethylase I and DMN demethylase II were determined in the ranges of 0.37 to 1.1 mM and 37 to 200 mM DMN, respectively.
[b] Not determined.
[c] Two separate determinations on two different batches of S-9.

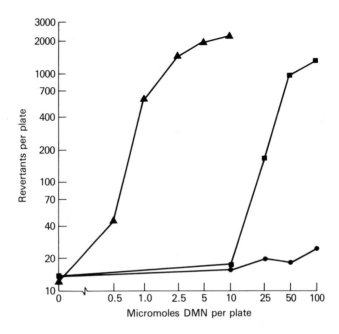

Figure 3. Mutagenic activity of DMN in the plate incorporation assay, using S-9 from Aroclor 1254-induced rat (●), mouse (■), or hamster (▲) liver.

should be a positive correlation between the DMN demethylase activities of various S-9 preparations and their ability to activate DMN to a mutagen. We tested the DMN demethylase activity of uninduced, Aroclor 1254-induced, and phenobarbital-induced rat, mouse, and hamster liver S-9 fractions, and also determined the ability of these same fractions to activate DMN to a mutagen in the plate incorporation assay with 3.7 mM DMN (10 μmoles per assay). At this DMN concentration, the enzyme activity is due primarily to DMN demethylase I. Table II presents the results obtained by averaging the data from two separate sets of experiments with two different batches of each type of S-9. The three types of rat liver S-9 had low enzyme activity and were unable to activate DMN to a mutagen. Conversely, the three types of hamster liver S-9 had high enzyme activities and activated DMN well. However, results with the uninduced and phenobarbital-induced mouse S-9 were unexpected, because the enzyme activities were similar to those obtained with hamster S-9 but no mutagenic activity was observed.

Since uninduced mouse and uninduced hamster liver S-9 fractions had approximately the same DMN demethylase activities but differed greatly in their ability to activate DMN to a mutagen, we performed all further experiments on S-9 fractions from uninduced animals. The use of uninduced animals

Table II
DMN Demethylase Activity of Different Types of S-9 and Their Ability to Activate DMN in the Plate Incorporation Assay[a]

Species	Inducer[b]	Revertants/plate	DMN demethylase activity (nmole/hr)
Rat	—	NSI[c]	129
	PB	NSI	140
	AR	NSI	46
Mouse	—	NSI	264
	PB	NSI	303
	AR	NSI	144
Hamster	—	≥ 2000	267
	PB	≥ 2000	352
	AR	≥ 2000	402

[a] Values are the averages of data from two separate sets of experiments with two separate preparations of each type of S-9. DMN concentration was 3.7 mM.
[b] PB, phenobarbital; AR, Aroclor 1254.
[c] No significant increase (less than 20) over spontaneous revertant count.

also has the advantage of minimizing complications due to the activity of DMN demethylase II, since both phenobarbital and Aroclor 1254 induce the high-K_m form of the enzyme.[7]

3.3. Effects of Mixing Liver Fractions from Different Species

Results given in Table II indicate that the differences in DMN demethylase activity cannot account for the ability of uninduced hamster liver S-9 to activate 3.7 mM DMN to a mutagen while uninduced mouse S-9 is unable to do so. One possibility we considered was that mouse liver S-9 contains some factor that interferes with the activation of DMN to a mutagen. To test this hypothesis, an experiment was conducted in which enzyme activity and DMN mutagenicity were determined using hamster S-9, mouse S-9, and a mixture of these two preparations. As shown in Table III, mouse S-9 interfered with the mutagenicity of DMN in the presence of hamster S-9. Use of twice the standard amount of hamster S-9 did not reduce the mutagenic activity of DMN, demonstrating that the mouse inhibition was not simply a nonspecific effect of an excess of S-9 or protein. Although mouse S-9 prevented detection of the mutagenic activity of DMN, it had no similar inhibiting effect on DMN demethylase activity.

We assumed that the inhibitor of DMN mutagenesis would be some soluble factor in the cytosol of the mouse liver S-9. However, when the S-9 was separated into microsomal and cytosolic fractions and each of these fractions was mixed with hamster liver S-9, mutagenesis was inhibited by the microsomal fraction but not by the cytosol (Table III).

To determine whether the inhibitor of mutagenesis found in the mouse

Table III
Effect of Mixing Hamster and Mouse Liver Fractions
on the Activation of DMN to a Mutagen and on DMN
Demethylase Activity[a]

Liver fraction(s) added	Revertants/plate[b]	DMN demethylase activity (nmole/hr)
Hamster S-9	2146	328
Mouse S-9	10	335
Hamster S-9 + mouse S-9	16	536
Hamster S-9 (300 µl)	2524	593
Hamster S-9 + mouse cytosol	1890	ND[c]
Hamster S-9 + mouse microsomes	14	576
Mouse microsomes	ND	241
Mouse microsomes (60°C)[d]	ND	2
Hamster S-9 + mouse microsomes (60°C)	17	330
Hamster S-9 + mouse microsomes (70°C)[e]	2209	ND

[a] All fractions were obtained from uninduced animals; 150 µl of each fraction, corresponding to 4.5 mg S-9 protein, was added unless otherwise indicated. The DMN concentration was 3.7 mM in all assays.
[b] Spontaneous revertant counts (8) not subtracted.
[c] ND, not determined.
[d] Microsomes heated to 60°C for 15 min.
[e] Microsomes heated to 70°C for 15 min.

microsomes might be mouse DMN demethylase, we heated mouse microsomes to 60°C for 15 min. This treatment destroyed the DMN demethylase activity of the preparation (Table III). When the heat-treated microsomal fraction was mixed with hamster liver S-9, it was still capable of inhibiting the mutagenic activity of DMN but had no effect on the DMN demethylase activity of hamster S-9. When the mouse microsomes were heated to 70°C, they lost their ability to inhibit DMN mutagenesis (Table III). Therefore, these microsomes contained an inhibitor of DMN mutagenesis that was not DMN demethylase and was destroyed by heating at 70°C but not at 60°C.

The failure of rat liver S-9 to activate DMN to a mutagen in the plate test might be explained by the low DMN demethylase activity of rat S-9 (Table II); however, the presence of an inhibitor in rat S-9 similar to that in mouse S-9 could also be an explanation. Therefore rat liver S-9, microsomes, and cytosol were tested in the same manner as the mouse liver fractions. Table IV shows that rat liver S-9 did, in fact, inhibit the mutagenicity of DMN in the presence of hamster S-9 and that it was similar to mouse S-9 in that the inhibitor was in the microsomal fraction, was stable at 60°C but was destroyed at 70°C, and did not inhibit hamster DMN demethylase.

The conditions under which the plate incorporation mutagenesis assay is performed are quite different from those of the DMN demethylase assay. The most important of these differences is probably the fact that the enzyme assay is

Table IV
Effect of Mixing Hamster and Rat Liver Fractions on the Activation of DMN to a Mutagen and on DMN Demethylase Activity[a]

Liver fraction(s) added	Revertants/plate[b]	DMN demethylase activity (nmole/hr)
Hamster S-9	2598	308
Rat S-9	12	70
Hamster S-9 + rat S-9	24	327
Hamster S-9 (300 µl)	2441	565
Hamster S-9 + rat cytosol	2417	ND[c]
Hamster S-9 + rat microsomes	107	381
Rat microsomes	ND	48
Rat microsomes (60° C)[d]	ND	0
Hamster S-9 + rat microsomes (60° C)	113	321
Hamster S-9 + rat microsomes (70° C)[e]	2589	ND

[a] All fractions were obtained from uninduced animals; 150 (µl) of each fraction, corresponding to 4.5 mg S-9 protein, was added unless otherwise indicated. The DMN concentration was 3.7 mM in all assays.
[b] Spontaneous revertant counts (7) not subtracted.
[c] ND, not determined.
[d] Microsomes heated to 60° C for 15 min.
[e] Microsomes heated to 70° C for 15 min.

terminated after no more than 40 min while the plates in the mutagenesis assay are incubated for 2 days. Therefore, we investigated whether mouse liver S-9 would interfere with DMN mutagenesis mediated by hamster S-9 under conditions more closely resembling those of the enzyme assay. A 40-min liquid suspension assay was performed with hamster S-9, mouse S-9, and a mixture of the two preparations. The results, shown in Table V, demonstrate that mouse S-9 does not interfere with the mutagenicity of DMN in the liquid suspension assay.

3.4. Cytosol Requirement for Mutagenicity of DMN

Since DMN demethylase is a microsomal enzyme, we performed an experiment to determine whether or not microsomes alone, in the absence of cytosol, would mediate the mutagenicity of DMN. As shown in Table VI, neither microsomes nor cytosol from uninduced hamster liver alone was capable of activating DMN to a mutagen in the plate incorporation assay, although the DMN demethylase activity of the S-9 was largely recovered in the microsomal fraction. When hamster liver microsomes were combined with cytosol, the ability to activate DMN was restored. Furthermore, the cytosol required for DMN mutagenesis could be derived from rat, mouse, or hamster liver S-9. The addition of cytosol to hamster microsomes did not greatly affect DMN demethylase activity. Heating the hamster cytosol to 60° C for 15 min partially des-

Table V

Mutagenic Activity of 3.7 mM DMN in Liquid Suspension Assays with Hamster Liver S-9, Mouse Liver S-9, and Hamster Plus Mouse Liver S-9 Fractions Combined[a]

Test chemical	S-9	his^+ cells/ml[b]	cfu/ml × 10^{-9} [b]	his^+ cells/cfu × 10^9
Water	Hamster	23.5	1.98	11.9
DMN	None	13.4	2.12	6.3
DMN	Hamster	340	1.81	188
DMN	Mouse	624	1.06	589
DMN	Hamster + mouse	3636	1.38	2634

[a] Values presented are means of results obtained on two suspension assay tubes of each type indicated.
[b] As determined in 0.8 ml resuspended culture volume.

Table VI

Requirement for Cytosol in the Activation of DMN to a Mutagen[a]

Liver fraction(s) added	Revertants/plate[b]	DMN demethylase activity (nmole/hr)
Hamster S-9	2297	214
Hamster microsomes	41	182
Hamster cytosol	21	16
Hamster microsomes + hamster cytosol	2270	188
Mouse cytosol	ND[c]	20
Hamster microsomes + mouse cytosol	2290	232
Rat cytosol	ND	5
Hamster microsomes + rat cytosol	2175	178
Hamster microsomes + hamster cytosol (60°C)[d]	749	ND
Hamster microsomes + hamster cytosol (70°C)[e]	67	ND
Hamster microsomes + 2.3 mg BSA[f]	38	ND
Hamster microsomes + 4.5 mg BSA	17	ND
Hamster microsomes (200 μl)	42	ND
Hamster microsomes (250 μl)	34	ND

[a] All fractions were obtained from uninduced animals; 150 μl of each fraction, corresponding to 4.5 mg S-9 protein, was added unless otherwise indicated. The DMN concentration was 3.7 mM in all assays.
[b] Spontaneous revertant counts (10) not subtracted.
[c] ND, not determined.
[d] Microsomes heated to 60°C for 15 min.
[e] Microsomes heated to 70°C for 15 min.
[f] Bovine serum albumin.

troyed its ability to assist in the activation of DMN, and almost all of this ability was destroyed at 70°C. We considered the possibility that the microsomal fraction was insufficient to activate DMN because some portion of the microsomes had been lost into the cytosol; thus, adding cytosol might simply serve to restore a sufficient quantity of microsomes to give mutagenic activity with DMN. To test this, the level of microsomal fraction was increased from 150 to 200 or 250 µl. Mutagenic activity was still not observed, as shown in Table VI.

To test whether the cytosolic activator of DMN mutagenesis is a protein, we incubated hamster liver cytosol with trypsin. After the incubation, the action of trypsin was stopped by adding either α-1-antitrypsin or soybean trypsin inhibitor. As controls, cytosol was incubated with trypsin, and either antitrypsin or trypsin inhibitor was added at the beginning of the trypsin treatment. As shown in Table VII, treatment of the cytosol with trypsin in the absence of α-1-antitrypsin or soybean trypsin inhibitor greatly reduced the ability of the cytosol to assist the microsomes in activating DMN to a mutagen. If antitrypsin or trypsin inhibitor was added at the same time as the trypsin, then no reduction in activity occurred. Furthermore, dialysis of 3.5 ml of hamster liver cytosol against 1 liter of homogenizing buffer for 24 hr did not eliminate the ability of the cytosol to stimulate mutagenicity of DMN in the presence of hamster liver microsomes (data not shown). Thus, the cytosolic activator of DMN mutagenesis contains protein.

3.5. Effects of Dimethylsulfoxide

We found that DMSO inhibits DMN mutagenesis in the plate incorporation assay when uninduced hamster liver S-9 is used (data not shown). Since DMN and DMSO are structurally similar (Figure 4), we hypothesized that the

Table VII
Effect of Tryspin Treatment on Activity of the Activator in Hamster Liver Cytosol

Cytosol pretreatment[a]	Added after pretreatment[b]	Revertants/plate[c]
Buffer	—	2551
Trypsin	AT	233
Trypsin + AT	—	2389
Trypsin	STI	316
Trypsin + STI	—	2256

[a] Cytosol was pretreated as indicated at 37°C for 12.5 hr. All mutagenesis assays were performed on *S. typhimurium* strain TA1530 using 3.7 mM DMN, 150 µl of uninduced hamster liver microsomes, and 150 µl of the pretreated, uninduced hamster liver cytosol.
[b] AT, α-1-antitrypsin; STI, soybean trypsin inhibitor.
[c] Spontaneous revertant counts (13) not subtracted.

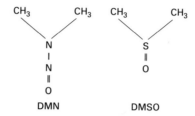

Figure 4. Structures of dimethylnitrosamine (DMN) and dimethylsulfoxide (DMSO).

inhibition of DMN mutagenicity may be due to the competitive inhibition of DMN demethylase.

To test this hypothesis, we performed DMN demethylase assays at various DMN concentrations, adding different amounts of DMSO to the reaction mixture. The lines in the Lineweaver–Burk plot shown in Figure 5 were generated from the raw data by using the same computer program used to calculate K_m values. The fact that the intercepts of the generated lines are quite close indicates that the inhibition of DMN demethylase by DMSO is predominantly competitive.

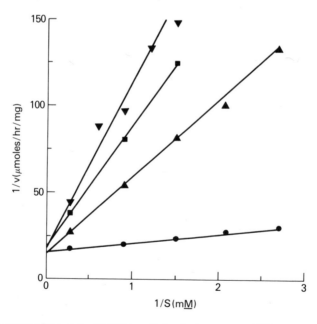

Figure 5. Lineweaver-Burk plots of DMN demethylase in the presence of different concentrations of DMSO. v= Rate of formation of formaldehyde; S = concentration of DMN. (●) No DMSO present; (▲) 26 mM DMSO; (■) 52 mM DMSO; (▼) 78 mM DMSO.

4. DISCUSSION

4.1. Apparent K_m Values and Regulation of DMN Demethylase

Our results obtained on the different K_m values of rodent liver DMN demethylase are in general agreement with the findings of others for the rat and mouse liver enzymes. The low apparent K_m in rat microsomes or S-9 has been reported as 0.32 mM[13]; 0.352 and 0.529 mM in fed and starved rats, respectively[14]; 0.22 mM[15]; 0.20 mM[10]; and 0.34 and 1.14 mM in uninduced and Aroclor-induced rats, respectively.[10] In the mouse, the low apparent K_m of the emzyme has been reported as 0.28[10]; 0.2[16]; 0.4[17]; and 0.23 and 0.35 mM in uninduced and Aroclor-induced mice, respectively.[10]

The high apparent K_m for this enzyme in rat liver S-9 and microsomes has been reported to be 35 mM[13]; 118 mM[18]; 51 mM[10]; and 87 and 129 mM for uninduced and Aroclor-induced rats, respectively.[10] In the mouse, the high apparent K_m has been reported to be 44 mM[10]; 40 mM[16,19]; and 43.5 and 56.1 mM in uninduced and Aroclor-induced mice, respectively.[10] The high apparent K_m values in our experiments were 158 mM for Aroclor-induced rats, 50 mM for Aroclor-induced mice, and 74 mM for Aroclor-induced hamsters (Table I). The fact that our results on the kinetics and the control[7] of DMN demethylase in rodent liver S-9 are consistent with those of others is important to the interpretation of our results, since we used only S-9 preparations that had been frozen while other investigators apparently used freshly prepared S-9 or microsome preparations.

4.2. Microsomal Inhibitor of DMN Mutagenesis

Our finding that mouse or rat liver S-9 or microsomes can interfere with the mutagenic activity of DMN in the plate assay is difficult to reconcile with two other observations we have made: (1) the microsomal inhibitor acts without affecting DMN demethylase activity, as determined in our enzyme assays, and (2) the inhibition does not appear to occur in liquid suspension assays. One possibility is that rat or mouse microsomes decrease the stability of hamster DMN demethylase during the prolonged incubation period of the plate incorporation assay. A more plausible explanation might be that the pathway of activation of DMN in the plate assay is different from that shown in Figure 1. Rat and mouse liver microsomes may be capable of trapping some intermediate in the plate test activation pathway that is not required for mutagenesis in suspension. Some difference in pathways between the suspension and plate assays would also be consistent with the observation that mouse liver S-9 is at least as efficient as hamster liver S-9 in activating DMN to a mutagen in suspension (Table V), but not on plates. Presumably the inhibitor in mouse microsomes that can interfere with DMN activation by hamster S-9 in the plate assay can also interfere with activation by mouse S-9 itself in this assay. In the suspension test, however, it does not appear to interfere with activation by either hamster or mouse liver S-9.

Hutton et al.[20] found that 1 mM DMN is mutagenic in the presence of hamster liver S-10 or microsomes, using a "preincubation" modification of the plate incorporation assay. They had previously found that this concentration of DMN was not sufficient to cause detectable mutagenesis in the presence of mouse liver S-10.[17] In the preincubation assay, the chemical is incubated with cells, liver extract, and cofactors in a liquid suspension for up to 1 hr and then the reaction mixture is plated in a soft agar overlay. After plating, the conditions are identical to those that occur in a plate incorporation assay. Although the results of Hutton et al.[20] appear to confirm our report[21] that 1 mM DMN is mutagenic in the plate incorporation assay with hamster S-9 but not with mouse S-9, they are actually difficult to reconcile with our findings. Hutton et al. have established that, under their test conditions, the mutagenic action of DMN is completed when the preincubation step is terminated and the reaction mixture poured on the agar plates. Thus, they are actually performing a liquid suspension assay with DMN. Contrary to their findings, we have observed that mouse liver S-9 is at least as effective as hamster S-9 in activating DMN is at a low concentration (3.7 mM) in liquid suspension (Table V).

4.3. Cytosolic Activator of DMN Mutagenesis

Czygan et al.[19] and Frantz and Malling[22] reported that DMN can be activated to a mutagen in liquid suspension assays by washed mouse liver microsomes. We found that cytosol is necessary for mutagenic activity under the conditions of the plate incorporation assay using washed hamster microsomes. It is possible that the cytosol stabilizes the DMN demethylase in the hamster microsomes during the long incubation period of the plate assays. Lake and co-workers[9] found that cytosol can stabilize DMN demethylase activity; however, these investigators also reported that cytosol caused a two- to four-fold increase in DMN demethylase activity of rat liver microsomes, but we did not observe this effect with hamster liver microsomes (Table VI).

An alternative to the hypothesis that cytosol stabilizes DMN demethylase in the plate assay is the previously discussed possibility that activation of DMN in the plate assay may involve a pathway other than that shown in Figure 1. Cytosol may be essential to this alternative route of activation. Others have found that cytosol increases DMN binding to protein[23] and to DNA.[24]

In the case of at least one N-nitroso compound, N-nitrosopyrrolidine, the presence of cytosol greatly enhances mutagenesis in liquid suspension.[25] Hecker et al.[26] postulated that the cytosol acts by removing the product of the first step in the pathway of activation of this compound, thereby accelerating its activation.

4.4. Inhibition by Dimethylsulfoxide

Since DMSO is the most widely used solvent for test chemicals in the *Salmonella* assay, it has been a matter of some concern that at least one potent carcinogen, DMN, could not be detected as a mutagen in the presence of this solvent. The finding of competitive inhibition of DMN demethylase by DMSO implies that DMSO is not acting by a general inactivation of the enzyme, such as by denaturation. Competitive inhibition implies that DMSO is likely to be displacing DMN in the active site of the enzyme, probably due to its similarity in structure to DMN. Thus the structural similarity of DMN and DMSO and our finding that DMSO is a competitive inhibitor of DMN demethylase may be the basis for the high degree of specificity of the inhibition of mutagenesis by DMSO. We have discussed above the possibility that DMN demethylase and the pathway in Figure 1 may not be sufficient to account for the mutagenic activity of DMN in the plate incorporation assay. Perhaps some other enzyme involved in the activation of DMN may also be inhibited by DMSO. To our knowledge, DMN and the structurally related diethylnitrosamine[2] are the only compounds whose mutagenicity is inhibited by DMSO.

4.5. Screening Chemicals with Hamster Liver S-9

The most widely used and recommended[1] type of S-9 for screening chemicals in the *Salmonella* assay is Aroclor 1254-induced rat liver S-9. It appears, however, that at least for many N-nitroso compounds, hamster liver S-9 is superior to that derived from rats. For example, Bartsch and co-workers[27] found that bis(2-hydroxy-*n*-propyl)-nitrosamine and methyl-*n*-propylnitrosamine are mutagenic when phenobarbital-induced hamster liver S-9, but not rat liver S-9, is used. We also found that diethylnitrosamine and N-nitrosodi-(*n*-butyl)-amine are more mutagenic in the plate assay when hamster liver S-9 is used than when mouse S-9 is used.[21] Rat liver S-9 was even less effective than mouse S-9. On the basis of these findings, we recommend that any N-nitroso compound that is not mutagenic in the presence of rat liver S-9 should be tested using hamster S-9.

We also found that benzidine is far more active as a mutagen when hamster S-9 is used than when rat S-9 is used.[28] Rat liver S-9 inhibited the hamster S-9-mediated mutagenic activity of benzidine[28] as it does for DMN mutagenicity (Table IV). Weinstein *et al.*,[29] on the other hand, did not find any inhibition of hamster S-9-mediated mutagenesis of phenacetin by rat liver S-9, even though the activity of this chemical could not be detected with rat S-9 alone.

However, in view of our findings of inhibition with DMN and benzidine, we cannot recommend that S-9 fractions from different species be mixed for screening of chemicals, as has been suggested by Weinstein and co-workers.[29]

5. REFERENCES

1. B. N. Ames, J. McCann, and E. Yamasaki, Methods for detecting carcinogens and mutagens with the *Salmonella*/mammalian-microsome mutagenicity test, *Mutat. Res. 31*, 347-367 (1975).
2. T. Yahagi, M. Nagao, Y. Seino, T. Matsushima, T. Sugimura, and M. Okada, Mutagenicities of N-nitrosamines on *Salmonella, Mutat. Res. 48*, 121-130 (1977).
3. H. Bartsch, A. Camus, and C. Malaveille, Comparative mutagenicity of N-nitrosamines in a semi-solid and in a liquid incubation system in the presence of rat or human tissue fractions, *Mutat. Res. 37*, 149-162 (1976).
4. *IARC Monographs on the Evaluation of the Carcinogenic Risk of Chemicals to Humans*, Vol. 17, pp. 125-175, International Agency for Research on Cancer, Lyon, France (1978).
5. H. Druckrey, R. Preussmann, S. Ivankovic, and D. Schmähl, Organotrope carcinogene Wirkungen bei 65 verscheidenen N-Nitroso-Verbindungen an BD-Ratten, *Z. Krebsforsch. 69*, 103-201 (1967).
6. A. E. Pegg, Metabolism of N-nitrosodimethylamine, in: *Molecular and Cellular Aspects of Carcinogen Screening Tests* (R. Montesano, H. Bartsch, and L. Tomatis, eds.), IARC Scientific Publications No. 27, pp. 3-22, International Agency for Research on Cancer, Lyon, France (1980).
7. M. J. Prival and V. D. Mitchell, Influence on microsomal and cytosolic fractions from rat, mouse, and hamster liver on the mutagenicity of dimethylnitrosamine in the *Salmonella* plate incorporation assay, *Cancer Res. 41*, 4361-4367 (1981).
8. H. J. Vogel and D. M. Bonner, Acetylornithinase of *Escherichia coli*: Partial purification and some properties, *J. Biol. Chem. 218*, 97-106 (1956).
9. B. G. Lake, J. C. Phillips, C. E. Heading, and S. D. Gangolli, Studies on the *in vitro* metabolism of dimethylnitrosamine by rat liver, *Toxicology 5*, 297-309 (1976).
10. J. C. Arcos, D. L. Davies, C. E. L. Brown, and M. F. Argus, Repressible and inducible enzymic forms of dimethylnitrosamine demethylase, *Z. Krebsforsch. 89*, 181-199 (1977).
11. B. H. J. Hofstee, Graphical analysis of single enzyme systems, *Enzymologia 17*, 273-278 (1956).
12. W. W. Cleland, Statistical analysis of enzyme kinetic data, *Methods Enzymol. 63* (Part A), 103-138 (1979).
13. B. G. Lake, C. E. Heading, J. C. Phillips, S. D. Gangolli, and A. G. Lloyd, Some studies on the metabolism *in vitro* of dimethylnitrosamine by rat liver, *Biochem. Soc. Trans. 2*, 610-612 (1974).
14. N. Venkatesan, J. C. Arcos, and M. F. Argus, Amino acid induction and carbohydrate repression of dimethylnitrosamine demethylase in rat liver, *Cancer Res. 30*, 2563-2567 (1970).
15. N. Venkatesan, M. F. Argus, and J. C. Arcos, Mechanism of 3-methylcholanthrene-induced inhibition of dimethylnitrosamine demethylase in rat liver, *Cancer Res. 30*, 2556-2562 (1970).
16. I. G. Sipes, M. L. Slocumb, and G. Holtzman, Stimulation of microsomal dimethylnitrosamine-N-demethylase by pretreatment of mice with acetone, *Chem.-Biol. Interact. 21*, 155-166 (1978).
17. J. J. Hutton, J. Meier, and C. Hackney, Comparison of the *in vitro* mutagenicity and metabolism and of dimethylnitrosamine and benzo[*a*]pyrene in tissues from inbred mice treated with phenobarbital, 3-methylcholanthrene or polychlorinated biphenyls, *Mutat. Res. 66*, 75-94 (1979).

18. I. Y. Chau, D. Dagani, and M. C. Archer, Kinetic studies on the hepatic microsomal metabolism of dimethylnitrosamine, diethylnitrosamine, and methylethylnitrosamine in the rat, *J. Natl. Cancer Inst. 61*, 517–521 (1978).
19. P. Czygan, H. Greim, A. J. Garro, F. Hutterer, F. Schaffner, H. Popper, O. Rosenthal, and D. Y. Cooper, Microsomal metabolism of dimethylnitrosamine and the cytochrome P-450 dependency of its activation to a mutagen, *Cancer Res. 33*, 2983–2986 (1973).
20. J. J. Hutton, C. Hackney, and J. Meier, Mutagenicity and metabolism of dimethylnitrosamine and benzo[*a*]pyrene in tissue homogenates from inbred Syrian hamsters treated with phenobarbital, 3-methylchloanthrene or polychorinated biphenyls, *Mutat. Res. 64*, 363–377 (1979).
21. M. J. Prival, V. D. King, and A. T. Sheldon, Jr., The mutagenicity of dialkyl nitrosamines in the *Salmonella* plate assay, *Environ. Mutagenesis 1*, 95–104 (1979).
22. C. N. Frantz and H. V. Malling, Factors affecting metabolism and mutagenicity of dimethylnitrosamine and diethylnitrosamine, *Cancer Res. 35*, 2307–2314 (1975).
23. H. M. Godoy, M. I. Diaz Gomez, and J. A. Castro, Mechanism of dimethylnitrosamine metabolism and activation in rats, *J. Natl. Cancer Inst. 61*, 1285–1289 (1978).
24. D. Y. Lai, S. C. Myers, Y. T. Woo, E. J. Greene, M. A. Friedman, M. F. Argus, and J. C. Arcos, Role of dimethylnitrosamine demethylase in the metabolic activation of dimethylnitrosamine, *Chem.-Biol. Interact. 28*, 107–126 (1979).
25. L. I. Hecker, R. K. Elespuru, and J. G. Farrelly, The mutagenicity of nitrosopyrrolidine is related to its metabolism, *Mutat. Res. 62*, 213–220 (1979).
26. L. I. Hecker, J. G. Farrelly, J. H. Smith, J. E. Saavedra, and P. A. Lyon, Metabolism of the liver carcinogen N-nitrosopyrrolidine by rat liver microsomes, *Cancer Res. 39*, 2679–2686 (1979).
27. H. Bartsch, C. Malaveille, and R. Montesano, The predictive value of tissue-mediated mutagenicity assays to assess the carcinogenic risk of chemicals, in: *Screening Tests in Chemical Carcinogenesis* (R. Montesano, H. Bartsch, and L. Tomatis, eds.), IARC Scientific Publications No. 12, pp. 467–486, International Agency for Research on Cancer, Lyon, France (1976).
28. M. J. Prival and V. D. Mitchell, Analysis of a method for testing azo dyes for mutagenic activity in *Salmonella typhimurium* in the presence of flavin mononucleotide and hamster liver S-9, *Mutat. Res. 92*, 103–116 (1982).
29. D. Weinstein, M. Katz, and S. Kazmer, Use of rat/hamster S-9 mixture in the Ames mutagenicity assay, *Environ. Mutagenesis 3*, 1–9 (1981).

9

The Relationship between Metabolism and Mutagenicity of Two Cyclic Nitrosamines

JAMES G. FARRELLY AND LANNY I. HECKER

1. INTRODUCTION

Animal studies with a large number of nitrosamines have demonstrated that different members of this chemical family can induce tumors in a remarkable range of mammalian tissues. It also appears that, to exert a biological effect, these carcinogens must first be activated by an appropriate enzyme system. Convincing evidence has accrued that indicates that the most important pathway for such activations is α-hydroxylation.[1-14] When the α-hydrogens of a nitrosamine are replaced with methyl groups, both carcinogenicity[3,4,6] and mutagenicity[13,15,17] are drastically reduced. Replacement of the α-hydrogens with deuterium also results in a reduction in carcinogenic potency.[5,18,19]

The α-hydroxynitrosamine formed metabolically is an unstable species that decomposes to form a series of reactive intermediates that can interact with cellular nucleophiles (Figure 4). In the case of heterocyclic nitrosamines, the hydroxylated carbon spontaneously cleaves from the ring nitrogen and is converted to an aldehyde. Concomitantly, the remainder of the molecule is believed

---

Abbreviations used in this chapter: NPYR, nitrosopyrrolidine; NHEX, nitrosohexamethyleneimine; THF, tetrahydrofuran; HPLC, high-pressure liquid chromatography; HEPES, N-2-hydroxyethylpiperazine-N'-2-ethanesulfonic acid; NMCE, nonmethylene chloride extractable.

JAMES G. FARRELLY AND LANNY I. HECKER • Chemical Carcinogenesis Program, NCI-Frederick Cancer Research Facility, Frederick, Maryland 21701.   By acceptance of this article, the publisher or recipient acknowledges the right of the U. S. Government to retain a nonexclusive, royalty-free license in and to any copyright covering the article.

to form a series of reactive intermediates, such as a diazohydroxide, or a diazonium or carbonium ion. It is not yet clear which product is the carcinogenic or mutagenic species. However, it would be possible for any of them to react with cellular macromolecules, be they nucleic acid, protein, carbohydrate, or lipid. Although an alteration in DNA by these intermediates is thought most likely to produce mutagenic lesions, the site of carcinogenic action is not known.

Since α-hydroxynitrosamines are too unstable to be isolated, α-acetoxy nitrosamines have been used as models for studies of activated nitrosamines.[10,12,20-22] In an aqueous environment, or when acted upon by esterases, α-acetoxynitrosamines are hydrolyzed to α-hydroxy compounds. In other words, these compounds are direct-acting mutagens in the Ames assay, unlike the simple dialkyl and cyclic nitrosamines, which must first be activated by microsomal enzymes.[9,12,23] A number of studies[7,8] have shown that α-acetoxynitrosamines induce tumors in animals at or near the site of administration. In contrast, the parent compounds, which do not contain the α-acetoxy group, usually produce a spectrum of tumors not influenced by the site or route of administration.

Of course, α-hydroxylation is not the only pathway by which cyclic nitrosamines are metabolized. When animals are given nitrosopyrrolidine (NPYR), β-hydroxynitrosopyrrolidine is found in the urine.[24,25] Nitrosopiperidine can be oxidized on the γ-carbon by rat liver microsomes, and in this report we will show that rat liver microsomes hydroxylate nitrosohexamethyleneimine (NHEX) on both the β- and γ-carbons. Although studies on urinary metabolites are informative in showing what pathways can occur in the animal, the relationship of this to carcinogenic events in the target organ has not been established.

To investigate this problem we studied the metabolism by liver microsomes of two cyclic nitrosamines, NPYR and NHEX, and attempted to relate the metabolic products and the biological effect that they produced.[26-28] We examined the role played by both the microsomal and supernatant fraction in metabolizing NPYR. A parallel study was made of the role these two fractions play in forming mutagens from this compound in bacterial systems.

The information presented in this paper shows the following: In our liver *in vitro* system, NO-PYR is metabolized only via α-hydroxylation. The major product of microsomal metabolism, 2-hydroxytetrahydrofuran (2-hydroxyTHF), can be further metabolized by the supernatant fraction to γ-hydroxybutyrate and 1,4-butanediol. Only in the presence of both microsomal and supernatant fractions does the reaction proceed fully. These studies showed that the presence of both fractions was necessary to produce a maximal mutagenic response. NHEX can be oxidized at the β- and γ-position, as well as at the α-position, by microsomes. When γ-hydroxyNHEX was added, it was not metabolized further no matter which liver fraction was used. Mutagenic studies with NHEX showed that only in the presence of an S-9 activation system was a significant mutagenic response elicited in the Ames system. Very few mutants were produced by γ-hydroxyNHEX under the same conditions. Thus, we found an excellent correlation between metabolism of the cyclic nitrosamines NPYR and NHEX (probably via oxidation at the α-carbon) and mutagenic potency in two bacterial mutagenesis systems.

2. MATERIALS AND METHODS

2.1. Chemicals

Unlabeled NPYR and NHEX were prepared as previously described.[29] The compounds labeled with ^{14}C in the α-carbon atoms were synthesized according to the method of Snyder et al.[29] and further purified by HPLC. [^{3}H]-2-hydroxy-THF was prepared according to Hecker et al.[27] The preparation of β- and γ-hydroxyNHEX was done according to the methods of Hecker and Saavedra.[28]

2.2. Preparation of Microsomes

Ten-week-old male Fischer rats or male Sprague–Dawley rats from the Frederick Cancer Research Center rat colony induced with either Aroclor 1254 or phenobarbital were decapitated, and their livers were removed and washed in 0.025 M Tris (pH 7.5), 0.25 M sucrose, and 0.001 M sodium EDTA. The livers were minced briefly in a VirTis homogenizer in 3 volumes of buffer containing 0.025 M Tris (pH 7.5), 0.34 M sucrose, and 0.001 M sodium EDTA, and the cells were broken with three to four strokes of a motor-driven Potter Elvehjem homogenizer. The homogenate was filtered through four layers of cheesecloth and centrifuged three times at 8000 $\times$ g for 10 min. After each spin, the lipid layer was removed and the pellet discarded. Part of the 8000 $\times$ g supernatant (S-8 fraction) was stored at $-70°$C, and the remainder was centrifuged at 200,000 $\times$ g in a 50 Ti rotor for 1 hr to pellet the microsomes. The lipid layer was removed and the postmicrosomal supernatant (supernatant) was stored at $-20°$C. The microsomal pellet was resuspended in buffer containing 0.01 M Tris (pH 7.5) and 0.1 M KCl and then pelleted at 200,000 $\times$ g in a 50 Ti rotor for 1 hr. The final microsomal pellet was resuspended in as many milliliters of the Tris-KCl buffer as there were grams of liver used, and stored at $-70°$C. The protein content of cellular fractions was determined on 5% trichloroacetic acid-precipitated material by the method of Lowry et al.[30] using bovine serum albumin as the standard. Animals were induced as follows: Aroclor 1254 (500 mg/kg) was injected intraperitoneally 5 days before sacrifice. Phenobarbital

(100 mg/kg) was injected intraperitoneally on each of 3 successive days. The animals were sacrificed on the fourth day.

2.3. Assay for Nitrosopyrrolidine Metabolism

The metabolism of NPYR was followed in reaction mixtures containing 0.05 M Tris (pH 7.5), an NADPH-generating system (1 mM NADP, 5 mM glucose-6-phosphate, and 0.2 units/ml glucose-6-phosphate dehydrogenase), 5 mM MgCl$_2$, 0.1 mM MnCl$_2$, 0.154 M KCl, [^{14}C]-NO-PYR at either 52 μM or 1.04 mM (containing 1 or 2 μCi/ml, respectively), and microsomes and supernatant (1.5 mg/ml each).

Microsomal reactions contained, in addition to microsomes, 0.05 M potassium phosphate buffer (pH 7.5), an NADPH-generating system, 1 mM sodium EDTA (pH 7.5), 0.154 M KCl, and NPYR (see above).

Reactions were initiated by the addition of NPYR and incubated for 1–5 hr at 37°C. Reactions were carried out in volumes of from 0.15–1.5 ml. Reactions were followed by the withdrawal of 150 to 500 μl aliquots, which were diluted two- or three-fold with water and extracted twice with equal volumes of methylene chloride. Aliquots of the aqueous phase were counted for ^{14}C in PCS (Amersham–Searle, Arlington Heights, Illinois) in a Packard liquid scintillation counter, or Millipore-filtered and analyzed by HPLC, using either a Waters μ Bondapak C$_{18}$ or a Dupont Zorbax-ODS column.

2.4. Large-Scale Production of Metabolites

2.4.1. Nitrosopyrrolidine

The major NPYR metabolites were isolated by the following methods. Three 20-ml reactions, containing both microsomes and supernatant and 1 mM [^{14}C]-NPYR were incubated at 37°C in 250 ml Erlenmeyer flasks in a shaking water-bath for 3–6 hr. The reactions were extracted twice with 20 ml of methylene chloride. The aqueous phase was precipitated with 3 volumes of ethanol at 0°C to remove nucleic acids and proteins that were pelleted by centrifugation for 10 min at 10,000 $\times$ g. The ethanol supernatant was reduced to 1–3 ml by air at 4°C, or a stream of nitrogen, ethanol-precipitated, reconcentrated and ethanol-precipitated once more. The final concentrated extract (1.5 ml) was applied to a Sephadex G-10 column (2.6 $\times$ 100 cm) and eluted with a buffer containing 1 mM Tris (pH 7.5) and 2 mM NaCl. After Sephadex G-10 chromatography, the individual fractions were pooled, concentrated, the pH was adjusted to 4.5 with HCl, and the concentrate chromatographed on a Sephadex LH-20 column (2.6 $\times$ 100 cm) eluted with 2 mM sodium acetate buffer (pH 4.5).

A combination of chromatography on DEAE-cellulose, Waters C$_{18}$, and Dupont Zorbax-ODS columns[27] was used to purify γ-hydroxybutyrate.

To purify 1,4-butanediol, a combination of chromatography on benzoylated DEAE cellulose, Waters C_{18}, Dupont Zipax-SCX, and Dupont Zorbax-ODS columns[27] was used.

To prepare 2-hydroxytetrahydrofuran we incubated a microsomal reaction with [2,5-^{14}C]-NPYR for 3 hr at 37°C. The reaction mixture was extracted twice with methylene chloride, and the aqueous phase, after concentration to 1–1.5 ml, was then passed through a small Sephadex G-10 column (2.6 × 21 cm). Six 1.5-ml fractions were pooled from a peak, the main component of which was [^{14}C]2-hydroxyTHF.

2.4.2. Nitrosohexamethyleneimine

Large-scale purification of radioactive metabolites of NHEX was essentially similar to the procedure used for NPYR, with the following changes. The reaction for NHEX was 0.05 M N-2-hydroxyethylpiperazine-N'-2-ethanesulfonic acid (HEPES) pH 7.5, 5 mM MgCl$_2$, 1mM, MnCl$_2$, an NADPH-generating system, 3 mM[^{14}C]-NHEX, microsomes and supernatant. A 30-ml reaction was run for 4 hr at 38°C and stopped by the addition of 10 ml Ba(OH)$_2$ and 10 ml ZnSO$_4$ according to the method of Somogyi.[31] Precipitated materials were removed by centrifugation, and the remaining supernatant was concentrated and ethanol-precipitated twice. A combination of chromatography on Sephadex G-10 and a 9.4 mm × 25 cm Whatman ODS-2 column was used to purify β- and γ-hydroxyNHEX.

2.5. Mutagenesis Assay of NPYR with *E. coli* WU 3610 (*tyr, leu*)

A 5% inoculum of stationary cells were grown in 10 ml of 0.8% Difco nutrient broth containing 0.5% NaCl for 1 hr at 37°C to early log phase. The cells were pelleted and resuspended in 7.5 ml 0.1 M NaCl. Aliquots of bacteria (187.5 μl) were incubated for 1 hr at 37°C in a shaking water-bath in a total of 750 μl of solution containing 450 μl of cell fractions, NO-PYR (or other suspected mutagens), a reaction mixture with a final concentration of 0.05 M Tris-HCl (pH 7.5), an NADPH-generating system [2 mM NADP, 0.5 mM NADPH, 10 mM glucose-6-phosphate, and 0.4 units/ml glucose-6-phosphate dehydrogenase (Sigma)], 5 mM MgCl$_2$, 0.1 mM MnCl$_2$, and 0.1 M KCl. Reactions were stopped by chilling on ice and 100 μl aliquots were plated in duplicate as previously described[32] for the detection of *tyr*$^+$ and *leu*$^+$ revertants. Aliquots of the reaction mixture were diluted and plated in duplicate on Difco nutrient agar to obtain cell counts.

2.6. Mutagenesis Assay of Nitrosohexamethyleneimine

The mutagenicity of NHEX and its hydroxylated metabolites was tested by the method of Ames *et al.*[33] using Aroclor-induced S-9 fractions.

3. RESULTS AND DISCUSSION

Most nitrosamines are not mutagenic unless activated by microsomes and do not cause tumors at the site of administration. These facts, together with the pronounced organotropy of these compounds, would lead to the assumption that the most active enzyme systems that are responsible for α-hydroxylation of specific nitrosamines must be present in the target tissue, or at least in a tissue sufficiently close for diffusion of the active metabolite to occur. Perhaps the reason that NPYR causes hepatocellular tumors in rats is that it is in the hepatocytes that adequate levels of α-hydroxylation occurs. In the case of NHEX the compound may be rapidly metabolized by the Küpfer cells (or adjacent hepatocytes) since this compound induces both reticuloendothelial and hepatocellular tumors in the liver.[34]

We selected the two cyclic nitrosamines NPYR and NHEX for our study. NPYR induces hepatocellular tumors in rats. NHEX, which has a seven-membered ring containing two more methylene groups than NPYR, causes both liver and esophageal tumors in three strains of rat.[6,34,42] Unlike NPYR, NHEX induces endothelial tumors, as well as some hepatocellular tumors.[34] By doing a comparative study of the metabolism and mutagenesis we hoped to gain some insight into the mechanism of action of these two liver carcinogens.

Early evidence suggested that many cyclic nitrosamines were only poorly metabolized to CO_2 when fed to rats.[35-37] However, we have shown that, when ^{14}C-labeled NPYR and NHEX were fed to rats by gavage at 2–4 mg per animal, 77% and 43%, respectively, of the radioactivity present in the α-carbon atoms were recovered as CO_2 within 24 hr. At higher doses (144 and 160 mg), CO_2 production fell to 14% and 4% of the administered radioactivity. At both dose levels, most of the remaining radioactivity was secreted in the urine as either unchanged nitrosamine or metabolites.

Although NPYR and NHEX can be metabolized to a great extent by the rat, relatively few studies of their metabolism have appeared. Both Krüger and Bertram[24] and Cottrell et al.[25] have detected 3-hydroxy-N-nitrosopyrrolidine (a β-oxidation product) in the urine of rats fed NPYR. Cottrell et al. also detected the following radioactive products of NPYR: 2-pyrrolidone, succinic semialdehyde, and γ-hydroxybutyric acid, as well as dimethylamine and urea, in the urine. Hecht and co-workers[38] confirmed the α-hydroxylation scheme for cyclic nitrosamines proposed by Krüger[39] by demonstrating that NPYR may be metabolized by rat liver microsomes to 2-hydroxyTHF, the product to be expected from α-hydroxylation of the parent nitrosamine (see Figure 4). A similar pathway for microsomal metabolism of N-nitrosopiperidine was also demonstrated by Leung et al.[40] Grandjean[35] found 11 major products, including ε-aminocaproic acid in the urine of rats fed NHEX, and concluded that many of the urinary products were the result of α-hydroxylation. Ross and

Mirvish[41] isolated a small amount of 1,6-hexanediol bound to rat liver nucleic acids after feeding [^{14}C]-NHEX, and interpreted this to indicate that the reactive intermediates that bind to nucleic acids result from a pathway involving α-hydroxylation.

3.1. Nitrosopyrrolidine

We studied the metabolism of both NPYR and NHEX by examining products that are not extracted with methylene chloride (NMCE products). Presumably, products arising from α-, β-, or γ-oxidation are considerably less soluble in methylene chloride than the substrate and will, therefore, be water extractable. No metabolites of NPYR were extracted into the methylene chloride fraction, but a fraction (approximately 20%–25%) of the somewhat less polar metabolites of NHEX did appear in the organic phase. Since the proportion of these partitioning into each fraction is constant, it was still possible to follow the extent of the reaction. For both nitrosamines the metabolites were extracted and analyzed.

Three very different spectra of metabolites are obtained depending on whether NPYR is metabolized by microsomes, supernatant, or by microsomes + supernatant (Figure 1). The metabolites produced by supernatant alone (Figure 1B) are similar to those produced by microsomes + supernatant (Figure 1C), but the quantity produced is much smaller. In Figure 1C the amount of supernatant is only one fourth that in Figure 1B. Although γ-hydroxybutyrate is the major product under the first peak, it should be noted that one other product is present as a shoulder. We believe this product to be γ-aminobutyrate.

The reaction of NPYR with microsomes proceeds for only 30–60 min (Figure 2) and the major product is 2-hydroxyTHF (Figures 1A, 3A, 3B), which arises from the spontaneous decomposition of α-hydroxyNPYR. It therefore seems possible that the accumulation of 2-hydroxyTHF inhibits any further microsomal metabolism of NPYR. If, at 60 or 120 min, supernatant is added to the microsomal reaction, metabolism starts up again and NMCE products are produced for several hours more (Figure 2). From Figure 3 it can be seen that, upon the addition of supernatant together with a fresh NADPH-generating system, the main microsomal product (2-hydroxyTHF) disappears with a concomitant rise in γ-hydroxybutyrate and 1,4-butanediol, products characteristic of the microsomal + supernatant reaction (see Figure 1C). With prolonged incubation, γ-hydroxybutyrate predominates, the level of 1,4-butanediol remains relatively constant, and 2-hydroxyTHF reappears (Figure 3D). Unless a fresh NADPH-generating system is added at 60 min together with the supernatant, then at 120 min γ-hydroxybutyrate is the major peak, and the tendency of γ-hydroxybutyrate to predominate increases with the time of incubation (Figure 3E). It should be noted that the microsomal reaction was carried out in the

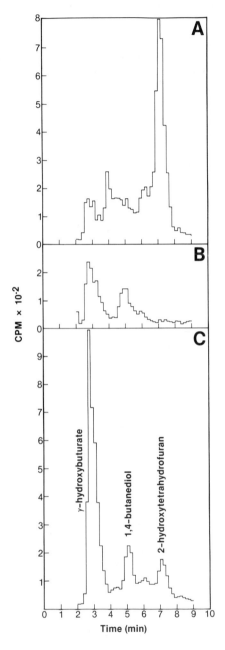

Figure 1. HPLC chromatograms of NPYR NMCE metabolites. (A) Microsomes alone; (B) supernatant alone; (C) microsomes plus supernatant. Reactions were incubated 3 hr with 52 μM NPYR. Aqueous fractions were applied to a Waters μ Bondapak C_{18} column and eluted with 0.01 M sodium acetate buffer pH 5.5 at 1.5 ml/min.[27]

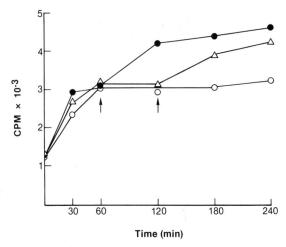

Figure 2. Microsomal reaction time course. Supernatant added at times designated by arrows. [^{14}C]-NMCE metabolites were measured versus time of incubation. ○, microsomes alone; ●, supernatant added at 60 min; △, supernatant added at 120 min; NPYR concentration is 1.04 mM.[27]

presence of Mg^{2+} and Tris buffer, conditions that are optimal for the microsomal + supernatant reaction, but not for the reaction with microsomes alone. The main product produced under these conditions is, however, that characteristic of the microsomal reaction (Figures 3A and 3B).

Since the product distribution was the same whether supernatant was added at 60 or 120 min, it appears that the accumulation of γ-hydroxybutyrate and 1,4-butanediol is related to the disappearance of 2-hydroxyTHF. Reactions that contained both microsomes + supernatant proceeded for up to 6 hr and produced an amount of metabolic products greater than the sum from the microsomal and supernatant reactions alone. At a concentration of 1.04 mM NPYR, 8%–10% of the initial substrate was metabolized after 4 hr by an amount of microsomes obtained from one quarter of a rat liver.

From our data we devised a scheme for the metabolism of NO-PYR by rat liver microsomes and supernatant (Figure 4). The initial step is α-hydroxylation followed by decomposition of the α-hydroxyNPYR to 4-hydroxybutanal, which, if isolated, exists predominantly as 2-hydroxyTHF.[38] Enzymes in the supernatant rapidly convert 4-hydroxybutanal to γ-hydroxybutyrate and 1,4-butanediol, with the most abundant product being γ-hydroxybutyrate. This compound, if fed to rats or humans, can be rapidly metabolized to CO_2 through a pathway involving β-oxidation to glycolate and acetate.[43,44] In mammalian brains, γ-hydroxybutyrate occurs in small quantities[45–47] and may be metabolized through succinate.[48–49] Such pathways could adequately account for the rapid metabolism of NPYR to CO_2 in the rat.[29]

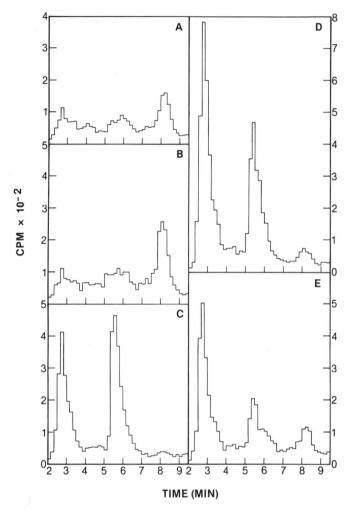

Figure 3. (A) Microsomes at 60 min; (B) microsomes at 240 min. with a fresh NADPH-generating system added at 60 min; (C) 120 min with supernatant plus a fresh NADPH-generating system added at 60 min; (D) 240 min with supernatant plus fresh NADPH-generating system added at 60 min, (E) 240 min with supernatant alone at 60 min. Aqueous fractions were applied to a Dupont Zorbax-ODS column and eluted with 5% methanol in 0.01 M sodium acetate buffer at 1.5 ml/min.

With an understanding of the major pathways by which NPYR is metabolized, we were able to examine the role of α-hydroxylation in the production of bacterial mutations without resorting to the use of acetoxy compounds. In these studies we employed *E. coli* WU 3610 (*tyr, leu*) for our tests[32]; both *tyr*[+] and *leu*[+] revertants produced in the presence of NPYR were scored. Since concentrations

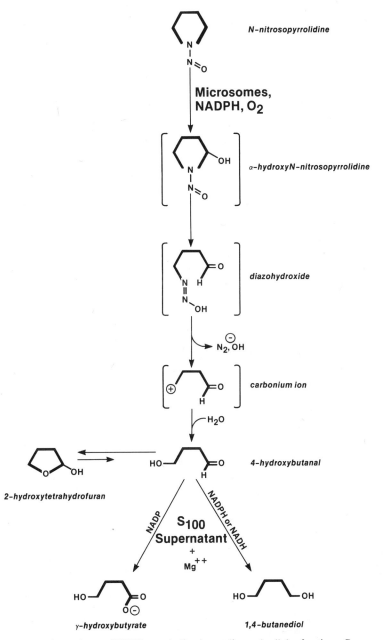

Figure 4. The major pathway of NPYR metabolism by rat-liver subcellular fractions. Compounds in parentheses are postulated unstable intermediates in the formation of 2-hydroxyTHF after α-hydroxylation of NPYR by microsomes.[38] In the presence of S-100 supernatant, 2-hydroxyTHF, which is in equilibrium with 4-hydroxybutanal, is rapidly converted to either γ-hydroxybutyrate or 1,4-butanediol.[27]

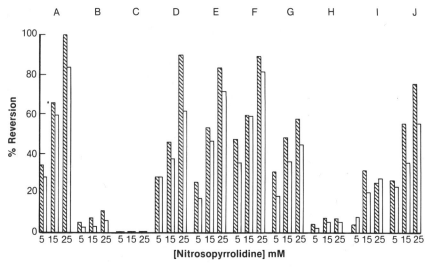

Figure 5. Relative mutagenicity vs. NPYR concentration. The number of tyr^+ revertants of *E. coli* WU 3610 (*tyr, leu*) produced by incubation of bacteria and uninduced S-8 fraction with 25 mM NPYR equals 100% mutagenicity. Hatched bars represent tyr^+ revertants and open bars represent leu^+ revertants. Bacteria were incubated for 1 hr at 37° C with the following cell fractions and plated on selective media.[32] (A) Uninduced S-8 from male Fischer rats; (B) uninduced microsomes; (C) uninduced S-100 supernatant; (D) uninduced microsomes + S-100 supernatant (1:1 v/v); (E) uninduced microsomes + S-100 supernatant (1:3); (F) phenobarbital-induced microsomes + S-100 supernatant (1:1) from male Sprague–Dawley rats; (G) Aroclor-induced S-8 from male Sprague–Dawley rats; (H) uninduced microsomes incubated 1 hr at 37° C bacteria and fresh NADPH-generating system added and incubated another hour; (I) same as (H) except bacteria + S-100 supernatant fraction added; (J) same as (H) except bacteria, S-100 supernatant fraction, and an NADPH-generating system added.[14]

of NPYR above 25 mM proved toxic, this was the highest level we used. There were approximately 400 revertants/10^8 survivors when uninduced S-8 fraction and 25 mM NPYR were incubated with 3×10^7 bacteria. The number of tyr^+ revertants using 25 mM NPYR and uninduced S-8 was set at 100%; other reversion rates were reported relative to this value. All results are the average from 2–6 duplicate experiments.

The percentage of revertants obtained in this system when NPYR was incubated with S-8 fraction from uninduced male Fischer rats is shown in Figure 5A. Neither of the two components of the S-8 fraction (microsomes of S-100 supernatant) incubated separately produced a significant number of revertants (Figures 5B and 5C). When microsomes and S-100 supernatant were combined in various proportions to produce a reconstituted S-8 fraction, most of the mutagenic activity was regained (Figures 5D and 5E). Phenobarbital-induced microsomes and S-100 fraction were no more effective than uninduced components, while Aroclor 1254-induced S-8 fractions were less effective than those from uninduced animals (Figures 5F and 5G). Since our metabolic experiments were carried out with cellular fractions from uninduced rats, it seemed more relevant to use uninduced enzyme fractions in the mutagenic experiments as well. Figures 5F and 5G show that there is no advantage in using induced microsomes in this system.

We next investigated the effect of preincubation of NPYR with microsomes before bacteria and S-100 were added. Microsomes were incubated with NPYR for 1 hr; after bacteria and a fresh NADPH-generating system were added the system was reincubated for a second hour (Figure 5H). The level of revertants was very low. However, Figure 5I shows the level of revertants obtained increased if bacteria and the S-100 fraction were added to the preincubation mixture after 1 hr. An almost maximal number of revertants was obtained upon addition of bacteria, S-100, and a fresh NADPH-generating system after the first hour (Figure 5J). Heat-denatured supernatant did not produce this effect. This is additional evidence that microsomes remain stable for the full time of incubation, and that even though microsomes alone produced relatively small numbers of revertants, they still retained their mutagenic potential as long as both supernatant and an NADPH-generating system were present. These results are paralleled by the metabolic reactions in which microsomal metabolism of NPYR lasts for only 30–60 min, but the microsomes have not lost their potential for activity (Figure 2). As we have demonstrated above, when S-100 is added to a microsomal reaction that has apparently stopped, the main product, 2-hydroxyTHF, is removed (with the concomitant production of γ-hydroxybutyrate and 1,4-butanediol), and the microsomal α-hydroxylation of NO-PYR then proceeds. Thus, 2-hydroxyTHF (or 4-hydroxybutanal) appears to inhibit the microsomal reaction.

These mutation studies lead us to suggest two possible explanations for the production of a mutagenic agent from NPYR. Either mutagenic agents are produced by α-hydroxylation of NPYR and the subsequent breakdown to 2-hydroxyTHF, or the metabolism of 2-hydroxyTHF to γ-hydroxybutyrate or 1,4-butanediol is responsible for these effects. If the latter is the case, the microsomal products should cause bacterial mutations when incubated with S-100 supernatant. This did not prove to be so; directly isolated total reaction products of the microsomal reaction showed no effect when incubated with bacteria and S-100. Purified samples of the principal products of NPYR metabolism (2-hydroxyTHF, γ-hydroxybutyrate, and 1,4-butanediol) also failed to produce revertants when incubated with either S-8 or S-100 supernatants over a range of substrate concentrations from 0.1 to 25 mM. We have, therefore, separated NPYR metabolism into two distinct phases: (1) microsomal α-hydroxylation with the production of a species highly reactive towards nucleophiles, including water (to produce 4-hydroxybutanal), and (2) subsequent oxidation and reduction of the initial product by the S-100 supernatant. The second phase of NPYR metabolism produced no reversions, but was indirectly responsible for the efficacy of the overall reaction by removing 2-hydroxyTHF and allowing the initial metabolism of NPYR to proceed at an accelerated rate. Thus, our best interpretation is that α-hydroxylation is responsible for the mutagenic activity of NPYR, although we have not ruled out the possibility that some minor pathways may be involved.

3.2. Nitrosohexamethyleneimine

The metabolism of NHEX was studied in much the same manner as was NPYR. After incubation of liver microsomal fractions with $[\alpha\text{-}^{14}C]$-NHEX, the reaction mixture was extracted with methylene chloride. We have found that every labeled metabolite of NHEX appears in the aqueous phase. Optimal conditions for the metabolism of NO-HEX are similar to those for NPYR. The K_m for the total metabolism of NO-HEX (0.7 mM) is close to that (0.34 mM) found for NPYR.[27] However, the percentage of 1 mM NHEX metabolized by liver microsomes + supernatant in a 4-hr reaction is ~16%, which is about double that for the same concentration of NPYR. We have shown that for the NPYR, the only detectable oxidative pathway of *in vitro* metabolism by microsomes obtained from either the liver (a target organ) or the lungs (a nontarget organ)[26] is α-oxidation. We were unable to detect microsomal oxidation of NPYR at the β-carbon. Other investigators have demonstrated the presence of β-hydroxyNPYR in the urine of rats[24,25] after *in vivo* administration. The source of this metabolite and its relationship, if any, to the mutagenic or carcinogenic action of NPYR is not known. Stolz and Sen have demonstrated that β-hydroxyNPYR may be mutagenic without activation,[50] but this work needs to be repeated using HPLC-purified compounds. However, as the ring-size of the cyclic nitrosamines increases, microsomal oxidations at carbons other than α-C become possible. Thus, both Hsieh *et al.*[51] and Rayman *et al.*[52] demonstrated that liver microsomes could metabolize N-nitrosopiperidine to its γ-hydroxy derivative. Studies on this compound and β-hydroxynitrosopiperidine showed that they were not mutagenic without microsomal activation.[15,16] Both β- and γ-hydroxynitrosopiperidines were somewhat less mutagenic than their parent. It is not known to what extent either of these hydroxylated compounds are metabolized by rat liver microsomes. It is not surprising that rat liver microsomes can hydroxylate NHEX at either the α-, β-, or γ-carbon. α-HydroxyNHEX, like α-hydroxyNPYR, is unstable and spontaneously decomposes to 6-hydroxy-*n*-hexanal, which may be further metabolized by microsomes and/or supernatant. The β- and γ-hydroxy derivatives of NHEX are stable and can easily be isolated and identified. When NHEX is incubated for 4 hr in the presence of liver microsomes + supernatant, approxi-

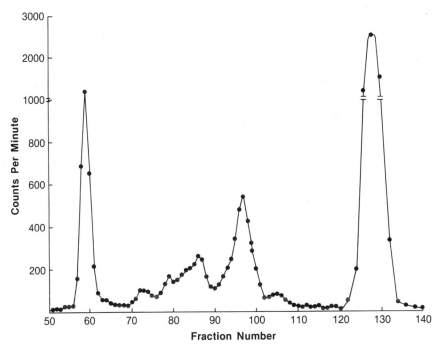

Figure 6. Sephadex G-10 (2.6 × 100 cm column) chromatography of a large-scale microsome + supernatant reaction using NHEX. Fractions are 4 ml each; ●, ^{14}C-radioactivity determined on 40-μl aliquots.

mately one-third of all detected metabolites are the stable β- and γ-hydroxylated nitrosamines (Figure 6). After precipitation of proteins by the method of Somogyi,[31] and removal of nucleic acids and large carbohydrates by ethanol precipitation, a concentrate of the reaction mixture was chromatographed on Sephadex G-10. The asymmetric peak of radioactivity falling between fractions 91–102, and the small trailing shoulder occurring from fractions 102–110, account for one-third of the radioactivity present in all the metabolites (fractions 56–110). These fractions (91–110) contain all the β- and γ-hydroxylated NO-HEX derivatives and no others. Further purification was achieved with a Whatman ODS-2 column (Figure 7). The first two peaks are the *anti* and *syn* conformers of γ-hydroxyNHEX, and the third and fourth are the *anti* and *syn* conformers of β-hydroxyNHEX. The ratio of γ: β-isomers is 3:1. Since the major stable metabolite from NHEX is γ-hydroxyNHEX, we sought to determine the contribution of this metabolite to the mutagenic activity of NHEX.

When a low concentration of NHEX (0.1 mM) was incubated for 4 hr with liver microsomes + supernatant, approximately half the starting material was

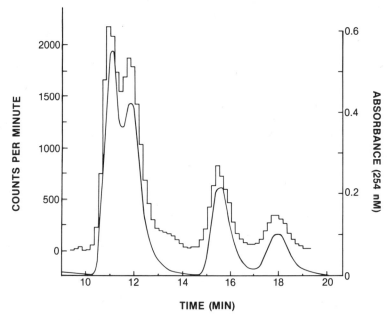

Figure 7. Whatman ODS-2 chromatography of fractions 91–110 from Figure 6. The column (9.4 mm × 25 cm) was run at 4.5 ml per min in 17% methanol/H_2O. Ten-microliter aliquots were counted for ^{14}C-radioactivity.

metabolized. Even with a concentration of 3 mM NHEX, 10% of the starting material was metabolized (Figure 6). In this illustration, all of the radioactivity before fraction 110 represents metabolites. Under the same reaction conditions, 0.1 mM γ-hydroxyNHEX is metabolized to about only 1%–2% (Figure 8). In this case, analyses were done on an analytical Whatman ODS-2 column (0.46 × 25 cm) with a program that did not distinguish the separate conformers of γ-hydroxyNHEX. The major peak, appearing between 8½ and 10 min, is γ-hydroxyNHEX, and it remained virtually unchanged after 4 hr. A small quantity of polar metabolites appeared at early elution times. Since γ-hydroxyNHEX is not metabolized to an appreciable extent, we would expect that it will not be mutagenic, unless it is direct-acting. γ-HydroxyNHEX was tested in the Ames system using *S. typhimurium* and Aroclor-induced S-9 fractions. We selected the Aroclor-induced fraction to increase the number of *his*[+] revertants so that low levels of mutation could be scored. Experiments are underway to compare the results of metabolic experiments with the mutagenic activity of NHEX in the presence of uninduced, Aroclor- and phenobarbital-induced S-9 fractions. The results from the Ames test (Figure 9) confirmed our expectations. In the presence of S-9, NHEX was mutagenic. Without a liver-

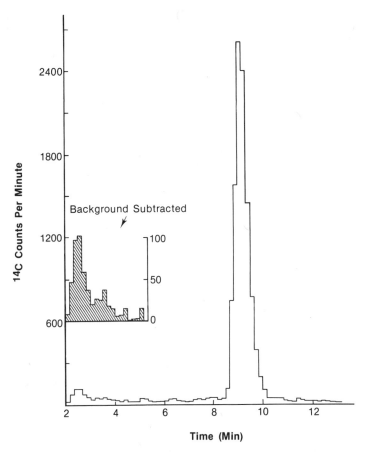

Figure 8. Whatman ODS-2 chromatography (0.46 x 25 cm) of 0.1 mM γ-hydroxyNHEX after a 4-hr reaction with microsomes and supernatant. The column was run at 1.5 ml per min using the following protocol: 6% methanol/water for 15 sec followed by a 2-min linear gradient to 28% methanol/water. The hatched area shows the first 5 min of elution on an expanded scale with a zero time point incubation subtracted.

activating system, the mutagenicity of NHEX was greatly reduced. The presence of a small number of reversions above background in the absence of S-9 may mean either that NHEX is a weak direct-acting mutagen, or that in this case the bacteria themselves can activate the nitrosamine. In contrast, γ-hydroxyNHEX was not mutagenic without the presence of liver subcellular fractions. In the presence of S-9, γ-hydroxyNHEX was about 2%–3% as mutagenic as NHEX itself. Thus, we found an excellent correlation between the level of metabolism and the mutagenicity of γ-hydroxyNHEX.

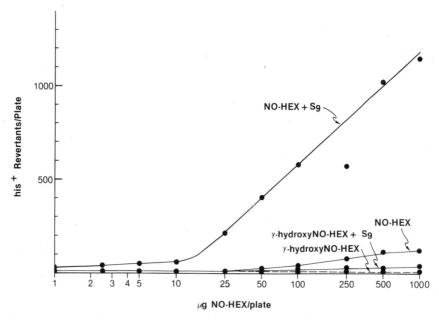

Figure 9. Mutagenicity in the Ames *Salmonella* reversion system of NHEX and γ-hydroxyNHEX in the presence and absence of S-9. Numbers of his^+ revertants per plate are compared over a 1000-fold concentration range of nitrosamines.

4. CONCLUSION

Before they can exert a mutagenic effect, most nitrosamines must first be metabolically activated. The mutagenic event probably follows hydroxylation of the carbon α to the nitroso group, which yields a species that can interact with nucleic acid. In most of the published work, this metabolic activation has been performed by liver S-9 fractions. *In vitro* studies have shown that the metabolism of nitrosopyrrolidine occurs via two separate stages. The compound is first α-hydroxylated by microsomes with the formation of 2-hydroxytetrahydrofuran (2-hydroxyTHF), which is then further metabolized by enzymes of the postmicrosomal supernatant. That is, in the presence of S-9 (microsomes + supernatant) 2-hydroxyTHF exists only as a transient entity. The mutagenic studies with nitrosopyrrolidine were made using a preincubation system with *E. coli* (*leu, tyr*). Only a small number of revertants were found using microsomes or supernatant alone. However, in the presence of S-9, which contains both fractions, the number of revertants was much higher. Thus, in this case, the removal of the product of microsomal α-hydroxylation by the enzymes of the supernatant was necessary before maximal rates of mutation were obtained.

In the case of nitrosohexamethyleneimine (NHEX), the γ-hydroxylated derivative is one of the major products of microsomal oxidation. *In vitro* studies have shown that the γ-hydroxylated compound (unlike the parent compound) is metabolized only to a small extent. Using the Ames assay, we have shown that the γ-hydroxylated derivative is 40- to 50-fold less mutagenic than is the parent nitrosamine. We conclude, therefore, that there is an excellent correlation between the extent to which this compound is metabolized and the mutagenic potency it exhibits in the Ames test.

ACKNOWLEDGMENTS. This work was supported by contract No. NO1-CO-75380 with the National Cancer Institute, NIH, Bethesda, Maryland 20205.

5. REFERENCES

1. P. N. Magee and J. M. Barnes, Carcinogenic nitroso compounds, *Adv. Cancer Res. 10*, 164–246 (1967).
2. H. Druckrey, S. Preussmann, S. Ivankovic, and D. Schmähl, Organotrope carcinogene Wirkungen bei 65 verschiedenen N-nitroso-Verbindungen an BD-Ratten, *Z. Krebsforsch. 69*, 103–201 (1967).
3. W. Lijinsky and H. W. Taylor, Carcinogenicity of methylated nitrosopiperidines, *Int. J. Cancer 16*, 318–322 (1975).
4. W. Lijinsky and H. W. Taylor, The effect of substitutes on the carcinogenicity of N-nitrosopyrrolidine in Sprague–Dawley rats, *Cancer Res. 36*, 1988–1990 (1976).
5. W. Lijinsky, H. W. Taylor, and L. K. Keefer, Reduction of rat liver carcinogenicity of 4-nitrosomorpholine by α-deuterium substitution, *J. Natl. Cancer Inst. 57*, 1311–1313 (1976).
6. W. Lijinsky and H. W. Taylor, Carcinogenicity of methylated derivatives of N-nitrosodiethylamine and related compounds in Sprague–Dawley rats, *J. Natl. Cancer Inst. 62*, 407–410 (1979).
7. M. Wiessler and D. Schmähl, Zur carcinogenen Wirkung von N-nitroso Verbindungen, 5. Mitteilung: Acetoxymethyl-Nitrosamine, *Z. Krebsforsch. 85*, 47–49 (1976).
8. S. R. Joshi, J. M. Rice, M. L. Wenk, P. P. Roller, and L. K. Keefer, Selective induction of intestinal tumors in rats by methyl(acetoxymethyl)nitrosamine, an ester of the presumed reactive metabolite of dimethylnitrosamine, *J. Natl. Cancer Inst. 58*, 1531–1535 (1977).
9. M. Okada, E. Suzuki, T. Anjo, and M. Mochizoki, Mutagenicity of α-acetoxydiakylnitrosamines: Model compounds for an ultimate carcinogen, *Gann 66*, 457–458 (1975).
10. J. E. Baldwin, W. E. Branz, R. F. Gomez, P. L. Kraft, A. J. Sinskey, and S. R. Tannenbaum, Chemical activation of nitrosamines into mutagenic agents, *Tetrahedron Lett. 5*, 333–336 (1976).
11. O. G. Fahmy and M. J. Fahmy, Mutagenic selectivity of carcinogenic nitroso compounds. III. N, α-acetoxymethyl-N-methylnitrosamine, *Chem.-Biol. Interact. 14*, 21–35 (1976).
12. A. M. Camus, M. Wiessler, C. Malaveille, and H. Bartsch, High mutagenicity of N-(α-acyloxy)alkyl-N-alkylnitrosamines in *S. typhimurium*: Model compounds for metabolically activated N,N-dialkylnitrosamines, *Mutat. Res. 49*, 187–194 (1978).
13. F. W. Larimer, D. W. Ramey, W. Lijinsky, and J. L. Epler, Mutagenicity of methylated N-nitrosopiperidines in *Saccharomyces cerevisiae*, *Mutat. Res. 57*, 155–161 (1978).
14. L. I. Hecker, R. K. Elespuru, and J. G. Farrelly, The mutagenicity of nitrosopyrrolidine is related to its metabolism, *Mutat. Res. 62*, 213–220 (1979).

15. C. E. Nix, B. Brewen, R. Wilkerson, W. Lijinsky, and J. L. Epler, Effects of N-nitrosopiperidine substitutions on mutagenicity in *Drosophila melanogaster*, *Mutat. Res. 67*, 27–38 (1979).
16. T. K. Rao, A. A. Hardigree, J. A. Young, W. Lijinsky, and J. L. Epler, Mutagenicity of N-nitrosopiperidines with *Salmonella typhimurium*/microsomal activation system, *Mutat. Res. 56*, 131–145 (1977).
17. T. K. Rao, J. A. Young, W. Lijinsky, and J. L. Epler, Mutagenicity of N-nitrosopiperazine derivatives in *Salmonella typhimurium*, *Mutat. Res. 57*, 127–134 (1978).
18. L. Keefer, W. Lijinsky, and H. Garcia, Deuterium isotope effect on the carcinogenicity of dimethylnitrosamine in rat liver, *J. Natl. Cancer. Inst. 51*, 299–302 (1973).
19. W. Lijinsky, J. E. Saavedra, M. D. Reuber, and B. N. Blackwell, The effect of deuterium labeling on the carcinogenicity of nitroso-2,6-dimethylmorpholine in rats, *Cancer Lett. 10*, 325–331 (1980).
20. O. G. Fahmy, M. J. Fahmy, and M. Wiessler, α-Acetoxy-dimethylnitrosamine: A proximate metabolite of the carcinogenic amine, *Biochem. Pharmacol. 24*, 1145–1148 (1975).
21. P. P. Roller, D. R. Shimp, and L. K. Keefer, Synthesis and solvolysis of methyl-(acetoxymethyl) nitrosamine, solution chemistry of the presumed carcinogenic metabolite of dimethylnitrosamine, *Tetrahedron Lett. 25*, 2065–2068 (1975).
22. P. L. Skipper, S. R. Tannenbaum, J. E. Baldwin, and A. Scott, Alkylation by α-acetoxy-N-nitrosamines: Models for N-nitrosamine metabolites, *Tetrahedron Lett. 49*, 4269–4272 (1977).
23. S. R. Tannenbaum, P. Kraft, J. Baldwin, and S. Branz, The mutagenicity of methylbenzylnitrosamine and its α-acetoxy derivatives, *Cancer Lett. 2*, 305–310 (1977).
24. F. W. Krüger and B. Bertram, Metabolism of nitrosamines *in vivo*. IV. Isolation of 3-hydroxyl-1-nitrosopyrrolidine from rat urine after application of 1-nitrosopyrrolidine, *Z. Krebsforsch. 83*, 255–260 (1975).
25. R. C. Cottrell, D. G. Walters, P. J. Young, J. C. Phillips, B. G. Lake, and S. D. Gangolli, Studies of the urinary metabolites of N-nitrosopyrrolidine in the rat, *Toxicol. Appl. Pharmacol. 54*, 368–376 (1980).
26. L. I. Hecker, *In vitro* metabolism of N-nitrosopyrrolidine by rat lung subcellular fractions: α-Hydroxylation in a non-target tissue, *Chem.-Biol. Interactions 30*, 57–65 (1980).
27. L. I. Hecker, J. G. Farrelly, J. H. Smith, J. E. Saavedra, and P. A. Lyon, Metabolism of the liver carcinogen N-nitrosopyrrolidine by rat liver microsomes, *Cancer Res. 39*, 2679–2686 (1979).
28. L. I. Hecker and J. E. Saavedra, *In vitro* formation and properties of β- and γ-hydroxy-N-nitrosohexamethyleneimine, *Carcinogenesis 1*, 1017–1025 (1980).
29. C. M. Snyder, J. G. Farrelly, and W. Lijinsky, Metabolism of three cyclic nitrosamines in Sprague–Dawley rats, *Cancer Res. 37*, 3530–3532 (1977).
30. O. H. Lowry, N. J. Rosebrough, A. L. Farr, and R. J. Randall, Protein measurement with the folin phenol reagent, *J. Biol. Chem. 193*, 265–275 (1951).
31. M. Somogyi, Determination of blood sugar, *J. Biol. Chem. 160*, 69–73 (1945).
32. R. K. Elespuru and W. Lijinsky, Mutagenicity of cyclic nitrosamines in *Escherichia coli* following activation with rat liver microsomes, *Cancer Res. 36*, 4099–4101 (1976).
33. B. N. Ames, J. McCann, and E. Yamasaki, Methods for detecting carcinogens and mutagens with the *Salmonella*/mammalian-microsome mutagenicity test, *Mutat. Res. 31*, 347–364 (1975).
34. W. Lijinsky and M. D. Reuber, Carcinogenic effect of nitrosopyrrolidine, nitrosopiperidine, and nitrosohexamethyleneimine in Fischer rats, *Cancer Lett. 12*, 99–103 (1981).
35. C. J. Grandjean, Metabolism of N-nitrosohexamethyleneimine, *J. Natl. Cancer Inst. 57*, 181–185 (1976).
36. F. W. Krüger, B. Bertram, and G. Eisenbrand, Metabolism of nitrosamines *in vivo*. Investigation on $^{14}CO_2$ exhalation, liver RNA labeling, and isolation of two metabolites from urine after administration of [2,5-^{14}C] dinitrosopiperazine to rats, *Z. Krebsforsch. 85*, 125–134 (1976).

37. B. W. Stewart, P. F. Swann, J. W. Holsman, and P. N. Magee, Cellular injury and carcinogenesis. Evidence for the alkylation of rat liver nucleic acids *in vivo* by N-nitrosomorpholine, *Z. Krebsforsch. 82*, 1–12 (1974).
38. S. S. Hecht, C. B. Chen, and D. Hoffmann, Evidence for metabolic α-hydroxylation of N-nitrosopyrrolidine, *Cancer Res. 38*, 215–218 (1978).
39. F. W. Krüger, New aspects in metabolism of carcinogenic nitrosamines, in: *Proceedings of the Second International Symposium of the Princess Takamatsu Cancer Research Fund* (W. Nakahara, S. Takayama, T. Sugimura, and S. Otashima, eds.), pp. 213–235, Univ. of Tokyo Press, Tokyo (1972).
40. K. H. Leung, K. K. Park, and M. C. Archer, Alphahydroxylation in the metabolism of N-nitrosopiperidine by rat liver microsomes: Formation of 5-hydroxypentanal, *Res. Commun. Chem. Pathol. Pharmacol. 19*, 201–211 (1978).
41. A. E. Ross and S. S. Mirvish, Metabolism of N-nitrosohexamethyleneimine to give 1,6-hexanediol bound to rat liver nucleic acids, *J. Natl. Cancer Inst. 58*, 651–655 (1977).
42. C. M. Goodall, W. Lijinsky, and L. Tomatis, Tumorigenicity of N-nitrosohexamethyleneimine, *Cancer Res. 28*, 1217–1222 (1968).
43. C. R. Lee, Evidence for the β-oxidation of orally administered 4-hydroxybutyrate in humans, *Biochem. Med. 17*, 284–291 (1977).
44. S. S. Walkenstein, R. Wiser, C. Gudmundsen, and H. Kimmel, Metabolism of γ-hydroxybutyric acid, *Biochem. Biophys. Acta 86*, 640–642 (1964).
45. N. J. Giarman and R. H. Roth, Differential estimation of gammabutyrolactone and gammahydroxybutyric acid in rat blood and brain, *Science 145*, 583–584 (1964).
46. R. H. Roth and N. J. Giarman, Preliminary report on the metabolism of γ-butyrolactone and γ-hydroxybutyric acid, *Biochem. Pharmacol. 14*, 177–178 (1965).
47. R. H. Roth and N. J. Giarman, Conversion *in vivo* of γ-aminobutyric to γ-hydroxybutyric acid in the rat, *Biochem. Pharmacol. 18*, 247–250 (1969).
48. J. D. Doherty, R. W. Stout, and R. H. Roth, Metabolism of [1-^{14}C]-γ-hydroxybutyric acid by rat brain after intraventricular injection, *Biochem. Pharmacol, 24* 469–474 (1975).
49. W. N. Fishbein and S. P. Bessman, γ-Hydroxybutyrate in mammalian brain. Reversible oxidation by lactic dehydrogenase, *J. Biol. Chem. 239*, 357–360 (1964).
50. D. R. Stolz and N. P. Sen, Mutagenicity of five cyclic N-nitrosamines: Assay with *Salmonella typhimurium, J. Natl. Cancer Inst. 58*, 393–394 (1977).
51. S.-T. Hsieh, P. L. Kraft, M. C. Archer, and S. R. Tannenbaum, Reaction of nitrosamines in the Udenfriend system: Principal products and biological activity, *Mutat. Res. 35*, 23–28 (1976).
52. M. P. Rayman, B. C. Challis, P. J. Cox, and M. Jarman, Oxidation of N-nitrosopiperidine in the Udenfriend model system and its metabolism by rat liver microsomes, *Biochem. Pharmacol. 24*, 621–626 (1975).

10

Structure–Activity Relations in Carcinogenesis by N-Nitroso Compounds

WILLIAM LIJINSKY

1. INTRODUCTION

Since 1956 when the carcinogenic activity of the first N-nitroso compound, nitrosodimethylamine, was reported by Magee and Barnes,[1] there has been growing interest in the biological activity of this group of compounds. Extensive studies of the carcinogenic activity of N-nitroso compounds of various structures have been carried out with the aim of using the differences in activity to suggest possible mechanisms of their carcinogenic action. This approach, which is common in pharmacology, has proved very useful.

The first experiments along these lines were carried out by Magee and his colleagues, while at the same time very extensive studies were conducted by Druckrey, Preussmann, and their associates. Reviews of their work[2,3] appeared more than a decade ago, and the focus of their interest was the probability that N-nitroso compounds acted through conversion to agents that alkylated DNA and other macromolecules, a suggestion first made by Magee and Farber.[4] The necessary truth of this idea, to which there are many exceptions, has been questioned,[5] even though it would unify the biological actions of these compounds as carcinogens and mutagens. More recently we have examined a large number of N-nitroso compounds in experiments designed to shed light on carcinogenic mechanisms, even possible ones not involving alkylation of DNA. In Tables I and II are given summaries of these experiments utilizing mouse skin

---

WILLIAM LIJINSKY • Chemical Carcinogenesis Program, NCI-Frederick Cancer Research Facility, Frederick, Maryland 21701.

Table I
Carcinogenicity of Nitrosoalkylamides in Mouse Skin
(20 Female Swiss Mice in Each Group)

		Concentration in acetone (mM)	Length of treatment (weeks)	Total dose (μmoles)	Time of first tumor (weeks)	No. of mice with skin tumors	Average latent period (weeks)
	Nitroso-						
Ia	Methylurea	40	40	80	23	18	34
Ib	Ethylurea	40	40	80	28	12	50
Ic	n-Propylurea	40	50	100	65	5	73
Ie	Allylurea	40	50	100	—	0	—
In	iso-Propylurea	40	50	100	65	4	80
Id	n-Butylurea	40	50	100	60	3	75
	iso-Butylurea	40	50	100	—	0	—
	sec-Butylurea	40	50	100	—	0	—
If	n-Amylurea	40	50	100	40	11	63
Ig	n-Hexylurea	40	50	100	55	11	66
Ih	n-Undecylurea	40	50	100	48	4	60
Ii	n-Tridecylurea	40	50	100	45	2	73
Io	Phenylurea	40	50	100	70	1	70
Ip	Cyclohexylurea	40	50	100	70	2	78

	Compound						
Iq	Benzylurea	40	50	100	—	0	—
Ir	2-Phenylethylurea	40	50	100	—	0	—
Ik	Hydroxyethylurea	40	45	90	42	15	48
Il	2-Hydroxypropylurea	40	40	80	31	16	45
Im	3-Hydroxypropylurea	40	45	90	35	16	50
Ij	Fluoroethylurea	10	40	20	40	15	54
	Oxazolidone	40	50	100	50	10	54
	5-Methyloxazolidone	40	50	100	43	19	47
IIIa	Methylurethane	40	60	120	55	1	—
IIIb	Ethylurethane	40	80	160	—	0	72
IIIc	Carbaryl	40	50	100	60	8	53
IIa	Methylnitro-guanidine	20	30	30	30	14	82
		8	30	12	60	4	47
IIb	Ethylnitro-guanidine	20	30	30	30	20	60
		8	30	12	35	14	—
	Cimetidine	20	110	110	—	1	—
		8	110	45	—	0	—
	Control (acetone)	—	—	—	—	0	—

Table II

Relative Carcinogenicity of N-Nitroso Compounds in Rats (Chronic Treatment)

	Compound	Dose rate (μmoles/week) Administration W=Drinking water G=Gavage in oil	Length of treatment (weeks)	Total dose per rat (μmoles)	Relative carcinogenicity		Organs in which tumors appear (time of death at 50% animals with tumors: weeks) SD=Sprague–Dawley rats F=Fischer rats W=Wistar rats
	Nitrosopiperidines:						
LIII	Nitroso-piperidine	87 W	44	3.9	+++	(SD)	Esophagus, nasal turbinates (42,38)
		87 W	28	2.5		(F)	Esophagus (30)
LXXXIX	2-Methyl-	87 W	50	4.4	++	(SD)	Esophagus, nasal turbinates (82,80)
XCII	3-Methyl-	87 W	50	4.4	+++	(SD)	Esophagus, nasal turbinates (55)
XCIII	4-Methyl-	87 W	40	3.5	+++	(SD)	Esophagus, nasal turbinates (40,42)
XC	2,6-Dimethyl-	87 W	50	4.4	0	(SD)	
	3,5-Dimethyl-	44 W	40	1.8	+	(SD)	Esophagus (100)
		87 W	50	4.4		(F)	Esophagus (61)
XCI	2,2,6,6-Tetramethyl-	87 W	50	4.4	0	(SD)	
LXXVIII	4-Phenyl-	87 W	50	4.4	++	(F)	Esophagus, liver (62)
XCIV	4-Cyclohexyl-	87 W	50	4.4	0	(F)	
LXXIX	4-t-Butyl-	87 W	40	3.5	−	(SD)	
CVI	3-Hydroxy-	87 W	60	5.3	+	(F)	Esophagus (71)
CVII	4-Hydroxy-	89 W	35	3.2	+++	(SD)	Esophagus, nasal turbinates, liver (hepatocellular) (43)
		89 W	35	3.2	+++	(SD)	Nasal turbinates, liver (hepatocellular) (44)
CVIII	4-Keto-	90 W	35	3.2	+++	(SD)	Nasal turbinates, liver (hepatocellular) (45)
CIX	4-Keto-3-Methyl-	90 W	30	2.7	+++	(F)	Esophagus, liver (45)
XCVIII	2-Carboxy-	90 W	26	2.3	0	(F)	Esophagus (27)
XCIX	4-Carboxy-	89 W	25	4.4	0	(W)	
		89 W	50	4.4	0	(SD)	

ID	Compound	Dose		Result	Strain	Tumor sites (refs)
CII	3-Chloro-	17 W	30	0.5	(F)	Esophagus, tongue, forestomach (40)
CIII	4-Chloro-	34 W	27	0.9	(SD)	Esophagus, nasal turbinates, liver (hepatocellular and sarcoma) (41)
CIV	3,4-Dichloro-	17 W	30	0.5	(F)	Esophagus, tongue, forestomach (37)
		34 W, 7 W	15,30	0.5, 0.2	(SD)	Esophagus (20,52)
CV	3,4-Dibromo-	17 W	21	0.35	(F)	Esophagus (24)
XCV	-1,2,3,6-Tetrahydro-pyridine	37 W	27	1.0	(SD)	Esophagus (36)
		89 W	26	2.3	(SD)	Liver (hepatocellular) (28)
		89 W, 36 W, 14 W	25,25,25	2.2, 0.9, 0.36	(F)	Esophagus, liver (sarcoma) (29,41,77)
	-1,2,3,6-Tetrahydro-pyridine-3-methyl	89 W	25	2.2	(F)	Esophagus (27)
XCVI	1,2,3,4-Tetrahydropyridine	89 W	25	2.2	(F)	Esophagus (30)
C	-Guvacoline	89 W	50	4.4	(SD)	
CI	-Methylphenidate	91 G	50	4.6	(SD)	
	Nitrosopyrrolidines:					
LII	Nitroso-pyrrolidine	200 W	50	10.0	(SD)	Liver (hepatocellular) (78)
		90 W	50	4.5	(F)	Liver (101)
LXXX	2,5-Dimethyl-	200 W	50	10.0	(SD)	
LXXXVII	3,4-Dichloro-	59 W	34	2.0	(SD)	Esophagus (44)
LXXXIII	2-Carboxy-	104 W	75	8.0	(W)	
LXXXIV	2-Carboxy-4-hydroxy-	94 W	75	7.0	(W)	
LXXXVI	-3-Pyrroline	100 W	50	5.0	(W)	Liver (hepatocellular) (80)
LXXXI	-Nornicotine	200 W	44	9.0	(SD)	Nasal turbinates (43)
	Nitrosomorpholines and other oxygen containing heterocycles:					
LVIII	Nitroso-morpholine	35 W, 7 W	30,30	1.0, 0.2	(SD)	Liver (hepatocellular) (75,98)
		35 W, 14 W	50,50	1.7, 0.7	(F)	Liver, esophagus (52,82)
LXII	2-Methyl	35 W	26	0.9	(F)	Liver, esophagus (27)
LXI	2,6-Dimethyl	35 W, 7 W	30,30	1.0, 0.2	(SD)	Esophagus, nasal turbinates (30,48)
		35 W, 14 W	22,30	0.75, 0.4	(F)	Esophagus (23,29)
	-Thiomorpholine	150 W, 38 W	38,38	5.8, 1.4	(W)	Esophagus (40,90)
LXV	-Phenmetrazine	87 W	50	4.4	(SD)	
LIX	Nitroso-1,3-oxazolidine	35 W	50	1.7	(F)	Liver (53)

Result column (+ notation):
- CII: +++
- CIII: +++
- CIV: ++++
- CV: ++++
- XCV: ++++, +++, +++
- XCVI: +++
- C: 0
- CI: 0
- LII: ++, 0
- LXXX: 0
- LXXXVII: ++++
- LXXXIII: 0
- LXXXIV: 0
- LXXXVI: ++
- LXXXI: +++
- LVIII: +++, ++++
- LXII: ++++
- LXI: ++++, ++
- LXV: 0
- LIX: +++

(*continued*)

Table II

	Compound	Dose rate (μmoles/week) Administration W=Drinking water G=Gavage in oil	Length of treatment (weeks)	Total dose per rat (μmoles)	Relative carcinogenicity	Organs in which tumors appear (time of death at 50% animals with tumors: weeks) SD=Sprague-Dawley rats F=Fischer rats W=Wistar rats
LXIII	-5-Methyl	35 W	50	1.7	++	(F) Liver (85)
LXIV	-2-Methyl	35 W	50	1.7	++	(F) Liver (88)
LX	Nitrosotetrahydro-1,3-oxazine	35 W	50	1.7	+++	(F) Liver (54)
	Nitrosopiperazines:					
LXVI	Dinitroso-piperazine	70 W	50	3.5	+++	(SD) Esophagus, nasal turbinates, liver (hepatocellular) (51)
LXVII	2-Methyl-	70 W	33	2.3	++++	(SD) Esophagus, nasal turbinates (33)
LXIX	2,5-Dimethyl-	70 W	50	3.5	+++	(SD) Esophagus, nasal turbinates (46)
LXVIII	2,6-Dimethyl-	70 W, 14 W	30,30	2.3, 0.4	++++	(SD) Esophagus, nasal turbinates (25,37)
		70 W, 14 W	20,20	1.4, 0.3		(F) Esophagus (27,41)
LXX	2,3,5,6-Tetramethyl-	70 W	50	3.5	0	(SD)
LXXI	-Homopiperazine	70 W	31	2.2	++++	(SD) Esophagus, nasal turbinates (36)
		70 W, 25 W, 11 W		2.1, 0.8, 0.35		(F) Esophagus (39,39,59)
LXXII	Nitroso-piperazine	170 W	60	10.5	0	(W)
LXXIII	4-Methyl-	190 W	50	9.7	0	(SD)
		1000 W	50	50	+	(F) Nasal turbinates (48)
LXXIV	3,5-Dimethyl-	70 W	29	2.0	+++	(F) Thymus Leukemia and lympho-sarcoma) (28)
LXXVII	4-Benzoyl-3,5-dimethyl	70 W	50	3.5	+	(F) Forestomach (114)
LXXVI	4-Acetyl-3,5-dimethyl	70 W	30	2.0	+++	(F) Esophagus (32)
LXXV	3,4,5-Trimethyl	70 W, 28 W, 11 W	26,30,30	1.8, 0.8, 0.3	+++	(F) Thymus, nasal turbinates (27,42,62)

	Cyclic nitrosamines:					
LI	Nitrosoazetidine	200 W, 67 W	50, 30	10.0, 2.0	+++	(SD) Liver (hepatocellular) (53,100)
LIV	Nitrosohexamethyleneimine	87 W		2.6	++++	(SD) Liver (sarcoma), esophagus, nasal turbinates (28)
LV	Nitrosoheptamethyleneimine	87 W		2.6		(F) Esophagus, liver (sarcoma) (31)
		70 W	21	1.5	++++	(SD) Lung, esophagus, nasal turbinates (25)
		70 W, 20 W, 11 W	13,25,25	0.9, 0.7, 0.3		(F) Esophagus (21,26,43)
LVI	Nitrosooctamethyleneimine	120 W	33	4.3	+++	(W) Lung, esophagus, nasal turbinates (43)
LVII	Nitrosododecamethyleneimine	230 G	50	11.0	+	(SD) Liver (hepatocellular) (92)
	Aliphatic nitrosamines:					
V	Nitrosodimethylamine	54 W	30	1.6	++++	(SD) Liver (sarcoma) (25)
		44 W, 18 W, 7 W	30,30,30	1.3, 0.5, 0.2		(F) Liver (sarcoma) (31,79,91)
VI	Nitrosodiethylamine	70 W	29	2.0	++++	(SD) Liver (hepatocellular), esophagus, nasal turbinates (30)
		110 W, 44 W, 17 W, 7 W	17,22,30,30	1.9, 1.0, 0.5, 0.2		(F) Esophagus, liver (18,26,35,45)
XXIX	Bis-(2-chloro)-	23 G	30	0.4	+++	(SD) Esophagus, forestomach, liver (hepato cellular) (84)
XXX	Bis-(2-cyano)-	70 W	50	3.5	0	(SD)
XIV	Bis-(2-hydroxy)-	2000 W, 750 W, 300 W	45,50,50	90, 37, 15	+	(F) Liver (hepatocellular) (47,61,80)
XXVII	Bis-(2-methoxy)-	70 W	50	3.5	+++	(SD) Esophagus, nasal turbinates, liver (hepatocellular) (53)
XXVIII	Bis-(2-ethoxy)-	70 W	50	3.5	+++	(SD) Esophagus, nasal turbinates, liver (hepatocellular) (92)
XXXIII	Bis-(2,2-diethoxy)-	70 W	50	3.5	0	(SD)
XXXIV	Bis-(2-trifluoro)-	17 W	30	0.5	0	(F)
VIIb	Nitrosodiisopropylamine	460 W, 70 W	40,50	18.5, 3.5	+	(SD) Nasal turbinates (55)
VIIa	Nitrosodi-n-propylamine	70 W	30	2.1	+++	(SD) Esophagus, nasal turbinates, liver (hepatocellular) (50)
		35 W	30	1.0		(F) Esophagus, forestomach (31)
		67 G, 34 G	30,30	2.0, 1.0		(F) Liver, esophagus, nasal turbinates, lung (34,54)

(continued)

Table II (*continued*)

	Compound	Dose rate (μmoles/week) Administration W=Drinking water G=Gavage in oil	Length of treatment (weeks)	Total dose per rat (μmoles)	Relative carcinogenicity	Organs in which tumors appear (time of death at 50% animals with tumors: weeks) SD=Sprague-Dawley rats F=Fischer rats W=Wistar rats
XXVI	Bis-(2-hydroxy)-	70 W	50	3.3	++	(SD) Nasal turbinates, liver (hepatocellular) (77)
XXXIX	Bis-(2-oxo)-	135 W	42	5.7		(F) Esophagus (42)
		70 W	30	2.0	+++	(SD) Liver (hepatocellular) (55)
		32 G	30	1.0		(F) Liver (hepatocellular) (88)
		35 W	50	1.7		(F) Liver (61)
XXXVIII	Bis-(2-chloro)-	70 G	20	1.4	++	(SD) Stomach
	Nitrosodi-isobutylamine	70 W	30	2.1	0	(SD)
		280 W	50	14.0	+	(F) Trachea, nasal turbinates (55)
	Nitrosodi-sec-butylamine	70 W	50	3.5	0	(SD)
VIII	Nitrosodi-n-butylamine	67 G	30	2.0	++	(F) Liver, lung, bladder, forestomach (94)
IX	Nitrosodi-n-octylamine	220 G	50	11.0	0	(SD)
XVII	Nitroso-methoxymethyl-amine	280 W	50	14.0	0	(SD)
Xa	-Methylethylamine	170 W,34 W,7 W	30,30,30	5.0,1.0,0.2	+++	(F) Liver, esophagus (31,63,105)
Xb	-Methyl-n-propylamine	14 W	23	0.3	++++	(F) Esophagus (28)
Xc	-Methyl-n-butylamine	14 W,5.5 W	20,23	0.3,0.12	++++	(F) Esophagus (24,30)
Xe	-Methyl-n-hexylamine	220 G,88 G,35 G	16,28,30	3.5,2.5,1.0	++++	(F) Esophagus, liver (17,31,63)
Xf	-Methyl-n-heptylamine	220 G,88 G	28,30	6.1,2.6	+++	(F) Liver, lung, trachea (27,51)
Xg	-Methyl-n-octylamine	220 G	30	6.5	+++	(F) Liver, bladder, lung, nasal turbinates (35)
Xh	-Methyl-n-nonylamine	220 G	30	6.5	+++	(F) Liver, lung (43)
Xi	-Methyl-n-decylamine	220 G	30	6.5	++	(F) Bladder (81)
Xj	-Methyl-n-undecylamine	220 G	30	6.5	++	(F) Lung, liver (48)

Code	Compound	Dose	Value	Value2	Rating	Sex	Organ (ref)
Xk	-Methyl-n-dodecylamine	220 G	30	6.5	++	(SD)	Bladder (62)
Xl	-Methyl-n-tetradecylamine	220 G,110 G	30,30	6.5,3.3		(F)	Bladder (53,101)
XXIV	-Methyl-2-phenylethyl-amine	220 G	30	6.5	++	(F)	Bladder (84)
		16 W,.5 W	30,32	0.5,0.2	++++	(F)	Esophagus (31,33)
XXIII	-Methyl-neopentylamine	17 W	22	0.4	++++	(F)	Esophagus (32)
XXII	-Methyl-trifluoroethyl-amine	44 W,17 W	28,30	1.2,0.5	+++	(F)	Esophagus (36,90)
XIII	-Methyl-cyclohexylamine	140 W,35 W	52,52	7.5,1.8	++++	(W)	Esophagus (58,84)
		35 W,.9 W	30,30	1.1,0.3		(F)	Esophagus (29,34)
XIIa	-Methyl-phenylamine	70 W	52	3.7	+++	(W)	Esophagus (66)
XIId	4-Fluoro-	37 W	50	1.8	+++	(F)	Esophagus (78)
XIIe	4-Nitro-	37 W	50	1.8	++	(F)	Esophagus (96)
		37 W	50	1.8	0	(F)	
XXV	-Methyl-benzylamine	9 W	22	0.2	++++	(F)	Esophagus (24)
XIIc	-Phenylbenzylamine	2000 Diet	70	140	+	(F)	Esophagus (69)
XIX	-Methyl-3-carboxypropyl-amine	200 W	57	11.5	++	(F)	Bladder (66)
XLIII	-Ethanolisopropanol-amine	135 W	50	6.7	++	(F)	Liver, esophagus (59)
XLIVa	-Hydroxypropyloxo-propylamine	135 W,35 W	21,40	2.9,1.4	+++	(F)	Esophagus, liver (24,41)
XLIV	-Ethanoloxopropylamine	35 W	50	1.7	+	(F)	Liver, tongue (113)
XLV	-Dihydroxypropyliso-propanolamine	135 W,35 W,.9 W	26,24,37	3.5,0.8,0.4	+++	(F)	Esophagus (27,24,63)
XLVI	-Dihydroxypropyloxo-propylamine	35 W,.9 W	24,30	0.8,0.3	++++	(F)	Esophagus (26,49)
XLVII	-Dihydroxypropylethanol-amine	135 W	50	6.7	+	(F)	· Liver, bladder (120)
XLVIII	-Allylisopropanolamine	135 W	50	6.7	++	(F)	Liver, esophagus (50)
L	-Allylethanolanamine	135 W	50	6.7	++	(F)	Liver, esophagus, nasal turbinates (54)
	-Allyldihydroxypropyl-amine	135 W	50	6.7	++	(F)	Esophagus (47)

(continued)

Table II (continued)

	Compound	Dose rate (μmoles/week) Administration W=Drinking water G=Gavage in oil	Length of Treatment (weeks)	Total dose per rat (μmoles)	Relative carcinogenicity		Organs in which tumors appear (time of death at 50% animals with tumors: weeks) SD=Sprague–Dawley rats F=Fischer rats W=Wistar rats
XLIX	-Allyloxopropylamine	35 W	50	1.8	+++	(F)	Liver (74)
XL	-Methyloxopropylamine	86 W, 22 W, 110 G	21,30, 14	1.8, 0.7, 1.5	+++	(F)	Esophagus, liver, trachea, nasal turbinates (23,40)
						(F)	Esophagus, nasal turbinates (14)
XLI	-Methylisopropanolamine	86 W	22	1.8	+++	(F)	Esophagus (29)
XLII	-Methyldihydroxypropyl-amine	86 W, 22 W	40	3.4	++	(F)	Esophagus (44,84)
	Nitrosamides:						
IVa	Nitroso-trimethylurea	38 W	50	1.9	+	(SD)	Nervous system
		95 W	47	4.5	++	(F)	Stomach, nervous system (44)
IVb	-Triethylurea	38 W	50	1.9	++	(SD)	Nervous system
		95 W	31	3.0	+++	(F)	Mammary gland (33)
	-1-Ethyl-3,3-dimethylurea	38 W	50	1.9	++	(SD)	Nervous system
IVc	-1-Methyl-3,3-diethylurea	38 W	50	1.9	+++	(SD)	Nervous system (46)
		95 W	33	3.1		(F)	Nervous system (32)
IIIa	Nitroso-methylurethane	22 G	10	0.2	++++	(SD)	Forestomach (65)
		11 G, 2.7 G	20,20	0.2, 0.05		(F)	Forestomach (88,77)
IIIb	-Ethylurethane	22 G	10	0.2	++++	(SD)	Forestomach (63)
		11 G, 2.7 G	20,20	0.2, 0.05		(F)	Forestomach (80,103)
IIIc	-Carbaryl	22 G	10	0.2	+++	(SD)	Forestomach (98)
	-Oxazolidone	60 G	40	2.3	+++	(F)	Forestomach (79)
	-5-Methyl	60 G	40	2.3	+++	(F)	Forestomach, duodenum (56)
	-Hydroxyethylurea	105 G, 40 G	25,30	2.6, 1.2	+++	(F)	Lung, forestomach, colon, bone (50,51)
	-2-Hydroxypropylurea	105 G, 40 G	25,30	2.6, 1.2	+++	(F)	Thymus, forestomach (28,32)

painting or oral administration of the compounds (numbering more than 140) to rats. Considerable effort was devoted to treatment of the rats with nearly equimolecular doses of the compounds belonging to one major group, so that the effects could be considered pharmacologically comparable. Several strains of rat have been used and the compounds have all been prepared in our laboratory to rigorous standards of purity.

2. NITROSOALKYLAMIDES

The early hypotheses of mechanisms of N-nitroso compound carcinogenesis relied heavily on the results of experiments with the supposedly directly acting nitrosoalkylamides, typified by nitrosomethylurea (**Ia**) and nitrosoethylurea (**Ib**). Nitrosomethylurea is a directly acting alkylating agent, methylating

$$H_2N-C-N{\overset{R}{\underset{NO}{}}}$$
$$\overset{\|}{O}$$

I

(a) R = CH_3-
(b) R = CH_3CH_2-
(c) R = $CH_3CH_2CH_2-$
(d) R = $CH_3[CH_2]_2CH_2-$
(e) R = $CH_2=CHCH_2-$
(f) R = $CH_3[CH_2]_3CH_2-$
(g) R = $CH_3[CH_2]_4CH_2-$
(h) R = $CH_3[CH_2]_9CH_2-$
(i) R = $CH_3[CH_2]_{11}CH_2-$
(j) R = FCH_2CH_2-
(k) R = $HOCH_2CH_2-$
(l) R = $CH_3CH(OH)CH_2-$
(m) R = $HOCH_2CH_2CH_2-$
(n) R = $CH_3CH(CH_3)-$
(o) R = C_6H_5-
(p) R = $C_6H_{11}-$
(q) R = $C_6H_5CH_2-$
(r) R = $C_6H_5CH_2CH_2-$

DNA and other macromolecules *in vivo* or *in vitro*. When administered orally to rats nitrosomethylurea induces tumors of the forestomach,[3] applied intrarectally it gives rise to tumors of the colon.[6] When given intravenously a large variety of tumors are induced,[7] and transplacentally it gives rise to tumors of the nervous system.[8] When painted continuously on the skin nitrosomethylurea induces papillomas and carcinomas,[9,10] although it does not seem to be a good initiator for two-stage carcinogenesis in experiments involving single application followed by croton oil treatment.[11] While the corresponding nitrosoalkylnitroguanidines, methyl (**IIa**) and ethyl (**IIb**), are more potent complete carcinogens when painted on mouse skin[11] than are the nitrosoalkylureas, nitrosomethylnitroguanidine is a relatively poor initiator (Hecker and Lijinsky, unpublished data). On the other hand, on mouse skin the two nitrosoalkylcarbamates, methyl (**IIIa**) and ethyl (**IIIB**), were noncarcinogenic on continuous painting (Lijinsky, unpublished data); nitrosocarbaryl (**IIIc**) is weakly carcinogenic by skin painting.[10] The ability of these compounds to alkylate target macromolecules directly, without metabolic activation, is well known,[2] and contrasts sharply with the relatively weak activity as carcinogens they show in mouse skin, compared with that of polycyclic aromatic hydrocarbons, which

$$O_2N-NH-\underset{\underset{NH}{\|}}{C}-N\genfrac{}{}{0pt}{}{R}{NO}$$

II

(a) R = CH_3^-
(b) R = $CH_3CH_2^-$

$$R_2O-\underset{\underset{O}{\|}}{C}-N\genfrac{}{}{0pt}{}{R_1}{NO}$$

III

(a) $R_1 = CH_3^-$; $R_2 = CH_3CH_2^-$
(b) $R_1 = CH_3CH_2^-$; $R_2 = CH_3CH_2^-$
(c) $R_1 = CH_3^-$; R_2 = 1-NAPHTHYL ($C_{10}H_7$)

require metabolic activation. However, when administered orally or intravenously, all of the nitrosoalkylamides are very potent carcinogens inducing in rats a great variety of tumors. The nitrosoalkylureas induce tumors of the nervous system, breast, stomach, pancreas, and other organs.[7] The nitrosoalkylnitroguanidines induce tumors of the glandular stomach,[13] and the nitrosoalkylcarbamates induce tumors of the forestomach at very low doses.[14] Therefore, the organs and cells affected seem to play at least as large a role in the number and type of tumors that appear as does the nature and reactivity of the carcinogenic nitrosamide applied. It seems likely that the explanations so far offered of the mechanism of carcinogenic action of these simple compounds, which are all strong directly acting mutagens, are incomplete and perhaps in error.

Further evidence that simple alkylation of macromolecules by these compounds is incomplete as a mechanism of carcinogenesis is offered by the failure of nitrosomethylurea to induce liver tumors in intact adult rats, even though it produces extensive alkylation of rat liver DNA when appropriately administered.[7] Furthermore, the nitrosotrialkylureas (IV) are very much more stable

$$\genfrac{}{}{0pt}{}{R_2}{R_2'}{\diagdown}H-\underset{\underset{O}{\|}}{C}-N\genfrac{}{}{0pt}{}{R_1}{NO}$$

IV

(a) $R_1 = R_2 = CH_3^-$
(b) $R_1 = R_2 = CH_3CH_2^-$
(c) $R_1 = CH_3^-$; $R_2 = CH_3CH_2^-$

compounds, requiring metabolic activation for their mutagenic and carcinogenic action and not converted to diazoalkanes at alkaline pH; in fact, they are stable in alkali.[15] Yet these nitrosoalkylureas are effective carcinogens in rats, giving rise to tumors of the central nervous system and malignant mammary tumors when administered chronically[3,15,16] (Table 2). Since the major chemical reactions undergone by nitrosotrialkylureas are so different from those of nitrosomonoalkylureas, yet their tumorigenic effects when administered systemically have great similarities, it can be surmised that the latter effects have much in common, and that possibly similar minor pathways account for the carcinogenic action of both types of nitrosamide. Unfortunately, the biochemistry of the nitrosotrialkylureas has not been investigated, but some indication that oxidation of the alkyl group on the same nitrogen as the nitroso group is involved has been given by the lesser carcinogenic activity of nitrosomethyl-d_3-diethylurea (**IVc**) compared with the unlabeled compound, although the target remains the nervous system.

The potency of the nitrosoalkylureas as skin carcinogens when applied in acetone solution to the skin of female Swiss mice was compared by measuring the latent period of the tumors and the number of animals with tumors in each treatment group.[10] These differences in potency (summarized in Table I) are considerable and do not seem to follow any particular pattern. For example, nitrosomethylurea and nitrosoethylurea are both very potent, whereas nitroso-n-propylurea (**Ic**) and nitroso-n-butylurea (**Id**) are much less potent and nitrosoallylurea (**Ie**) failed to induce tumors. Assuming that alkylation of a key macromolecule is involved in carcinogenesis by these compounds, this suggests that there is a considerable difference in the extent of alkylation or in stability of the alkylated macromolecule produced by the compounds. Nitrosoamylurea (**If**) and nitrosohexylurea (**Ig**) are considerably more potent than the propyl and butyl compounds, while nitrosoundecylurea (**Ih**) and nitrosotridecylurea (**Ii**) are weaker. No pattern relating increased or decreased liposolubility or stability of the nitrosoureas with carcinogenicity can be discerned in these results. Among the substituted alkylnitrosoureas, nitroso-2-fluoroethylurea (**Ij**) is highly carcinogenic, as potent as nitrosoethylurea, as is nitroso-2-hydroxyethylurea (**Ik**). Nitroso-2-hydroxypropylurea (**Il**) and nitroso-3-hydroxypropylurea (**Im**), on the other hand, are equally potent skin carcinogens as nitrosohydroxyethylurea and much more potent than nitroso-n-propylurea or nitroso-isopropylurea (**In**). Perhaps because of the bulk of the hydrocarbon residue (which might inhibit insertion of it into DNA), all of the cyclic nitrosoalkylureas were either noncarcinogenic or very weakly so on mouse skin. These compounds included phenyl- (**Io**), cyclohexyl- (**Ip**), benzyl- (**Iq**), and 2-phenylethyl-nitrosourea (**Ir**); phenylnitrosourea and benzylnitrosourea were carcinogenic by other routes.[17,18] It seems that the explanation of carcinogenesis by nitrosoalkylureas as simple alkylating agents is inadequate to explain the differences in carcinogenicity among them.

3. ALKYLATION BY NITROSAMINES

There was considerable support in the beginning for the hypothesis that formation from a nitrosamine of an agent that alkylated nucleic acids was the primary reaction in carcinogenesis by these compounds. Implicit in the hypothesis was the concept that metabolism of the nitrosamine *in vivo* was necessary to form the alkylating moiety. Thus most of the carcinogenic nitrosodialkylamines, nitroso-dimethylamine (**V**), -diethylamine (**VI**), -dipropylamine (**VIIa**), and methyl-n-butylamine (**Xc**), could be converted to alkylating agents and,

$$\begin{array}{cccc}
\text{CH}_3\!\diagdown & \text{CH}_3\text{CH}_2\!\diagdown & \text{CH}_3\text{CH}_2\text{CH}_2\!\diagdown & \text{CH}_3\text{CH}(\text{CH}_3)\!\diagdown \\
\text{N-NO} & \text{N-NO} & \text{N-NO} & \text{N-NO} \\
\text{CH}_3\!\diagup & \text{CH}_3\text{CH}_2\!\diagup & \text{CH}_3\text{CH}_2\text{CH}_2\!\diagup & \text{CH}_3\text{CH}(\text{CH}_3)\!\diagup \\
\mathbf{V} & \mathbf{VI} & \mathbf{VIIa} & \mathbf{VIIb}
\end{array}$$

$$\begin{array}{cc}
\text{CH}_3[\text{CH}_2]_2\text{CH}_2\!\diagdown & \text{CH}_3[\text{CH}_2]_6\text{CH}_2\!\diagdown \\
\text{N-NO} & \text{N-NO} \\
\text{CH}_3[\text{CH}_2]_2\text{CH}_2\!\diagup & \text{CH}_3[\text{CH}_2]_6\text{CH}_2\!\diagup \\
\mathbf{VIII} & \mathbf{IX}
\end{array}$$

$$\begin{array}{c}
\text{CH}_3\!\diagdown \\
\text{N-NO} \\
\text{R}\!\diagup \\
\mathbf{X}
\end{array}$$

(a) R = CH_3CH_2-
(b) R = $\text{CH}_3\text{CH}_2\text{CH}_2$-
(c) R = $\text{CH}_3[\text{CH}_2]_2\text{CH}_2$-
(d) R = $\text{CH}_3[\text{CH}_2]_3\text{CH}_2$-
(e) R = $\text{CH}_3[\text{CH}_2]_4\text{CH}_2$-
(f) R = $\text{CH}_3[\text{CH}_2]_5\text{CH}_2$-
(g) R = $\text{CH}_3[\text{CH}_2]_6\text{CH}_2$-
(h) R = $\text{CH}_3[\text{CH}_2]_7\text{CH}_2$-
(i) R = $\text{CH}_3[\text{CH}_2]_8\text{CH}_2$-
(j) R = $\text{CH}_3[\text{CH}_2]_9\text{CH}_2$-
(k) R = $\text{CH}_3[\text{CH}_2]_{10}\text{CH}_2$-
(l) R = $\text{CH}_3[\text{CH}_2]_{12}\text{CH}_2$-

indeed, were found in most cases that were examined to give rise to 7-alkylguanines as a major product in DNA. Although the presence of this particular alkylated base is now known not to be related to carcinogenesis or mutagenesis, its formation is an index of alkylating activity of the nitrosamine. An important support for this alkylation mechanism of carcinogenesis was the failure of nitrosomethyl-tertiary-butylamine (**XI**) to induce tumors or to be a methylating agent *in vivo*.[19] However, there have been discrepancies that are more difficult to accommodate within a comprehensive hypothesis relating alkylation of DNA with carcinogenesis. Principal among these is nitrosomethylaniline (**XIIa**), which theoretically cannot form an alkylating agent—and does not in practice—yet is a carcinogen in rats, inducing tumors of the esophagus.[3,20-22] A very similar compound, nitrosomethylcyclohexylamine (**XIII**),

$$\begin{array}{c}
\text{CH}_3\!\diagdown \\
\text{CH}_3\!\diagdown\text{N-NO} \\
\text{C} \\
\text{CH}_3\!\diagup\text{CH}_3 \\
\mathbf{XI}
\end{array}$$

$$\text{R}_2\!-\!\!\langle\text{O}\rangle\!-\!\text{N-NO}$$
$$|$$
$$\text{R}_1$$
XII

(a) $R_1 = \text{CH}_3$-; $R_2 = \text{H}$
(b) $R_1 = \langle\text{O}\rangle$- ; $R_2 = \text{H}$
(c) $R_1 = \langle\text{O}\rangle\text{-CH}_2$-; $R_2 = \text{H}$
(d) $R_1 = \text{CH}_3$-; $R_2 = \text{F}$
(e) $R_1 = \text{CH}_3$-; $R_2 = \text{NO}_2$

XIII: Phenyl-N(NO)-CH₃

also has the rat esophagus as its target[21,23] and both theoretically and experimentally gives rise to methylated nucleic acids through metabolism.[5] However, while *in vivo* this compound produces extensive methylation in rat liver nucleic acids, there was no detectable methylation in the nucleic acids of the esophagus;[24] this nitrosamine does not induce liver tumors in rats, but it is an effective esophageal carcinogen. The possibility arises that the extensive methylation of liver nucleic acids observed with several nitrosamines is related to liver toxicity, but not to carcinogenesis. It seems probable that liver cell toxicity is essential to liver carcinogenesis by nitrosamines, but that induction of other changes are necessary for tumor induction, and few nitrosamines are able to induce those changes. Several nitrosamines are now known to be tumorigenic with little or no apparent genotoxicity. These include, in addition to nitrosomethylaniline, the rat bladder carcinogen nitrosodiphenylamine (**XIIb**),[25] the rat liver carcinogen nitrosodiethanolamine (**XIV**),[3,26] and the rat esophageal carcinogen nitrosophenylbenzylamine (**XIIc**), none of which has been mutagenic in any system. On the other hand, the potent directly acting mutagen nitroso-2-acetoxydibenzylamine (**XV**) has been at most very weakly carcinogenic in rats after chronic oral administration (Lyle and Lijinsky, unpublished), although it is an analog of nitroso-2-acetoxydimethylamine (**XVI**) and similar aliphatic nitrosmaines, which are both potent mutagens and potent carcinogens.[27]

XIV: (HOCH₂CH₂)₂N-NO

XV: (PhCH₂)(PhCH(O-C(=O)-CH₃))N-NO

XVI: (CH₃)(CH₃C(=O)-O-CH₂)N-NO

On the other hand, nitrosomethoxymethylamine (nitroso-O,N-dimethylhydroxylamine (**XVII**), which is not a typical nitrosamine, could give rise through oxidation only to a methoxylating agent and is not carcinogenic,[3,28] although it is mutagenic.[29] In contrast, the homologous diethyl compound, nitroso-O,N-diethylhydroxylamine (**XVIII**), is a potent liver carcinogen in rats,[30] although alpha oxidation would give rise only to an ethoxylating agent. Similarly, blockage of alpha oxidation of nitrosodiethyamine by methyl substitution, as in nitrosodiisopropylamine (**VIIb**), reduces carcinogenic potency enormously.[31]

$$\begin{array}{cc} \mathrm{CH_3O} \\ \phantom{\mathrm{CH_3}}{\diagdown}\mathrm{N{-}NO} \\ \mathrm{CH_3}{\diagup} \end{array} \qquad \begin{array}{cc} \mathrm{CH_3CH_2O} \\ \phantom{\mathrm{CH_3CH_2}}{\diagdown}\mathrm{N{-}NO} \\ \mathrm{CH_3CH_2}{\diagup} \end{array}$$

XVII XVIII

4. NITROSOMETHYLALKYLAMINES (X)

It seems, therefore, that the mechanisms of carcinogenesis are less simple than formation of a methylating or ethylating agent, and this is shown clearly by the large differences in carcinogenicity between the members of the homologous series of nitrosomethylalkylamines. These differences are in potency, as well as in the organs in which tumors are induced. The simplest member of this series is nitrosomethylethylamine (**Xa**), which gives rise to liver tumors in rats but is much less potent than either nitrosodimethylamine or nitrosodiethylamine,[3,32] of which it could be considered a hybrid. It is hard to reconcile this weaker carcinogenicity with the fact that, if nitrosomethylethylamine acts through formation of an alkylating intermediate, it must form an ethylating or methylating agent (or both), which is true of nitrosodimethylamine and nitrosodiethylamine. At high doses, nitrosomethylethylamine induces some esophageal tumors in rats, as well as liver tumors.[32]

The next two homologs, nitrosomethyl-*n*-propylamine (**Xb**) and nitrosomethyl-*n*-butylamine (**Xc**), induce only tumors of the esophagus[33] (and related parts of the upper gastrointestinal tract, such as tongue, pharynx, and forestomach), but no liver tumors (except that nitrosomethylbutylamine was reported to induce liver tumors in MRC rats.[34] These two asymmetric nitrosamines are very much more potent than nitrosomethylethylamine, very small doses leading to death of the rats with esophageal carcinomas within a short time (3 to 4 months). Nitrosomethyl-*n*-pentylamine (**Xd**) and nitrosomethyl-*n*-hexylamine (**Xe**) are also esophageal carcinogens of high potency in rats. The nitrosomethylalkylamines with 3, 4, 5, or 6 carbon atoms in the long chain are very similar in potency. The next higher homolog, nitrosomethyl-*n*-heptylamine (**Xf**), was formerly reported as noncarcinogenic when given by subcutaneous injection,[3] but in the Fischer rat it is a reasonably potent liver carcinogen, but induced no esophageal tumors unless given in drinking water. It is unclear how one additional carbon atom in the alkyl chain could make so dramatic a change in target organ, and it is difficult to reconcile this with formation of a methylating agent from these compounds (which is demonstrable) as the common mechanism of action.

As the length of the alkyl chain in this series increases, the carcinogenic potency tends to decrease and with that appear further changes in organ specificity in rats. Thus, nitrosomethyl-*n*-octylamine (**Xg**) gives rise to liver tumors and to transitional cell carcinomas of the urinary bladder.[35] This is the same pattern shown by nitrosodi-*n*-butylamine (**VIII**) at somewhat higher doses.[3] The action of nitrosomethyloctylamine presents a dilemma if oxidation of the octyl group with release of a methylating agent is not involved in carcinogenesis. If the action of these compounds involves formation of an agent that alkylates nucleic acids, the alternative to methylation by nitrosomethyloctylamine is octylation, or alkylation by some moiety derived by metabolic shortening of the octyl chain. Presumably the same pathway should be followed by nitrosodi-*n*-octylamine (**IX**), yet this compound is not detectably carcinogenic in rats,[36] although it is mutagenic to *Salmonella*.[37]

The carcinogenic effectiveness of nitrosomethylalkylamines of higher molecular weight than the -octylamine shows an interesting periodicity, in conformity with earlier predictions of Okada.[38] Those homologs with an even number of carbon atoms in the chain are bladder carcinogens in rats, while the homologs with an odd number of carbon atoms in the chain do not induce bladder tumors in rats. Thus, nitrosomethyldecylamine (**Xi**), -dodecylamine (**Xk**), and -tetradecylamine (**Xl**) induce almost 100% incidence of transitional cell carcinoma of the urinary bladder, all with about equal potency. Nitrosomethylnonylamine (**Xh**) and nitrosomethylundecylamine (**Xj**) both induce liver tumors, the -nonyl compound being appreciably more potent then the -undecylamine.

As suggested by Okada, metabolism of the nitrosomethylalkylamines with long carbon chains might occur by a process similar to that of long-chain fatty acids (which they perhaps resemble), namely, a series of beta oxidations, yielding as end products nitrosomethylcarboxypropylamine (**XIX**) and nitrososarcosine (**XXI**) from those nitrosamines with even-numbered chains, and nitrosomethylcarboxyethylamine (**XX**) from those with odd-numbered chains.[38]

$$\begin{array}{ccc}
HOOCCH_2CH_2CH_2 & HOOCCH_2CH_2 & HOOCCH_2 \\
\diagdown N-NO & \diagdown N-NO & \diagdown N-NO \\
CH_3\diagup & CH_3\diagup & CH_3\diagup \\
XIX & XX & XXI
\end{array}$$

Nitrosomethylcarboxypropylamine is a bladder carcinogen, although less effective than the equivalent dose of a nitrosomethylalkylamine from which it could arise, and nitrososarcosine is a very weak esophageal carcinogen,[3] while nitrosomethylcarboxyethylamine appears not to have been tested for chronic toxicity. The activation scheme suggested by Okada appears, then, to be correct as far as it goes, but there are further steps needed to approach the true carcinogenic metabolite.

Nitrosomethylethylamine gives rise mainly to carcinomas of the liver when given to rats at a variety of doses, although, as already mentioned, it is a much weaker carcinogen than either nitrosodimethylamine or nitrosodiethylamine.

At high doses, nitrosomethylethylamine also induces esophageal tumors in reasonable incidence,[32] suggesting that more than one pathway of activation to a carcinogenic agent exists for this nitrosamine in rats. This is a conclusion that can also be drawn from studies of the effect of deuterium substitution at various positions in this molecule on carcinogenic activity.[39] Surprisingly, substitution of deuterium for hydrogen in the methyl portion of the ethyl group, a region of the molecule not usually considered to be important in carcinogenesis, leads to induction of esophageal tumors in most of the treated rats, although there is no significant effect on survival of the rats.[40] That the beta carbon atom of nitrosomethylamine is, indeed, important in carcinogenesis by this compound is shown by the profound effect of substitution on that carbon atom on carcinogenicity. With the exception of hydroxyl (which produces a liver carcinogen like the parent), every substituent in that position leads to abolition of liver carcinogenesis and an enhancement of esophageal carcinogenicity, often a very large enhancement.[40] For example, a methyl substituent forms nitrosomethyl-*n*-propylamine (**Xb**), a very potent esophageal carcinogen that induces no liver tumors. Nitrosomethyltrifluoroethylamine (**XXII**) and nitrosomethylneopentylamine (**XXIII**) (the analogous trimethyl compound derived from nitrosomethylethylamine) are both potent esophageal carcinogens, but induce no liver tumors in rats.[40] Replacement of one of the beta hydrogens in nitrosomethylethylamine with phenyl, as in nitrosomethyl-2-phenylethylamine (**XXIV**), forms an extremely potent esophageal carcinogen, similar in effectiveness to nitrosomethylbenzylamine (**XXV**), a very potent carcinogen.[40] Nitrosomethylphenyl-

CF$_3$-CH$_2$ \ N-NO / CH$_3$	CH$_3$ | CH$_3$-C-CH$_2$ \ | / N-NO CH$_3$ / CH$_3$	⟨O⟩-CH$_2$CH$_2$ \ N-NO / CH$_3$	⟨O⟩-CH$_2$ \ N-NO / CH$_3$
XXII	**XXIII**	**XXIV**	**XXV**

ethylamine is so potent that a total dose of 1.3 mg given to each of a group of rats during 33 weeks as a solution of 0.4 ppm in drinking water induces a 45% incidence of tumors of the upper gastrointestinal tract.[41]

All of these compounds have in common a tendency to induce tumors of the upper GI tract in rats and an oxidizable methyl group attached to nitrogen. They could also give rise to a methylating agent, through oxidation of the alpha carbon on the other side of the nitrogen, and this has been suggested by the experiments of Kleihues[42] as the mechanism by which they induce esophageal cancer in rats. On the other hand, nitrosomethylaniline is also an esophageal carcinogen in rats, although less potent than nitrosomethylbenzylamine, but cannot give rise to a methylating agent by a known mechanism and is not known to alkylate nucleic acids *in vivo*.[24] In the case of this compound it must be assumed that activation to a carcinogenic form involves oxidation of the methyl group, which could also, of course, occur with the analogous nitrosamines that are esophageal carcinogens in rats. Substitution of fluorine in the 4- position (**XIId**) of nitrosomethylaniline reduces carcinogenic activity considerably, with the esophagus being the target, whereas the analog with a nitro- group in the 4- position (**XIIe**) has failed to induce any tumors thus far. The saturated analog of nitrosomethylaniline, nitrosomethylcyclohexylamine, is an esophageal carcinogen of considerable potency in the rat,[21] but induces no liver tumors, although it gave rise to extensive methylation of nucleic acids of the liver *in vivo*; no methylation was detected in the nucleic acids of the esophagus,[24] even though this was the target organ. Nitrosophenylbenzylamine is also something of an anomaly, since it would be expected to be noncarcinogenic (by analogy with the inactive nitrosodibenzylamine[3]), but it induces esophageal tumors and hemangioendothelial tumors of the spleen on feeding to rats. Although it is carcinogenic only at relatively high doses,[43] it is difficult to understand its mechanism of action other than oxidation of the methylene group, releasing a phenyldiazonium ion, which could also be formed by oxidation of nitrosomethylaniline. Whether this could lead to interaction with cellular macromolecules and thence to cancer is not known.

5. SUBSTITUTED ACYCLIC NITROSAMINES

The effects of substitution of various functional groups on the carcinogenic activity of open-chain nitrosamines, both symmetrical and asymmetrical, fall in a pattern, although they are not always predictable. Considering the symmetrical nitrosamines, the effect of hydroxy substitution on the carcinogenicity of nitrosodiethylamine, for example, is to reduce the potency quite markedly, so that nitrosodiethanolamine (**XIV**) is a much weaker carcinogen than the parent compound, although not as weak as hitherto supposed.[3] This derivative induced almost entirely hepatocellular carcinomas (with a few tumors of the nasal cavity) in rats,[26] in contrast with the hemangioendothelial tumors of the liver and esophageal tumors, as well as hepatocellular carcinomas, induced by nitrosodiethylamine in rats.[44] Comparisons of potency might not be precise because so large a proportion of administered nitrosodiethanolamine is excreted unchanged in the urine.[45] Nitrosobis-(2-hydroxypropyl)amine (**XXVI**), analogous to nitrosodiethanolamine, is also a considerably weaker carcinogen than

$$CH_3CH(OH)CH_2 \diagdown$$
$$N-NO$$
$$CH_3CH(OH)CH_2 \diagup$$

XXVI

nitrosodi-*n*-propylamine (**VII**), and, like the parent compound, induced mainly esophageal tumors in rats,[46,47] but no liver tumors when given in drinking water; nitrosodi-*n*-propylamine induced liver tumors, as well as esophageal tumors when administered by gavage in oil.

Although nitrosodiethanolamine is a relatively weak carcinogen, the methyl and ethyl ethers of this compound [nitrosobis-(2-methoxyethyl)-amine (**XXVII**) and nitrosobis-(2-ethoxyethyl)-amine (**XXVIII**)] are quite potent carcinogens, inducing both liver tumors and tumors of the upper gastrointestinal tract in rats.[48] The increase in carcinogenic potency is probably related to the lesser ease of elimination from the body of the more lipophilic ethers than of the alcohol; a similar conclusion was drawn from experiments conducted long ago with the methyl ether of a carcinogenic polynuclear hydrocarbon, which was carcinogenic whereas the phenol was inactive.[49]

$$\begin{array}{cc} CH_3OCH_2CH_2 \\ \diagdown \\ N\text{-}NO \\ \diagup \\ CH_3OCH_2CH_2 \end{array} \qquad \begin{array}{cc} CH_3CH_2OCH_2CH_2 \\ \diagdown \\ N\text{-}NO \\ \diagup \\ CH_3CH_2OCH_2CH_2 \end{array}$$

XXVII   XXVIII

The carcinogenicity of a number of symmetrical beta-substituted derivatives of nitrosodiethylamine show interesting comparisons, in addition to the two ethers discussed above. The effect of two chlorine atoms, as in nitrosobis-(2-chloroethyl)amine (**XXIX**), is to reduce the potency of the nitrosamine compared with the parent compound, but the target organs remain the liver and the upper gastrointestinal tract.[48] The presence of two cyano groups on the beta carbon of nitrosodiethylamine, as in nitrosiminodipropionitrile (**XXX**), abolishes carcinogenic activity.[48] The lower homolog, nitrosiminodiacetonitrile (**XXXI**), is also noncarcinogenic,[3] and so is nitrosiminodiacetic acid (**XXXII**),[50] which is of additional interest because it could be formed by interaction of the detergent builder nitrilotriacetic acid (NTA) with nitrosating agents and thereby could represent a considerable human exposure.[51] It is possible that the effects of these substituents (cyano- and carboxyl-) on the carcinogenicity of nitrosamines is electronic; the same effect is observed in cyclic nitrosamines.[52]

$$\begin{array}{cccc} ClCH_2CH_2 & NC\text{-}CH_2\text{-}CH_2 & NCCH_2 & HOOCCH_2 \\ \diagdown & \diagdown & \diagdown & \diagdown \\ N\text{-}NO & NO & N\text{-}NO & N\text{-}NO \\ \diagup & \diagup & \diagup & \diagup \\ ClCH_2CH_2 & NC\text{-}CH_2\text{-}CH_2 & NCCH_2 & HOOCCH_2 \end{array}$$

XXIX   XXX   XXXI   XXXII

There has been a suggestion that nitrosodiethanolamine might exert its carcinogenic effect through formation of the aldehyde, following the normal pathway of alcohol oxidation. This is made less likely by the finding that nitrosobis-(2,2-diethoxyethyl)-amine (**XXXIII**) is noncarcinogenic in rats,[48] because this compound is the diethylacetal of the aldehyde, nitrosiminodiacetaldehyde, which can be expected to decompose easily to form the latter. It can, perhaps, be concluded that the effect of the carbonyls is to reduce or inhibit reactivity at the alpha carbon atoms of the nitrosamine. The same is probably true of nitrosobis-(2-trifluoroethyl)-amine (**XXXIV**), which also is noncarcinogenic[53] and nonmutagenic, in contrast with the relatively strong carcinogenicity of the asymmetric nitrosamine with the trifluoroethyl group on only one side of the molecule,[40] as in nitrosomethyltrifluoroethylamine (**XXII**). Similarly, there are several cases in which a nitrosomethylalkylamine is carcinogenic, whereas the symmetrical analogous nitrosodialkylamine is inactive. For example, nitrosomethylallylamine (**XXXV**) is carcinogenic, whereas nitrosodiallylamine (**XXXVI**) is inactive in rats.[3] Nitrososarcosine (**XXI**) is carcinogenic,[3] if weak, while nitrosiminodiacetic acid is inactive; nitrosomethylacetonitrile (**XXXVII**) is carcinogenic, while nitrosiminodiacetonitrile is only marginally active.[3]

As in the nitrosodiethylamine series, substitution in nitrosodi-*n*-propylamine of various groups changes carcinogenic activity. Nitrosobis-(2-hydroxypropyl)-amine (**XXVI**) (analogous to nitrosodiethanolamine) is a weaker carcinogen than nitrosodipropylamine and induces mainly tumors of the upper gastrointestinal tract.[46,47] Nitrosobis-(2-chloropropyl)-amine (**XXXVIII**) [analogous to nitrosobis-(2-chloroethyl)-amine] is quite a weak carcinogen, inducing only a low incidence of tumors of the forestomach, but it is a rather unstable compound, quite toxic to the rats.[46] On the other hand, the analog of the aldehyde corresponding to nitrosodiethanolamine, which is nitrosobis-(2-oxopropyl)-amine (**XXXIX**), is a potent carcinogen giving rise to hepatocellular carcinomas in rats,[46] as well as colon tumors in one report.[54] This ketone is also a potent inducer of carcinomas of the pancreatic duct and of liver tumors in Syrian hamsters.[55] The oxygenated derivatives of nitrosomethyl-*n*-propylamine corresponding in structure to the symmetrical oxygenated derivatives of nitrosodipropylamine are carcinogens of considerable potency, though less potent than the parent compound. Nitrosomethyl-2-oxopropylamine **XL** induces tumors of both the esophagus and liver, whereas nitrosomethyl-2-hydroxypropylamine **XLI** induces only esophageal tumors in rats, and both are of approximately equal potency. Nitrosomethyl-2-oxopropylamine is also a potent inducer of tumors of the pancreatic duct in Syrian hamsters.[56] Nitrosomethyl-2,3-dihydroxypropylamine **XLII** is considerably less potent than **XLI**, but yet induces only esophageal tumors in rats.

Examination of the carcinogenic activities of a number of asymmetrical oxygenated nitrosamines reveals important differences, perhaps discrepancies. For example, nitroso-2-hydroxypropylethanolamine (**XLIII**) is both a liver carcinogen and an esophageal carcinogen in rats, and intermediate in potency between nitrosodiethanolamine, which induces liver tumors, and nitrosobis-(2-hydroxypropyl)amine, which induces esophageal tumors. It is possible that oxidation of one side of the molecule leads to the formation of an active (alkylating) moiety having the esophagus as its target, while activation of the other side of the molecule leads to an active moiety having the liver as its target. However, nitroso-2-oxopropylethanolamine (**XLIV**) has induced only a few liver tumors in rats after two years, although both sides of this molecule are assumed to be activated to a liver carcinogen. Perhaps the pharmacodynamics of these molecules, which are both hydrophilic and lipophilic, plays a very large and unexplored role in their carcinogenic activity. Nitroso-2,3-dihydroxypropyl-2-hydroxypropylamine (**XLV**) is a more potent carcinogen for the rat esophagus than is nitrosobis-(2-hydroxypropyl)-amine, and the corresponding dihydroxypropyl-2-oxopropyl- compound (**XLVI**) is also a very potent inducer of esophageal tumors, although both compounds are highly hydrophilic. Nitrosodihydroxypropylethanolamine (**XLVII**), however, given at equimolar doses, has failed to induce any tumors in rats.

$$
\begin{array}{ccc}
\text{XLIII} & \text{XLIV} & \text{XLIVa} \\
\text{XLV} & \text{XLVI} & \text{XLVII}
\end{array}
$$

Because nitrosodiallylamine is noncarcinogenic in rats, it can be assumed that oxidation of the allyl group of a nitrosamine does not lead to a carcinogenic product. Therefore, an allyl group could be considered as inactivating that side of a nitrosamine molecule. However, in only one case did the presence of an allyl group increase the potency of a nitrosamine relative to its symmetrical counterpart. For example, nitrosoallyl-2-hydroxypropylamine (**XLVIII**) is less potent than nitrosobis-(2-hydroxypropyl)-amine, but induces liver tumors as well as esophageal tumors; nitrosoallyl-2-oxopropylamine (**XLIX**) is less potent than nitrosobis-(2-oxopropyl)-amine and induces only liver tumors. On the other hand, nitrosoallylethanolamine (**L**) is considerably more potent than nitrosodiethanolamine and induces tumors of the esophagus as well as liver tumors. These results indicate the difficulty in assessing the effect of changes in the structure of a nitrosamine on carcinogenic activity, because the alkyl groups on both sides of the nitrogen atom are, or can be, activated independently and the activation of one affects the activation of the other. The use of deuterium

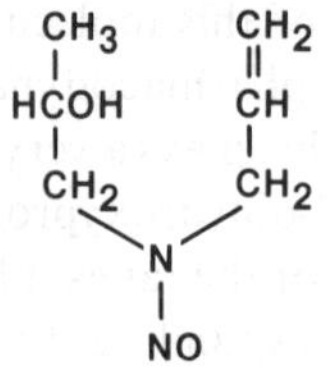

XLVIII: CH₃-HCOH-CH₂-N(NO)-CH₂-CH=CH₂

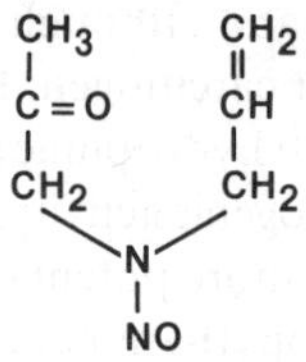

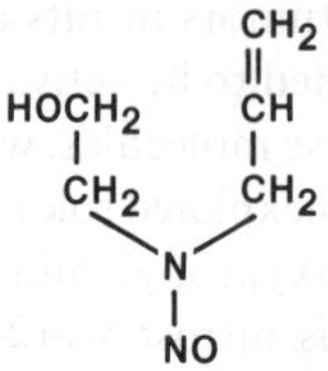

XLIX: CH₃-C(=O)-CH₂-N(NO)-CH₂-CH=CH₂

L: HOCH₂-CH₂-N(NO)-CH₂-CH=CH₂

substitution in selected parts of the molecule has already been mentioned as a means of overcoming this difficulty to some extent, and it appears from those experiments and from examination of the metabolites of a number of nitrosamines *in vivo* that almost every carbon atom in a nitrosamine molecule is susceptible to oxidation.[57,58]

6. CYCLIC NITROSAMINES

The problem of elucidating mechanisms of carcinogenesis from observations of the effects of changes in structure on carcinogenic activity becomes equally difficult in the case of cyclic nitrosamines. In this type of molecule there is an added complication in that configuration and conformation of the molecule seem to play an important role in carcinogenic activity. Many cyclic nitrosamines have flexible molecules, but the presence of substituents can make the molecule rigid so that configurations favorable or unfavorable for activation to carcinogenic forms can predominate. It has not yet been possible to identify likely proximate carcinogenic moieties derived from cyclic nitrosamines, analogous to the simple alkylating agents derived from simple nitrosodialkylamines. It is known that the extent to which cyclic nitrosamines give rise to products that can interact with nucleic acids is orders of magnitude smaller than that from the much-studied acyclic nitrosamines or nitrosoalkylamides.[5]

The simplest stable cyclic nitrosamine is nitrosoazetidine (**LI**), first prepared in the last century. Nitrosoazetidine is a liver carcinogen in rats, inducing a high incidence of hepatocellular carcinomas,[59,60] but only after administration of fairly high doses. The next higher homolog is nitrosopyrrolidine (**LII**), which is a compound of environmental importance, being present in cigarette smoke and in bacon when fried. Nitrosopyrrolidine is also a liver carcinogen in rats,[3] seemingly somewhat less potent than nitrosoazetidine.[61-63] A great deal of

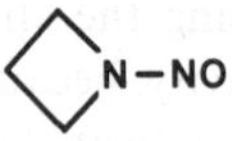

LI: azetidine N–NO

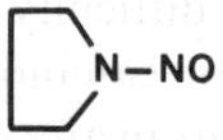

LII: pyrrolidine N–NO

investigation of the mechanism of carcinogenesis by nitrosopyrrolidine has been carried out, notably by Hecht and his associates[64] and by Farrelly and Hecker (see Chapter 9 of this volume), but the mechanism of carcinogenic action remains uncertain.

The next higher homolog is nitrosopiperidine (**LIII**), which has induced in most strains of rat mainly tumors of the upper gastrointestinal tract,[3,63] with few or no liver tumors. This is a sharp difference from the two lower homologs, which do not affect the upper gastrointestinal tract and are much less potent than nitrosopiperidine. The differences might be due to lack of activation in the unresponsive organs or to different pathways of metabolism in those organs. The metabolism of nitrosopiperidine has not been studied intensely, but the examination of the effect of various substituents on carcinogenicity has involved a larger number of derivatives of nitrosopiperidine than of any other nitrosamine. A great variety of derivatives of piperidine have been prepared and the carcinogenicity of their nitroso derivatives will be described and discussed later. Nitrosopiperidine itself, with its rather restricted carcinogenic action in rats, has not been of great interest. On the other hand, the next higher homolog, nitrosohexamethyleneimine (**LIV**), is a carcinogen equally potent with nitrosopiperidine, yet inducing both esophageal tumors (and other tumors of the upper GI tract) and liver tumors, which were hemangioendothelial sarcomas (similar to those induced in rats by nitrosodimethylamine) and hepatocellular carcinomas.[31,63,65] Some studies of the metabolism and activation of nitrosohexamethyleneimine have been carried out[66] and a nucleic acid adduct has been isolated from the livers of animals treated with the ^{14}C-labeled compound.[67] However, there has been no substantial insight into the mechanism by which this compound induces tumors of the liver or esophagus in rats.

⟨N–NO⟩ ⟨N–NO⟩

LIII   LIV

Nitrosoheptamethyleneimine (**LV**) and its next higher homolog, nitrosooctamethyleneimine (**LVI**), unlike nitrosohexamethyleneimine, do not induce liver tumors in rats, although both give rise to esophageal tumors and other tumors of the upper GI tract. In addition, both of the former compounds induce squamous carcinomas of the lung by oral administration,[68] not, as is the case with most other lung carcinogens, by inhalation or direct application to the lungs. This could be of importance because much of tobacco smoke, with its high content of amines, is swallowed. The implications of this have been discussed in Chapter 1. Nitrosoheptamethyleneimine is considerably more potent than nitrosooctamethyleneimine, leading to death of the treated animals much earlier after administration of equivalent doses.[68] The response of Fischer rats to induction of lung tumors by nitrosoheptamethyleneimine appears to be less than that of Sprague-Dawley[69] or Wistar rats,[70] although the response to the induction of upper GI tract tumors by this compound does not differ much between the strains. Alpha oxidation seems to be an important step in activation of nitrosoheptamethyleneimine to a carcinogenic form in rats, since replacement of hydrogen by deuterium in the alpha positions greatly reduces carcinogenic potency.[71] Few studies of the metabolism and activation of nitrosoheptamethyleneimine have been carried out, but there are suggestions that, unlike the lower homologous cyclic nitrosamines, the pattern of metabolism in rat lung differs markedly from that in rat liver, suggesting a unique metabolic pathway in the lung that might be related to the induction of tumors in that organ. Studies of Reznik-Schüller[72] indicate that only one or two types of the large variety of cells in the lung activate nitrosoheptamethyleneimine and those are the cells from which tumors eventually arise.

LV   LVI

The only cyclic nitrosamine larger than nitrosooctamethyleneimine that has been tested for carcinogenicity is nitrosododecamethyleneimine (**LVII**), which appears to be a weaker carcinogen than those homologs with 5, 6, 7, or 8 carbon atoms in the ring. It gives rise mainly to hepatocellular carcinomas[36] in rats and is of comparable potency with nitrosopyrrolidine. In mice, it is one of the few nitrosamines that induces tumors of the glandular stomach in rodents.[73]

LVII

A number of cyclic nitrosamines containing oxygen in the ring have been studied, some very intensely, especially nitrosomorpholine (**LVIII**) (which might be an environmentally important carcinogenic nitrosamine, because it has been found in rubber factories[74]; see Chapter 1). Morpholine was one of the first secondary amines shown to give rise to a carcinogenic nitrosamine *in vivo* after feeding to rats together with nitrite.[75] Nitrosomorpholine is a liver carcinogen and produces few tumors of the upper gastrointestinal tract, in contrast with its analog without the oxygen, nitrosopiperidine. The latter seems to be somewhat more potent than nitrosomorpholine. The lower homolog of nitrosomorpholine, nitrosooxazolidine (**LIX**), is a carcinogen of approximately equal potency with the former, and induces only hepatocellular carcinomas in rats. This is also true of an isomer of nitrosomorpholine, nitrosotetrahydro-1,3-oxazine (**LX**). Thus, although these three nitrosamines can be expected to be activated to rather different proximate carcinogens, their carcinogenic activities are almost identical, suggesting either a surprising coincidence or a mechanism of carcinogenesis presently unexplored. The latter is made plausible by a comparison of the carcinogenic activities of methyl-substituted derivatives of nitrosomorpholine and nitrosooxazolidine. The same is probably true of nitrosothiomorpholine, which is a weaker carcinogen than nitrosomorpholine and induces esophageal tumors in rats, but no liver tumors.[76]

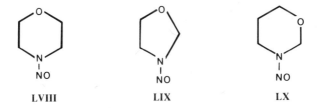

LVIII LIX LX

7. SUBSTITUTED CYCLIC NITROSAMINES

Nitroso-2,6-dimethylmorpholine (**LXI**) is a much more potent carcinogen than nitrosomorpholine and induces esophageal tumors in rats, as opposed to liver tumors induced by the latter.[77,78] In Syrian hamsters nitroso-2,6-dimethylmorpholine is one of those few carcinogens that gives rise to carcinomas of the pancreatic ducts,[79-81] and the effects of deuterium labelling in either the alpha or beta positions have been opposite in hamsters[81] and rat,[82] suggesting that the mechanisms of carcinogenesis by this compound in rats and hamsters are different. Similarly, the *cis* and *trans* isomeric forms in which nitroso-2,6-dimethylmorpholine exists show quite different relative carcinogenic activities in rats, Syrian hamsters, and guinea pigs. In rats the *trans* isomer is more potent than the *cis*,[83] whereas in hamsters the *cis* is more potent than the *trans*[81] and the *cis* is more potent than the *trans* in guinea pigs, in which angiosarcomas of the liver are induced.[84] These findings again suggest that the mechanisms of carcinogenesis by this compound are quite different in these species.

LXI

Nitroso-2-methylmorpholine (**LXII**) is intermediate in carcinogenic potency between nitrosomorpholine and nitroso-2,6-dimethylmorpholine, and induces both esophageal tumors and liver tumors in rats. This suggests the possibility that activation of one "side" of the molecule is related to induction of liver tumors, while activation of the other "side" is related to induction of esophageal tumors. The effect of methyl substitution in nitrosooxazolidine is quite different from the enhancing effect on carcinogenicity of nitrosomorpholine. Nitroso-5-methyloxazolidine (**LXIII**), which has been reported as a contaminant in synthetic cutting oils,[85] is a less potent carcinogen than the unmethylated parent, and nitroso-2-methyloxazolidine (**LXIV**) is also less potent; both methylated nitrosooxazolidines have induced only liver tumors in rats. It seems, therefore, that activation of the methylated nitrosomorpholines and methylated nitrosooxazolidines follow different pathways. There is evidence to suggest that metabolism of nitroso-2,6-dimethylmorpholine involves oxidation at the beta carbon atom, followed by ring opening[86] and, presumably, further metabolism. It is not certain, however, how important this beta oxidation is in carcinogenesis, since it seems to occur with comparable ease in rats, hamsters, and guinea pigs, as judged by the similar patterns of urinary metabolites excreted by the three species.[87] Therefore, it is difficult to discern differences in metabolism of this compound that will correlate with the very dramatic differences in carcinogenic activity between the three species, unless one or two of the metabolites, still unidentified,[87] are the end products of pathways of activation yet unknown.

LXII **LXIII** **LXIV**

Other factors than the availability of sites for activation are almost certainly involved in carcinogenesis by derivatives of nitrosomorpholine, because one such derivative, nitroso-2-phenyl-3-methylmorpholine (**LXV**), the nitroso derivative of the drug phenmetrazine,[88] is neither mutagenic[29] nor carcinogenic[89] when given to rats at doses equivalent to those of nitrosomorpholine that rapidly induce a very high incidence of tumors. Therefore, although one alpha and one beta carbon atom of nitrosophenmetrazine are available for reaction, it appears that the compound cannot be activated to a carcinogenic form, possibly because of the bulk of the molecule due to the benzene ring. This is a complication in the elucidation of mechanisms of nitrosamine carcinogenesis that is entirely unexplored.

LXV

A number of methylated derivatives of nitrosopiperazine have been investigated, and are of interest because this compound is quite a strong base and has a second nitrosatable amino group. The fully nitrosated derivative of piperazine, dinitrosopiperazine (**LXVI**), is a fairly potent carcinogen in rats, inducing both esophageal tumors and liver tumors.[3,90] Methylation of dinitrosopiperazine produces molecules that are much more rigid than the unmethylated compound, and was predicted to lead to compounds of increased carcinogenic potency compared with the parent compound.[91] This proved to be the case, so that 2-methyl- (**LXVII**), 2,6-dimethyl- (**LXVIII**), and 2,5-dimethyl-dinitrosopiperazine (**LXIX**) were all much more potent carcinogens than dinitrosopiperazine, the 2- and 2,6- derivatives being approximately equal and somewhat more potent than the 2,5- derivative.[90] All of these methylated derivatives induced esophageal tumors and tumors of the nasal cavity, but no liver tumors. As would be expected, the fully methylated derivative, 2,3,5,6-tetramethyldinitrosopiperazine (**LXX**), in which access to all of the carbon atoms is blocked, preventing metabolic activation, is noncarcinogenic.[28] The higher homolog of dinitrosopiperazine, dinitrosohomopiperazine (**LXXI**), is much more potent than the former and induces no liver tumors, but mainly tumors of the upper GI tract and nasal cavity.[90]

LXVI   **LXVII**   **LXVIII**

LXIX   **LXX**   **LXXI**

Another example of the profound differences between species in response to nitrosamines is that dinitroso-2,6-dimethylpiperazine, equally potent with nitroso-2,6-dimethylmorpholine in inducing esophageal tumors in rats, is a much weaker carcinogen than the latter in Syrian hamsters and has so far not induced pancreatic tumors in this species. Therefore, even though it seems likely that the two compounds would be metabolized and activated similarly, this is obviously not so in hamsters, although it might be so in rats. This again suggests that the mechanisms by which a given nitrosamine induces tumors might be quite different between one species and another.

Mononitrosopiperazine (**LXXII**) appeared to be a very weak carcinogen when first tested in rats.[76] It later seemed probable that it is not carcinogenic at all, but gave rise to tumors through transnitrosation forming dinitrosopiperazine *in vivo*.[92] This conforms with the lack of mutagenicity of mononitrosopiperazine in several systems.[93,94] N-methylnitrosopiperazine (4-methyl-1-nitrosopiperazine) (**LXXIII**) has been tested several times in rats and has been reported noncarcinogenic or very weakly carcinogenic.[3,28] Nitrosomethyl-piperazine has been nonmutagenic in most systems,[93,94] but has shown weak mutagenic activity in one report.[95] A recent test of nitrosomethylpiperazine at very high dose rates in rats has resulted in a high incidence of only one tumor, in the nasal cavity (H. Reznik-Schüller and W. Lijinsky, unpublished data); no other tumors were induced in significant incidence. Therefore, both nitrosopiperazine and N-methylnitrosopiperazine can be considered, at best, very weak carcinogens, perhaps because both are quite polar compounds and might be unable to cross cell membranes readily. The same stricture would seem to apply to the C-methyl derivatives of nitrosopiperazine and N-methylnitrosopiperazine. It was surprising, therefore, to find that both 3,5-dimethyl-1-nitrosopiperazine (**LXXIV**) and 3,4,5-trimethyl-1-nitrosopiperazine (**LXXV**) are potent carcinogens when given to rats in drinking water, giving rise to a high incidence of leukemias and lymphomas of the thymus,[96] a tumor very rarely induced by a nitrosamine, together with tumors of the nasal cavity. At lower dose rates trimethylnitrosopiperazine gave rise mainly to nasal cavity tumors.

It was thought that acylation of the imino group in 1-nitrosopiperazine would lend rigidity to the molecule, thereby making it a more potent carcinogen.

Nitroso-4-acetyl-3,5-dimethylpiperazine (**LXXVI**) was a potent carcinogen, inducing tumors in the rat esophagus, but was not more potent than the unacetylated amine, nitroso-3,5-dimethylpiperazine.[96] On the other hand, the analogous benzoylated compound, nitroso-4-benzoyl-3,5-dimethylpiperazine (**LXXVII**), was a very weak carcinogen, inducing only a few tumors of the forestomach, mainly benign papillomas, and reducing the lifespan of the animals hardly at all.[96] It is possible, of course, that the benzoyl substituent made the molecule too bulky to fit into the active site of the activating enzyme, but this is pure speculation since other molecules of considerable bulk, such as nitroso-4-phenylpiperidine (**LXXVIII**) and nitroso-4-tertiary-butylpiperidine (**LXXIX**), are carcinogens of considerable potency,[97] although somewhat weaker than nitrosopiperidine.

LXXVI   **LXXVII**   **LXXVIII**   **LXXIX**

Comparatively few derivatives of nitrosopyrrolidine have been examined for carcinogenic activity. Among them has been nitroso-2,5-dimethylpyrrolidine (**LXXX**), in which the presence of the methyl groups at both alpha carbon atoms reduced or abolished carcinogenic activity, since, at the dose given, no tumors were induced.[62] The finding that blockage of the alpha carbons by methyl reduced carcinogenic activity was an indication that oxidation at the alpha carbon atom was a key step in carcinogenesis by nitrosamines. This proposition has been supported by studies of the effect of alpha methylation on carcinogenesis by nitrosopiperidines[98] and aliphatic nitrosamines,[31] in which the result was invariably reduction or elimination of carcinogenic activity, as previously discussed.

LXXX

The effects of other substituents on carcinogenic activity of cyclic nitrosamines complements in most cases the observations in open chain nitrosamines. In the case of nitrosopyrrolidine, for example, substitution of a pyridine ring in the alpha position, as in nitrosonornicotine (**LXXXI**), increased the carcinogenic potency and changed the target organ from the liver to the nasal cavity.[99] Nitrosonornicotine has been reported to give rise to esophageal tumors as well as tumors of the nasal cavity in rats following treatment by a different protocol.[100] The analogous pyridinyl derivative of nitrosopiperidine, nitrosoanabasine (**LXXXII**), is a considerably weaker carcinogen than nitrosopiperidine and induces tumors of the upper respiratory tract in rats.[100] Nitrosoanabasine is, like nitrosonornicotine, a constituent of tobacco smoke.[101]

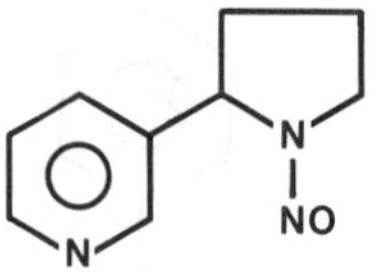

LXXXI 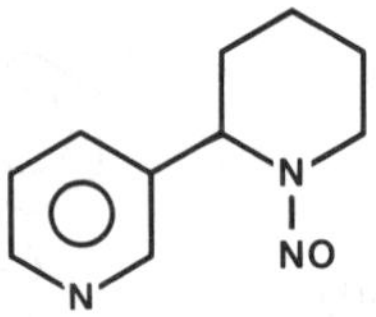   LXXXII

Nitrosoproline (**LXXXIII**) and nitrosohydroxyproline (**LXXXIV**) are nitrosated amino acids and can be considered carboxylic acid derivatives of nitrosopyrrolidine and nitroso-3-hydroxypyrrolidine (**LXXXV**), which itself is a somewhat weaker carcinogen than nitrosopyrrolidine in rats,[102] but similarly induces liver tumors. Both nitrosoproline and nitrosohydroxyproline appear to be quite inactive, failing to give rise to tumors even after administration of quite high doses to rats or mice.[52,103] This accords well with the failure to detect any appreciable binding of nitrosoproline to nucleic acids or proteins of rat liver.[104] Most of administered nitrosoproline appears to be excreted unchanged.[105] The failure of the carboxylic acid derivatives to induce tumors recalls similar findings with carboxylic acid derivatives of several aliphatic nitrosamines, which are inactive or are weak carcinogens, and is a result similar to that obtained with carboxylic acid derivatives of nitrosopiperidine, to be described later. It is possible that these strongly ionized molecules only cross cell membranes with difficulty.

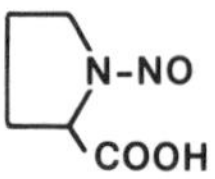

LXXXIII 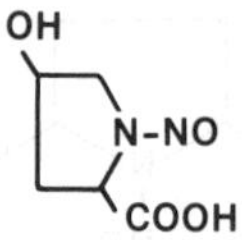   LXXXIV 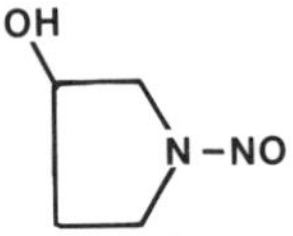   LXXXV

Unsaturation of nitrosopyrrolidine, as in nitroso-3-pyrroline (**LXXXVI**), did not greatly affect the carcinogenicity of the molecule, since the unsaturated compound had similar activity to nitrosopyrrolidine and also induced tumors of the liver.[70] The effect of substitution of chlorine in the beta positions of nitrosopyrrolidine was to increase greatly carcinogenic activity. Nitroso-3,4-dichloropyrrolidine (**LXXXVII**) induced exclusively tumors of the upper gastrointestinal tract, mainly the esophagus, and no liver tumors,[62] in sharp contrast with nitrosopyrrolidine. The analogous nitroso-3,4-dibromopyrrolidine (**LXXXVIII**) has not been tested by chronic administration to animals. However, unlike the dichloro compound, which is a potent mutagen in *Salmonella*, nitrosodibromopyrrolidine is not mutagenic,[106] although this does not necessarily mean that it is noncarcinogenic.

LXXXVI LXXXVII LXXXVIII

8. DERIVATIVES OF NITROSOPIPERIDINE

A large number of derivatives of nitrosopiperidine have been examined with quite a variation in effect on carcinogenicity of nitrosopiperidine in evidence. Methyl substitution at the alpha carbon atoms reduces or eliminates carcinogenic activity, so that 2-methylnitrosopiperidine (**LXXXIX**) is considerably weaker than nitrosopiperidine itself and induces a small incidence of liver tumors, especially in females.[98,107] Nitroso-2,6-dimethylpiperidine (**XC**) and nitroso-2,2,6,6-tetramethylpiperidine (**XCI**), in which both alpha carbon atoms were blocked to oxidation, were both noncarcinogenic. Substitution of a methyl group in the 3-position (**XCII**) did not significantly change carcinogenic activity, while nitroso-4-methylpiperidine (**XCIII**) seemed to be somewhat more carcinogenic than nitrosopiperidine.[98] On the other hand, nitroso-4-*t*-butylpiperidine (**LXXIX**) was considerably less carcinogenic than nitrosopiperidine itself and, like the latter, induced mainly tumors of the upper gastrointestinal tract.[97] Although the presence of the 4-substituents tends to make the molecule more rigid (and, therefore, probably more active as a carcinogen, as in the case of the nitrosopiperazines already discussed), the *t*-butyl group decreased carcinogenic activity. It is possible that the bulkiness of the substituent played a role in reducing the extent of metabolism of the compound. Nitroso-4-phenylpiperidine (**LXXVIII**) was a more potent carcinogen than the *t*-butyl analog, although less potent than nitrosopiperidine, and it induced liver tumors in rats as well as tumors of the upper GI tract.[97] There is no explanation at hand for the induction of liver tumors by nitroso-4-phenylpiperidine, while nitroso-4-*t*-butylpiperidine does not induce liver tumors. One possibility that arises from the failure of nitrosopiperidine itself to induce liver tumors is that it is so potent that all of the treated rats die early from esophageal tumors and liver tumors take longer to develop. However, the animals treated with the *t*-butyl compound also survived much better than nitrosopiperidine-treated rats, and yet gave rise to no liver tumors. Equally strange is that treatment of rats with an equimolar dose of nitroso-4-cyclohexylpiperidine (**XCIV**), the saturated analog of nitroso-4-phenylpiperidine, failed to induce tumors at all.[97] The contrast between this compound and the analogs with bulky substituents in the 4- position raises the question whether the bulkiness can be a factor in carcinogenesis by these derivatives of nitrosopiperidine. It is possible that comparisons of the metabolism of these compounds will reveal differences compatible with their differences in carcinogenicity.

LXXXIX　　　XC　　　XCI

XCII　　　XCIII　　　XCIV

The effect of unsaturation on carcinogenesis of nitrosopiperidine is in some ways similar to that in nitrosopyrrolidine. Thus, nitroso-1,2,3,6-tetrahydropyridine (nitroso-Δ^3-dehydropiperidine) (**XCV**) and nitroso-1, 2, 3, 4-tetrahydropyridine (nitroso-Δ^2-dehydropiperidine) (**XCVI**) are both approximately equally potent carcinogens with nitrosopiperidine itself.[108] However, while the Δ^2 compound induced only tumors of the upper GI tract (like nitrosopiperidine), nitroso-Δ^3-dehydropiperidine induced liver tumors as well in Fischer rats, and exclusively liver tumors in Sprague-Dawley rats. Furthermore, the liver tumors in SD rats were hepatocellular carcinomas,[109] while in

Fischer rats they were mainly angiosarcomas.[108] The reasons for these profound effects of unsaturation are not known, but it does not seem likely that formation of an epoxide is involved, since nitrosopiperidine-3,4-oxide (**XCVII**), administered by gavage, was a weak carcinogen, inducing only a few tumors of the liver and esophagus.[108] There is little or no effect of methyl substitution on the carcinogenicity of nitroso-Δ^3-dehydropiperidine.

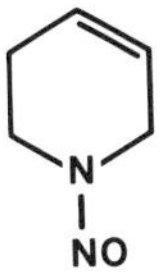

XCV 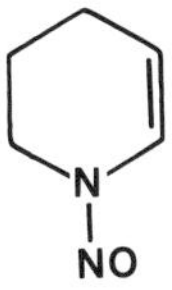 XCVI 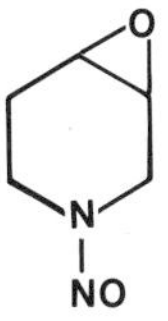 XCVII

As in the case of nitrosopyrrolidine, the effect of carboxyl substitution in nitrosopiperidine was to eliminate carcinogenic activity. Thus, nitrosopipecolic (**XCVIII**) acid and nitrosoisonipecotic (**XCIX**) acid are both noncarcinogenic[52,28] and are also nonmutagenic to bacteria.[110] The ester nitrosoguvacoline (**C**), derived from the acid guvacine, a carboxylic acid related to Δ^3-dehydropiperidine (and which is found in betel nut), is also noncarcinogenic[109] (although it is mutagenic to bacteria[110]). Another ester derived from piperidine is methylphenidate, a widely used drug, and this forms a nitroso derivative (**CI**) that is neither mutagenic[29,110] nor carcinogenic.[89] All of the nitroso derivatives of amino acids and their esters have available alpha carbon atoms for activation, yet they lack carcinogenic activity, illustrating the importance of the structure of the molecule as a whole in conferring carcinogenic activity.

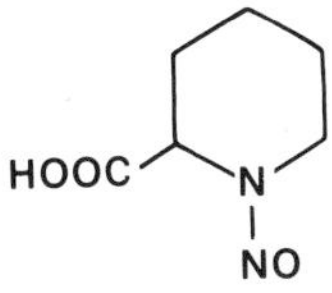

XCVIII 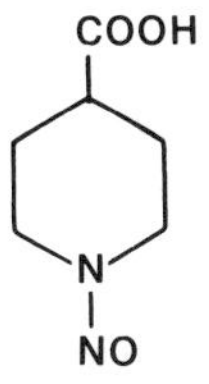 XCIX 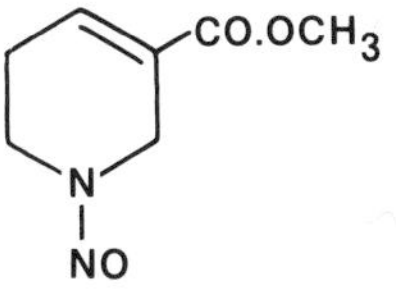 C

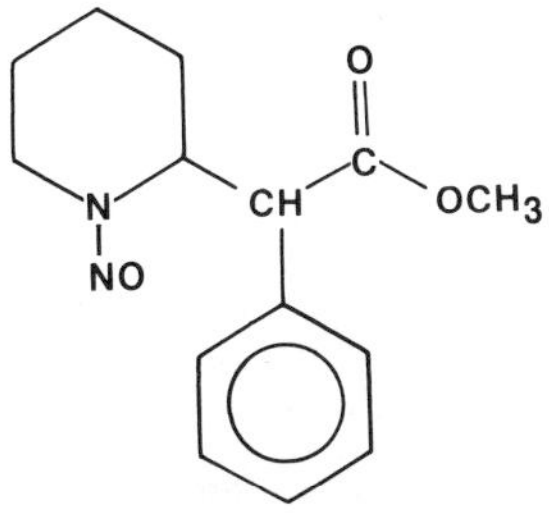

CI

The substitution of halogen atoms for hydrogen in nitrosopiperidine has large effects on carcinogenic potency, although the target organs usually remain the upper gastrointestinal tract. In Fischer rats 3-chloro- (**CII**) and 4-chloro-nitrosopiperidine (**CIII**) are more potent carcinogens than the parent compound, and 3,4-dichloronitrosopiperidine (**CIV**) is even more potent.[111] In Sprague–Dawley rats the same is true, except that 4-chloronitrosopiperidine induced liver tumors as well as tumors of the upper GI tract.[112] On the other hand, 3,4-dichloronitrosopiperidine induces only tumors of the upper GI tract in Sprague–Dawley rats, as does 3,4-dibromonitrosopiperidine (**CV**), but the dichloro compound is considerably more potent than the dibromo compound.[113] This effect of the halogen substitution in increasing carcinogenic potency without changing the target organs might be an effect of the substituents on the conformation of the molecule, as has been suggested previously, or there might be electronic effects on the susceptibility of the alpha carbon atoms to activation. These matters remain to be resolved through studies of metabolism of the compounds and comparisons of their interaction with macromolecules in target cells.

CII **CIII** **CIV** **CV**

One possible route of activation of cyclic nitrosamines is through oxidation at positions other than the alpha carbon atoms, for example at the 3- or 4-positions of nitrosopiperidine. As one way of assessing this, several oxygenated derivatives of nitrosopiperidine were prepared and their carcinogenicity compared with that of the parent compound. Nitroso-3-hydroxypiperidine (**CVI**) had very similar carcinogenic activity to that of nitrosopiperidine, inducing tumors of the esophagus and nasal cavity, together with a small incidence of liver tumors in Sprague–Dawley rats.[114] On the other hand, in Sprague–Dawley rats neither nitroso-4-hydroxypiperidine (**CVII**) nor the corresponding ketone, nitroso-4-piperidone (**CVIII**), induced any esophageal tumors, although they were of similar potency to nitrosopiperidine. Instead, both 4-substituted derivatives induced tumors of the nasal cavity and liver tumors (hepatocellular carcinomas).[114] On the other hand, in Fischer rats nitroso-4-piperidone gave rise to tumors of the upper GI tract, as did nitrosopiperidine, but also induced a high incidence of liver tumors. In this study nitrosopiperidone was a less potent carcinogen than nitrosopiperidine. The effect of methyl substitution on the carcinogenicity of nitroso-4-piperidone was to increase considerably carcinogenic potency and to induce only tumors of the upper GI tract. It is possible that nitroso-3-methyl-4-piperidone (**CIX**) failed to induce liver tumors as well because the rats died too early from the esophageal tumors for tumors of the liver to develop. However, in the absence of supporting evidence it is unwise to assume that this was the case, because there are many nitrosamines that induce liver tumors causing death of animals within 27 weeks of the beginning of treatment. The results of the tests of the oxygenated nitrosopiperidines do not support the concept that the activation of nitrosopiperidine involves formation of any of these compounds as an intermediate in the process of carcinogenesis.

CVI   **CVII**   **CVIII**   **CIX**

9. CONCLUSION

It is clear that there is no definite conclusion regarding mechanisms of carcinogenesis by N-nitroso compounds that can be drawn from the extensive series of structure–activity studies that have been described. As is shown elsewhere in this volume, many carcinogenic nitrosamines are not mutagenic to bacteria, and some are not active in any of the short-term tests presently in use. Even among those that are mutagenic and carcinogenic, there is no quantitative relationship between the two types of biological activity. Nevertheless, it cannot be assumed that interaction with macromolecules controlling mitosis, especially DNA, is not involved in the process of carcinogenesis. It might be that our methods of detecting and measuring genotoxic activity are insufficiently precise or sensitive to reveal the important interactions. Furthermore, so many metabolic events (not all of them necessarily in the liver) might be involved between administration of the carcinogen to the animal and the subsequent induction of the tumor, that it would be surprising if a mutagenic effect produced in bacteria or other sensitive cells by an agent formed through activation with rat liver microsomes were correlated with carcinogenesis. On the other hand, it is conceivable that not all carcinogenic activity occurs through interaction with DNA. Other types of effect on the delicate mechanism controlling mitosis might exist and should be sought with equal vigor.

ACKNOWLEDGMENT. The experimental work was supported by contracts from the National Cancer Institutes, United States Public Health Service.

REFERENCES

1. P. N. Magee and J. M. Barnes, The production of malignant primary hepatic tumours in the rat by feeding dimethylnitrosamine, *Brit. J. Cancer 10*, 114–122 (1956).
2. P. N. Magee and J. M. Barnes, Carcinogenic nitroso compounds, *Adv. Cancer Res. 10*, 163–246 (1967).
3. H. Druckrey, R. Preussman, S. Ivankovic, and D. Schmähl, Organotrope carcinogene Wirkungen bei 65 verschiedenen N-Nitroso-Verbindungen an BD-Ratten, *Z. Krebsforsch. 69*, 103–201 (1967).
4. P. N. Magee and E. Farber, Toxic liver injury and carcinogenesis. Methylation of rat-liver nucleic acids by dimethylnitrosamine *in vivo, Biochem. J. 83*, 114–124 (1962).
5. W. Lijinsky, Interaction with nucleic acids of carcinogenic and mutagenic N-nitroso compounds, *Prog. Nucleic Acid Res. 17*, 247–269 (1976).
6. T. Narisawa, C.-Q. Wong, R. R. Maronpot, and J. H. Weisburger, Large bowel carcinogenesis in mice and rats by several intrarectal doses of methylnitrosourea and negative effect of nitrite plus methylurea, *Cancer Res. 36*, 505–510 (1976).
7. W. Lijinsky, H. Garcia, L. Keefer, and J. Loo, Carcinogenesis and alkylation of rat liver nucleic acids by nitrosomethylurea and nitrosoethylurea administered by intraportal injection, *Cancer Res. 32*, 893–897 (1972).
8. Y. Ishida, M. Tamura, H. Kanda, and K. Okamoto, Histopathological studies of the nervous system tumors in rats induced by N-nitroso-methylurea, *Acta Path. Jpn. 25*, 385–401 (1975).
9. A. Graffi, F. Hoffmann, and M. Schütt, N-methyl-N-nitrosourea as a strong topical carcinogen when painted on skin of rodents, *Nature 214*, 611 (1967).
10. W. Lijinsky and C. Winter, Skin tumors induced by painting nitrosoalkylureas on mouse skin, *Cancer Res. Clinical Oncology 102*, 13–20 (1981).
11. H. B. Waynforth and P. N. Magee, The effect of various doses and schedules of administration of *N*-methyl-*N*-nitrosourea, with and without croton oil promotion, on skin papilloma production in BALB/c mice, *Gann Monogr. Cancer Res. 17*, 439–448 (1975).
12. S. Takayama, N. Kuwabara, Y. Azama, and T. Sugimura, Skin tumors in mice painted with N-methyl-N'-nitro-N-nitrosoguanidine and N-ethyl-N'-ethyl-N'-nitro-N-nitrosoguanidine, *J. Natl. Cancer Inst. 46*, 973–977 (1971).
13. T. Sugimura and S. Fujimura, Tumour production in glandular stomach of rat by N-methyl-N'-nitro-N-nitroso-guanidine, *Nature 216*, 943–944 (1967).
14. W. Lijinsky and H. W. Taylor, Carcinogenesis in Sprague–Dawley rats of N-nitroso-N-alkylcarbamate esters, *Cancer Lett. 1*, 275–279 (1976).
15. W. Lijinsky and H. W. Taylor, Induction of neurogenic tumors by nitrosotrialkylureas in rats, *Z. Krebsforsch. 93*, 315–321 (1975).
16. W. Lijinsky, M. D. Reuber, and B.-N. Blackwell, Carcinogenicity of nitrosotrialkylureas in Fischer rats, *J. Natl. Cancer Inst. 65*, 451–453 (1980).
17. R. Preussmann, H. Druckrey, and J. Bücheler, Carcinogene Wirkung von phenyl-nitrosoharnstoff, *Z. Krebsforsch. 71*, 63–74 (1968).
18. S. Ivankovic, Carcinogene Wirkung von N-benzyl-N-nitrosoharnstoff (BzNH) an BD-Ratten, *Z. Krebsforsch. 91*, 63–67 (1978).
19. P. N. Magee and K. Y. Lee, Cellular injury and carcinogenesis, *Biochem. J. 91*, 35–42 (1964).
20. E. Boyland, F. J. C. Roe, J. W. Gorrod, and B. C. V. Mitchley, The carcinogenicity of nitrosoanabasine, a possible constituent of tobacco smoke, *Br. J. Cancer 18*, 265–270 (1964).
21. C. M. Goodall, W. Lijinsky, L. Tomatis, and C. E. M. Wenyon, Toxicity and oncogenicity of

nitrosomethylaniline and nitrosomethylcyclohexylamine, *Toxicol. Appl. Pharmacol. 17*, 426-432 (1970).
22. N. Napalkov and K. M. Pozharisski, Morphogenesis of experimental tumors of the esophagus, *J. Natl. Cancer Inst. 42*, 927-940 (1969).
23. W. Lijinsky and M. D. Reuber, Carcinogenicity of deuterium-labeled N-nitroso-N-methylcyclohexylamine in rats, *J. Natl. Cancer Inst. 64*, 1535-1536 (1980).
24. W. Lijinsky, L. Keefer, J. Loo, and A. E. Ross, Studies of alkylation of nucleic acids in rats by cyclic nitrosamines, *Cancer Res. 33*, 1634-1641 (1973).
25. R. H. Cardy, W. Lijinsky, and P. Hildebrandt, Neoplastic and nonneoplastic urinary bladder lesions induced in Fischer 344 rats and B6C3F1 hybrid mice by N-nitrosodiphenylamine, *Ecotoxicol. Environ. Safety 3*, 29-35 (1979).
26. W. Lijinsky, M. D. Reuber, and W. B. Manning, Potent carcinogenicity of nitrosodiethanolamine in rats, *Nature 288*, 309-310 (1980).
27. M. Wiessler and D. Schmähl, Zur carcinogenen Wirkung von N-nitrosoverbindungen. V. Acetoxymethyl-methyl-nitrosamin, *Z. Krebsforsch. 85*, 47-49 (1976).
28. W. Lijinsky and H. W. Taylor, Carcinogenesis tests of nitroso-N-methylpiperazine, 2,3,5,6-tetramethyldinitrosopiperazine, nitrosoisonipecotic acid and nitrosomethoxymethylamine in rats, *Z. Krebsforsch. 89*, 31-36 (1977).
29. A. W. Andrews, L. H. Thibault, and W. Lijinsky, The relationship between mutagenicity and carcinogenicity of some nitrosamines, *Mutat. Res. 51*, 319-326 (1978).
30. M. Wiessler and D. Schmähl, Zur carcinogenen Wirkung von N-nitrosoverbindungen. 4. N-nitroso-O,N-diäthylhydroxylamin, *Z. Krebsforsch. 83*, 205-206 (1975).
31. W. Lijinsky and H. W. Taylor, Carcinogenicity of methylated derivatives of nitrosodiethylamine and related compounds in Sprague-Dawley rats, *J. Natl. Cancer Inst. 62*, 407-410 (1979).
32. W. Lijinsky and M. D. Reuber, Comparative carcinogenesis by some aliphatic nitrosamines in Fischer rats, *Cancer Lett. 14*, 297-302 (1981).
33. W. Lijinsky, M. D. Reuber, J. E. Saavedra, and B.-N. Blackwell, Effect of deuterium on the carcinogenicity of nitroso-methyl-n-butylamine, *Carcinogenesis 1*, 157-160 (1980).
34. D. F. Heath and P. N. Magee, Toxic properties of dialkylnitrosamines and some related compounds, *Br. J. Ind. Med. 19*, 276-282 (1962).
35. W. Lijinsky, J. Saavedra, and M. D. Reuber, Induction of carcinogenesis in Fischer rats by methylalkylnitrosamines, *Cancer Res. 41*, 1288-1292 (1981).
36. W. Lijinsky and H. W. Taylor, Carcinogenicity tests in rats of two nitrosamines of high molecular weight, nitrosododecamethyleneimine and nitrosodi-n-octylamine, *Ecotoxicol. Environ. Safety 2*, 407-411 (1978).
37. T. K. Rao, J. A. Young, W. Lijinsky, and J. L. Epler, Mutagenicity of aliphatic nitrosamines in *Salmonella typhimurium, Mutat. Res. 66*, 1-9 (1979).
38. M. Okada, E. Suzuki, and M. Mochizuki, Possible important role of urinary N-methyl-N-(3-carboxypropyl)nitrosamine in the induction of bladder tumors in rats by N-methyl-N-dodecylnitrosamine, *Gann 67*, 771-772 (1976).
39. W. Lijinsky and M. D. Reuber, Carcinogenicity in rats of nitrosomethylethylamines labeled with deuterium in several positions, *Cancer Res. 40*, 19-21 (1980).
40. W. Lijinsky, J. E. Saavedra, M. D. Reuber, and S. S. Singer, Esophageal carcinogenesis in Fischer 344 rats by nitrosomethylethylamines substituted in the ethyl group, *J. Natl. Cancer Inst. 68*, 681-684 (1982).
41. W. Lijinsky, M. D. Reuber, T. S. Davies, J. E. Saavedra, and C. W. Riggs, Dose-response studies in carcinogenesis by nitrosomethyl-2-phenylethylamine in rats and the effect of deuterium, *Food Cosmet. Toxicol. 20*, 393-399 (1982).
42. P. Kleihues, C. Veit, M. Wiessler, and R. M. Hodgson, DNA methylation by N-nitrosomethylbenzylamine in target and non-target tissues of NMRI mice, *Carcinogenesis 2*, 897-899 (1981).
43. W. Lijinsky and M. D. Reuber, Transnitrosation by nitrosamines *in vivo*, in: *N-Nitroso*

Compounds: Occurrence and Biological Effects (H. Bartsch, I. K. O'Neill, M. Castegnaro, and M. Okada, eds.), IARC Scientific Publications No. 41, pp. 625-631, International Agency for Research on Cancer, Lyon, France (1982).
44. W. Lijinsky, M. D. Reuber, and C. W. Riggs, Dose-response studies in rats with nitrosodiethylamine, *Cancer Res. 41*, 4997-5003 (1981).
45. R. Preussmann, G. Würtele, G. Eisenbrand, and B. Spiegelhalder, Urinary excretion of N-nitrosodiethanolamine administered orally to rats, *Cancer Lett. 4*, 207-209 (1978).
46. W. Lijinsky and H. W. Taylor, Comparative carcinogenicity of some derivatives of nitrosodi-*n*-propylamine in rats, *Ecotoxicol. Environ. Safety 2*, 421-426 (1978).
47. U. Mohr, G. Reznik, and P. Pour, Carcinogenic effects of diisopropanolnitrosamine in Sprague-Dawley rats, *J. Natl. Cancer Inst. 58*, 361-364 (1977).
48. W. Lijinsky and H. W. Taylor, Relative carcinogenic effectiveness of derivatives of diethylnitrosamine in rats, *Cancer Res. 38*, 2391-2394 (1978).
49. J. W. Cook and R. Schoental, Carcinogenic activity of metabolic products of 3:4-benzpyrene: Application to rats and mice, *Br. J. Cancer 6*, 400-406 (1952).
50. W. Lijinsky, M. Greenblatt, and C. Kommineni, Feeding studies of nitrilotriacetic acid and derivatives in rats, *J. Natl. Cancer Inst. 50*, 1061-1063 (1973).
51. W. Lijinsky, L. Keefer, E. Conrad, and R. Van de Bogart, The nitrosation of tertiary amines and some biologic implications, *J. Natl. Cancer Inst. 49*, 1239-1249 (1972).
52. H. Garcia and W. Lijinsky, Studies of the tumorigenic effect in feeding of nitrosamino acids and of low concentrations of amines and nitrite to rats, *Z. Krebsforsch. 79*, 141-144 (1973).
53. R. Preussmann, M. Habs, B. Pool, D. Stummeyer, W. Lijinsky, and M. D. Reuber, Fluorosubstituted N-nitrosamines. 1. Inactivity of N-nitroso-bis-(2,2,2-trifluoroethyl)amine in carcinogenicity and mutagenicity tests, *Carcinogenesis 2*, 753-756 (1981).
54. P. Pour, A new and advantageous model for colorectal cancer, *Cancer Lett. 4*, 293-298 (1978).
55. P. Pour, J. Althoff, F. W. Krüger, and U. Mohr, A potent pancreatic carcinogen in Syrian hamsters: N-nitrosobis (2-oxopropyl)amine, *J. Natl. Cancer Inst. 58*, 1449-1453 (1977).
56. P. Pour, R. Gingell, R. Langenbach, D. Nagel, C. Grandjean, T. Lawson, and S. Salmasi, Carcinogenicity of N-(nitrosomethyl-2-oxopropyl)amine in Syrian hamsters, *Cancer Res. 40*, 3588-3590 (1980).
57. L. Blattmann, N. Joswig, and R. Preussmann, Struktur von Metaboliten des Carcinogen Methyl-*n*-butyl-nitrosamin in Rattenurin, *Z. Krebsforsch. 81*, 71-73 (1974).
58. E. Suzuki and M. Okada, Metabolic fate of N,N-dipropylnitrosamine and N,N-diamylnitrosamine in the rat, in relation to their lack of carcinogenic effect on the urinary bladder, *Gann 72*, 552-561 (1981).
59. W. Lijinsky, K. Y. Lee, L. Tomatis, and W. H. Butler, Nitrosoazetidine—A potent carcinogen of low toxicity, *Naturwissenschaften 54*, 518 (1967).
60. W. Lijinsky and H. W. Taylor, Carcinogenicity of nitrosoazetidine and tetradeuteronitrosoazetidine in Sprague-Dawley rats, *Z. Krebsforsch. 89*, 215-219 (1977).
61. M. Greenblatt and W. Lijinsky, Nitrosamine studies: Neoplasms of liver and genital mesothelium in nitrosopyrrolidine treated MRC rats, *J. Natl. Cancer Inst. 48*, 1687-1696 (1972).
62. W. Lijinsky and H. W. Taylor, The effect of substituents on the carcinogenicity of N-nitrosopyrrolidine in Sprague-Dawley rats, *Cancer Res. 36*, 1988-1990 (1976).
63. W. Lijinsky and M. D. Reuber, Carcinogenic effect of nitrosopyrrolidine, nitrosopiperidine and nitrosohexamethyleneimine in Fischer rats, *Cancer Lett. 12*, 99-103 (1981).
64. S. S. Hecht, C. B. Chen, and D. Hoffmann, Evidence for metabolic α-hydroxylation of N-nitrosopyrrolidine, *Cancer Res. 38*, 215-218 (1978).
65. C. M. Goodall, W. Lijinsky, and L. Tomatis, Tumorigenicity of N-nitroso-hexamethyleneimine, *Cancer Res. 28*, 1217-1222 (1968).

66. C. J. Grandjean, The metabolism of N-nitrosohexamethyleneimine in rats. Identification of urinary metabolites, *J. Natl. Cancer Inst.* 57, 181-185 (1976).
67. A. E. Ross and S. S. Mirvish, Metabolism of N-nitrosohexamethyleneimine to give 1,6-hexanediol bound to rat liver nucleic acids, *J. Natl. Cancer Inst.* 58, 651-655 (1977).
68. W. Lijinsky, L. Tomatis, and C. E. M. Wenyon, Lung tumors in rats treated with N-nitrosoheptamethyleneimine and N-nitrosooctamethyleneimine, *Proc. Soc. Exper. Biol. Med.* 130, 945-949 (1969).
69. H. Schreiber, P. Nettesheim, W. Lijinsky, C. B. Richter, and H. E. Walburg, Induction of lung cancer in germ free, specific pathogen free and infected rats by N-nitrosoheptamethyleneimine: Enhancement by respiratory infection, *J. Natl. Cancer Inst.* 49, 1107-1114 (1972).
70. H. Garcia and W. Lijinsky, Tumorigenicity of five cyclic nitrosamines in MRC rats, *Z. Krebsforsch.* 77, 257-261 (1972).
71. W. Lijinsky, M. D. Reuber, T. S. Davies, and C. W. Riggs, Dose-response studies with nitrosoheptamethyleneimine and its alpha deuterium labeled derivative in F344 rats, *J. Natl. Cancer Inst.* 69, 1127-1133 (1982).
72. H. M. Reznik-Schüller and B. F. Hague, A morphometric study of the pulmonary clara cell in normal and nitrosoheptamethyleneimine-treated European hamsters, *Exp. Path.* 18, 366-376 (1980).
73. C. M. Goodall, W. Lijinsky, L. Keefer, and E. F. D'Ath, Oncogenic activity of N-nitrosododecamethyleneimine in liver, glandular stomach and other tissue of NZO/B1 mice, *Internat. J. Cancer* 11, 369-376 (1972).
74. J. M. Fajen, G. A. Carson, D. P. Rounbehler, T. Y. Fan, R. Vita, U. E. Goff, M. H. Wolf, G. S. Edwards, D. H. Fine, V. Reinhold, and K. Biemann, N-Nitrosamines in the rubber and tire industry, *Science* 205, 1262-1264 (1979).
75. J. Sander and G. Bürkle, Induktion maligner Tumoren bei Ratten durch gleichzeitige Verfütterung von Nitrit und Sekundaren Aminen, *Z. Krebsforsch.* 73, 54-66 (1969).
76. H. Garcia, L. Keefer, W. Lijinsky, and C. E. M. Wenyon, Carcinogenicity of nitrosothiomorpholine and 1-nitrosopiperazine in rats, *Z. Krebsforsch.* 74, 179-184 (1970).
77. W. Lijinsky and H. W. Taylor, Increased carcinogenicity of 2,6-dimethylnitrosomorpholine compared with nitrosomorpholine in rats, *Cancer Res.* 35, 2123-2125 (1975).
78. W. Lijinsky and H. W. Taylor, The change in carcinogenic effectiveness of some cyclic nitrosamines at different doses, *Z. Krebsforsch.* 92, 221-225 (1978).
79. U. Mohr, G. Reznik, E. Emminger, and W. Lijinsky, Induction of pancreatic duct carcinomas in the Syrian golden hamster with 2,6-dimethylnitrosomorpholine, *J. Natl. Cancer Inst.* 58, 429-432 (1977).
80. G. Reznik, U. Mohr, and W. Lijinsky, Carcinogenic effect of N-nitroso-2,6-dimethylmorpholine in Syrian golden hamsters, *J. Natl. Cancer Inst.* 60, 371-378 (1978).
81. M. S. Rao, D. G. Scarpelli, and W. Lijinsky, Carcinogenesis in Syrian hamsters by N-nitroso-2,6-dimethylmorpholine, its *cis* and *trans* isomers, and the effect of deuterium labeling, *Carcinogenesis* 2, 731-735 (1981).
82. W. Lijinsky, J. E. Saavedra, M. D. Reuber, and B.-N. Blackwell, The effect of deuterium labeling on the carcinogenicity of nitroso-2,6-dimethylmorpholine in rats, *Cancer Lett.* 10, 325-331 (1980).
83. W. Lijinsky and M. D. Reuber, Comparison of carcinogenesis by two isomers of nitroso-2,6-di-methylmorpholine, *Carcinogenesis* 1, 501-503 (1980).
84. W. Lijinsky and M. D. Reuber, Comparative carcinogenicity of two isomers of dimethylnitrosomorpholine in guinea pigs, *Cancer Lett.* 14, 7-11 (1981).
85. R. W. Stephany, J. Freudenthal, and P. L. Schuller, N-Nitroso-5-methyl-1,3-oxazolidine identified as an impurity in a commercial cutting fluid, *Recl. Trav. Chim. Pays-Bas* 97, 177-178 (1978).

86. R. Gingell, L. Wallcave, D. Magee, R. Kupper, and P. Pour, Common metabolites of N-nitroso-2,6-dimethylmorpholine and N-nitroso-bis-(2-oxopropyl)amine in the Syrian hamster, *Cancer Lett. 2*, 47–52 (1976).
87. B. Underwood and W. Lijinsky, Comparative metabolism of 2,6-dimethylnitrosomorpholine in rats, hamsters and guinea pigs, *Cancer Res. 42*, 54–58 (1982).
88. M. Greenblatt, V. Kommineni, E. Conrad, L. Wallcave, and W. Lijinsky, In vivo conversion of phenmetrazine into its N-nitroso derivative, *Nature New Biol. 236*, 25–26 (1972).
89. W. Lijinsky and H. W. Taylor, Carcinogenicity tests of N-nitroso derivatives of two drugs, phenmetrazine and methylphenidate, *Cancer Lett. 1*, 359–363 (1976).
90. W. Lijinsky and H. W. Taylor, Carcinogenicity of methylated dinitrosopiperazines in rats, *Cancer Res. 35*, 1270–1273 (1975).
91. R. E. Lyle, J. E. Saavedra, G. G. Lyle, M. M. Fribush, J. L. Marshall, W. Lijinsky, and G. M. Singer, Conformational stereospecificity in electrophilic reactions with cyclic anions, *Tetrahedron Lett.* 4431–4434 (1976).
92. L. A. Love, W. Lijinsky, L. K. Keefer, and H. Garcia, Chronic oral administration of 1-nitrosopiperazine at high doses to MRC rats, *Z. Krebsforsch. 89*, 69–73 (1977).
93. R. K. Elespuru and W. Lijinsky, Mutagenicity of cyclic nitrosamines in *E. coli* following activation with rat liver microsomes, *Cancer Res. 36*, 4099–4101 (1976).
94. T. K. Rao, J. A. Young, W. Lijinsky, and J. L. Epler, Mutagenicity of N-nitrosopiperazine derivatives in *Salmonella typhimurium*, *Mutat. Res. 57*, 127–134 (1978).
95. H. Bartsch, C. Malaveille, and R. Montesano, The predictive value of tissue-mediated mutagenicity assays to assess the carcinogenic risk of chemicals, in: *Screening Tests in Chemical Carcinogenesis* (R. Montesano, H. Bartsch, and L. Tomatis, eds.), IARC Scientific Publications No. 12, pp. 467–486, International Agency for Research on Cancer, Lyon, France (1976).
96. S. S. Singer, G. M. Singer, J. E. Saavedra, M. D. Reuber, and W. Lijinsky, Carcinogenesis by derivatives of 1-nitroso-3,5-dimethylpiperazine in rats, *Cancer Res. 41*, 1034–1038 (1981).
97. W. Lijinsky, G. M. Singer, and M. D. Reuber, The effect of 4-substitution on the carcinogenicity of nitrosopiperidine, *Carcinogenesis 2*, 1045–1048 (1981).
98. W. Lijinsky and H. W. Taylor, Carcinogenicity of methylated nitrosopiperidines, *Intern. J. Cancer 16*, 318–322 (1975).
99. G. M. Singer and H. W. Taylor, The carcinogenicity of N-nitrosonornicotine in Sprague-Dawley rats, *J. Natl. Cancer Inst. 57*, 1275–1276 (1976).
100. D. Hoffmann, R. Raineri, S. S. Hecht, R. Maronpot, and E. L. Wynder, Effects of N'-nitrosonornicotine and N'-nitrosoanabasine in rats, *J. Natl. Cancer Inst. 55*, 977–981 (1975).
101. D. Hoffmann, J. D. Adams, K. D. Brunnemann, and S. S. Hecht, Assessment of tobacco-specific N-nitrosamines in tobacco products, *Cancer Res. 39*, 2505–2509 (1979).
102. G. Eisenbrand, M. Habs, D. Schmähl, and R. Preussmann, Carcinogenicity of N-nitroso-3-hydroxypyrrolidine and dose–response study with N-nitrosopiperidine in rats, in: *N-Nitroso Compounds: Analysis, Formation and Occurrence* (E. A. Walker, L. Griciute, M. Castegnaro, and M. Börzsönti, eds.), IARC Scientific Publications No. 31, pp. 657–666, International Agency for Research on Cancer, Lyon, France (1980).
103. M. Greenblatt and W. Lijinsky, Null carcinogenic effect of feeding amino acids with nitrite to MRC rats, *J. Natl. Cancer Inst. 50*, 799–802 (1973).
104. C. Chu and P. N. Magee, Metabolic fate of nitrosoproline in the rat, *Cancer Res. 41*, 3653–3657 (1981).
105. H. Ohshima and H. Bartsch, Quantitative estimation of endogenous nitrosation in humans by monitoring N-nitrosoproline excreted in the urine, *Cancer Res. 41*, 3658–3662 (1981).
106. T. K. Rao, J. T. Cox, B. E. Allen, J. L. Epler, and W. Lijinsky, Mutagenicity of N-nitrosopyrrolidine derivatives in *Salmonella* (Ames) and *E. coli* K-12 (343/113) assays, *Mutat. Res. 89*, 35–43 (1981).

107. M. Wiessler and D. Schmähl, Zur carcinogenen Wirkung von N-n-nitrosoverbindungen. 2. S (+) and R (-)-Nitroso-2-methyl-piperidin, Z. Krebsforsch. 79, 118-122 (1973).
108. R. Kupper, M. D. Reuber, B. N. Blackwell, W. Lijinsky, S. R. Koepke, and C. J. Michejda, Carcinogenicity of the isomeric N-nitroso-Δ^2-piperidines in rats and the in vivo isomerization of the Δ^3- to the Δ^2-isomer, Carcinogenesis 1, 753-757 (1980).
109. W. Lijinsky and H. W. Taylor, Carcinogenicity of two unsaturated derivatives of nitrosopiperidine in Sprague-Dawley rats, J. Natl. Cancer Inst. 57, 1315-1317 (1976).
110. T. K. Rao, A. W. Hardigree, J. A. Young, W. Lijinsky, and J. L. Epler, Mutagenicity of N-nitrosopiperidines with Salmonella typhimurium/microsomal activation system, Mutat. Res. 56, 131-145 (1977).
111. W. Lijinsky, J. E. Saavedra, M. D. Reuber, and B.-N. Blackwell, Carcinogenicity of 3-chloro-, 4-chloro- and 3,4-dichloronitrosopiperidine in Fischer rats, Cancer Res. 40, 3325-3327 (1980).
112. W. Lijinsky and H. W. Taylor, Carcinogenicity of 4-chloronitrosopiperidine in Sprague-Dawley rats, Z. Krebsforsch. 92, 217-220 (1978).
113. W. Lijinsky and H. W. Taylor, Carcinogenicity of N-nitroso-3,4-dichloro- and N-nitroso-3,4-dibromopiperidine in rats, Cancer Res. 35, 3209-3211 (1975).
114. W. Lijinsky and H. W. Taylor, Tumorigenesis by oxygenated nitrosopiperidines, J. Natl. Cancer Inst. 55, 705-708 (1975).

11

Comparison of Mutagenic and Carcinogenic Properties

A Critique

W. LIJINSKY, J. L. EPLER, and T. K. RAO

1. INTRODUCTION

Because of the broad support of a somatic mutation theory to explain carcinogenesis, a number of short-term mutagenesis assays have been widely applied as prescreens for detecting carcinogens. These assays measure a wide spectrum of genetic damage such as gene mutations, DNA repair synthesis, chromatid or chromosomal abberations, and certain physiological or structural changes. The philosophy of using short-term assays and the merits and limitations of each test system have become topics of scientific investigation and discussion, especially because of the role they play in regulatory policies. The purpose of organizing this volume was an attempt to examine and compare the genotoxic properties of one extensively studied class of compounds, namely, N-nitroso compounds.

The preceding chapters have each detailed a specific aspect of the existing biological data on a large group of N-nitroso compounds. The underlying link has been the availability of these compounds at the Frederick Cancer Research Facility because of the research going on there. The selection and application of the various biological assays has facilitated the comparison of the carcinogenic

---

W. LIJINSKY • Chemical Carcinogenesis Program, NCI-Frederick Cancer Research Facility, Frederick, Maryland 21701.   J. L. EPLER AND T. K. RAO • Biology Division, Oak Ridge National Laboratory, Oak Ridge, Tennessee 37830.   Dr. Rao's present address is Environmental Health Research and Testing, Inc., Research Triangle Park, North Carolina 27709.   By acceptance of this article, the publisher or recipient acknowledges the U.S. Government's right to retain a nonexclusive, royalty-free license in and to any copyright covering the article.

and mutagenic end points; however, the array of investigators and various short-term assays came about through the investigators' mutual interest in the research, not solicitation because of their field of interest. Thus, gaps in the "battery" of short-term assays applied are obvious, especially in the area of DNA repair. The following critique attempts to integrate, compare, and observe the concordance among the data available with this chemical class.

The principal focus of the approach has been the testing of N-nitroso compounds in rapid, inexpensive, short-term mutagenicity assays or, perhaps, more generally, assays for the detection of genetic toxicants, since end points other than gene mutation were considered. These results have then been compared with the results obtained in whole-animal tests for carcinogenicity. Selection of the test compounds emphasized structure–activity and, when feasible, the metabolism of the N-nitrosamines under test. Metabolism was in most cases mediated by the addition of mammalian enzymes or, as in the *Drosophila* data cited,[1,2] through metabolism in the organism.

The potential predictive value of assays for genetic damage and the correlation with carcinogenic damage has been emphasized in a number of reports.[3,4] This critique focuses on this correlation or, rather, concordance. The bacterial test system developed by Ames[5] has been reported to be widely applicable for prescreening a large number of pure chemical carcinogens and noncarcinogens. Complex environmental mixtures,[6] effluents, or commercial and industrial products (e.g., smoke condensates, soots, hair dyes, synthetic oils, waste oils, cutting oils, etc.) have been examined and analytical chemical approaches have been coupled with the biological end points. The assumption has been made that the mutagenic end point will bear some direct relationship to the carcinogenicity of the material; this generalization might well be in error.[7] Different assays measure different classes of compounds to different degrees of sensitivity. Our present examination further reinforces this caution in interpretation. The importance of tests in additional genetic systems, along with the need to continue whole-animal testing for both mutagenic and carcinogenic end points, must be considered. Because of the intrinsic limitation of each genetic assay, testing with only one system has often led to faulty conclusions about specific compounds.[8] To overcome this potential shortcoming, the work summarized in this volume has utilized an array of mutagenicity assays.

The critique is composed of three related sections. First, a tabular summation and integration of the qualitative results from all of the chapters and allied published information. Second, a comparison of the qualitative agreement between the various mutagenic end points. Finally, a critique of the relationships, both qualitative and quantitative, between the short-term mutagenicity tests and the carcinogenic end points is presented. Emphasis in this final section is given to the *Salmonella* and phage induction assays, since the largest array of compounds tested for carcinogenicity were tested in these assays. Particular attention is paid to the important role of metabolic activation coupled to these systems. Generalizations alerting the reader to difficulties and precautions in interpreting the data, or extrapolating the results from one biological end point to another, are brought forth.

2. TABULAR SUMMATION AND INTEGRATION OF RESULTS

A large number of N-nitroso compounds that have been examined for carcinogenic activity were examined for genotoxic properties. Most of these studies were carried out independently, or in some cases as a collaborative effort, between eleven laboratories utilizing ten different assays. The studies examined a spectrum of genetic damage. The results are integrated from open literature or contributions in this volume or as a personal communication at the workshop on "Genotoxicology of Nitrosamines," July 14–15, 1980, Bethesda, Maryland, sponsored by the Subcommittee on Environmental Mutagenesis, DHEW-CCERP.

2.1. Selection of Nitroso Compounds for the Comparative Study

The nitroso compounds used in the comparative study were provided by Dr. Lijinsky, Frederick Cancer Research Facility, Frederick, Maryland. Their carcinogenic properties have been evaluated and the results have been tabulated.[9] A subjective assessment of carcinogenicity (based on dose, dose-rate, tumor incidence, and time taken for tumor development) is presented in the last column of Tables I to X.

2.2. *In Vitro* Assays

A number of *in vitro* genetic systems ranging from bacteria to mammalian cells in culture were used for determining genotoxicity. The bacterial mutagenesis assays utilized the well-publicized *Salmonella* histidine reversion system developed by Dr. B. N. Ames.[5] The studies incorporated several modifications of the standard procedures. Rao *et al.* (column a) used both Aroclor- and phenobarbital-induced rat liver S-9 and have noted[10] that phenobarbital-induced S-9 was specific for activating nitrosamines. Andrews *et al.* (column b)[11] have used both hamster and rat liver S-9. Several carcinogenic nitrosamines that were not mutagenic with rat liver S-9 were activated into a mutagenic form by the hamster liver S-9. Guttenplan (column c)[12] used a low-pH incubation medium and found that several nitrosamines that did not respond in the standard plate assay were mutagenic under these conditions. Zeiger *et al.*[13,14] (column d) have compared a conventional liquid suspension test with the standard pour-plate incorporation assay.

Escherichia coli has also been used for examining mutagenic nitrosamines.

Elespuru[15] used a phage induction assay and the reversion of *tyr* locus in the WP-2 strain[16] (columns a and b, respectively) while Rao *et al.*[17,18] have used a multipurpose strain (K-12, 343/113) that is capable of detecting both reversion (*arg* and *nad* loci) and forward mutations (*gal* locus) (column c).

Saccharomyces cerevisiae has been used[19,20] as another microbial system where reversion of a missense locus for histidine (*his*) and forward mutation for Canavanine resistance (CAN^R) were monitored. It is noteworthy that microbial assays utilized in this study were deficient for DNA repair. Rao *et al.*[18] have previously reported that the excision repair deficient mutants were more sensitive for nitrosamine-induced mutagenesis than other tester strains.

Mammalian cells in culture were also used in the mutagenic analysis of nitrosamines. Jones *et al.*[21] have used selection for 6-thioguanine (column a) and Ouabain resistance (column b) as genetic markers in Chinese hamster cells (V79). Ho *et al.*[22] have selected for HGPRT-deficient mutants in the Chinese hamster ovary cell (column c). In a parallel study, Ho *et al.* have also examined the induction of sister-chromatid exchanges and chromosomal aberrations in human leukocytes. As in the microbial assays, the mammalian cell systems have incorporated exogenous metabolic systems by the addition of rat liver homogenates.

2.3. *In Vivo* Assay

Drosophila melanogaster, the common fruit fly, is the only *in vivo* genetic assay system that was considered in this comparative study. Induction of sex-linked recessive lethals[1,2] was used as a genetic end point. A wide spectrum of genetic damage, such as point mutation or chromosomal breaks, could be attributed as the causative factors in the induction of these mutations. In addition to normal wild-type flies, a stock of flies deficient for DNA repair (excision repair mutants) were shown to be extremely useful in identifying mutagenic nitrosamines (C. Nix, personal communication).

3. COMPARISON OF MUTAGENIC END POINTS

A large effort has gone into generating comparative mutagenesis data with these compounds and the results are a product of several years of investigation by several researchers. The empirical correlations attempted in this chapter are of practical importance. These comparisons are meant for assessing the importance of modifications in genetic testing methodologies and not to recommend any particular assay or battery of assays in genetic toxicological studies.

A large number of nitrosoalkylamides (guanidines, carbamates, and ureas) were examined for mutagenicity in *Salmonella* and *E. coli* (Table I). The concurrence between several mutagenicity assays is notable. The mutagenicity

Table I
Mutagenicity of Nitroso Alkylamides in Bacterial Assays[a]

Nitroso compound	Mutagenicity		Carcinogenicity
	Salmonella b	E. coli a	
Nitroso-guanidines			
Methylnitroguanidine	+	+	+
Ethylnitroguanidine	+	+	+
Cimetidine	+	+	0
Nitroso-carbamates			
Methylurethane	+	+	+
Ethylurethane	+	+	+
Carbaryl	+	+	+
Methomyl	+	+	+
Carbofuran	+	+	+
Landrin	+	+	+
Aldicarb	+	+	+
Buxten	+	+	+
Methylphenylcarbamate	+	+	+
Baygon	+		+
Oxazolidone	+		+
5-Methyloxazolidone	+		+
Nitroso-ureas[b]			
Methylurea	+	+	+
Ethylurea	+	+	+
n-Propylurea	+	+	+
Allylurea	+	+	0
Isopropylurea	+	+	+
n-Butylurea	+	+	+
Isobutylurea	+	0	0
sec-Butylurea	+		0
n-Amylurea	+	+	+
n-Hexylura	+	±	+
Undecylurea	+	0	+
Tridecylurea	+	0	±
Phenylurea	+	+	0
Cyclohexylurea	+	+	±
Benzylurea	+	+	0
Phenylethylurea	+	+	0
Hydroxyethylurea	+		+
Fluoroethylurea	+	+	+
Isopropanolurea	+		+
3-Hydroxypropylurea	+		+

[a] See footnote a in Table II for the description of assays.
[b] Carcinogenicity in mouse skin painting.

Table II

Comparative Mutagenicity of Symmetrical Acyclic Nitrosamines[a]

Nitroso compound	Salmonella				Escherichia coli			Saccharo-myces	Cultured hamster cells			Drosophila	Carcinogenicity
	a	b	c	d	a	b	c		a	b	c		
Dimethylamine	+	+	+	+	+	+		+	+	+	+	+	+
Diethylamine	+	+	+		+	+		+	+	+	+	+	+
Di-n-propylamine	+	+	+		+		+	+	+	+		+	+
Di-iso-propylamine	0		+		±		+					+	+
Di-n-butylamine	+	+	+		±				+	+		+[b]	+
Di-iso-butylamine	+	+										0	+
Di-n-octylamine	+	+							0	0		0	0
Di-allylamine	+				+		+					+	0

[a] *Salmonella* assay: *his* reversion; a. Rao; b. Andrews; c. Guttenplan; d. Zeiger. *E. coli*: a. Elespuru (phage induction); b. Elespuru (*trp* reversion); c. Rao (*arg* reversion; 343/113). *Saccharomyces*: Larimer (*his* reversion; CAN^R). Cultured hamster cells: a. Jones (V79/OUA^R); b. Jones (V79/6TG^R); c. Ho (CHO/6TG^R). Drosophila: Nix (SLRL), SCE/CA: Ho (human leukocytes). Carcinogenicity: Lijinsky (rat).
[b] Repair deficient only.

Table III
Comparative Mutagenicity of Substituted Aliphatic Nitrosamines in Bacterial Assays[a]

Nitroso compound	Salmonella a	Salmonella b	E. coli a	Carcinogenicity
Ethylethanolamine	+			+
Bis-(2-cyanoethyl)amine	0			0
Bis-(2-methoxyethyl)amine	0			+
Bis-(2-ethoxyethyl)amine	+		0	+
Bis-(2,2-diethoxyethyl)amine	0		0	0
Bis-(2-hydroxypropyl)amine	0	+	0	+
Bis-(2-oxopropyl)amine	0	+	0	+
Bis-(2-chloropropyl)amine	+		+	+
Diethanolamine	0	0	0	+
Bis-(2-chloroethyl)amine	+		+	+
Bis-(-trifluoroethyl)amine	0			0
Methyl-2-oxopropylamine		+		+
Dihydroxypropyl-oxo-propylamine		+		+
2,3-Dihydroxypropylallylamine		+		+
Allylhydroxypropylamine		+		+
Dihydroxypropylethanolamine		+		±
Allylethanolamine		+		+
Ethanolisopropanolamine		0		+
Hydroxyethyloxopropylamine		+		+
Hydroxypropyloxopropylamine		+		+

[a] See footnote a in Table II for the description of assays.

of a group of symmetrical acyclic nitrosamines with varying chain length was analyzed. The concurrence between mutagenicity assays is again excellent (Table II). Nitroso-diallylamine, a noncarcinogen in rats, was mutagenic in a number of test systems. This compound was shown to be carcinogenic in hamsters.[23] A group of substituted aliphatic nitrosamines was also investigated for mutagenicity in *Salmonella* and *E. coli* (Table III). Unlike other nitroso compounds, these aliphatic nitrosamines failed to follow any consistent pattern when mutagenicity and carcinogenicity are compared. The *Salmonella* assay with rat liver S-9 (column a) and with rat or hamster liver S-9 (column b) exhibited some differences. Hamster liver S-9 appeared to activate both carcinogenic and noncarcinogenic nitrosamines. Table IV illustrates the results obtained with a group of nitrosomethylalkylamines in *Salmonella* and *E. coli*, which exhibited a greater degree of correlation than the aliphatic compounds in Table III. Another group of asymmetric carcinogenic nitrosamines was examined for mutagenicity in a larger number of test systems (Table V). Nitrosomethylethylamine, which was not mutagenic in *Salmonella*, was found to be

Table IV
Mutagenicity of Nitrosomethyl Alkylamines in Bacterial Assays[a]

Nitroso compound	Mutagenicity		Carcinogenicity
	Salmonella b	E. coli a	
Methyl-			
Methoxyamine	+	+	0
2-Hydroxypropylamine	+		+
Dihydroxypropylamine	+		+
2-Oxopropylamine	+		+
Carboxypropylamine	0	±	+
n-Hexylamine	+	+	+
n-Heptylamine	+	+	+
n-Octylamine	+	+	+
n-Nonylamine	+	+	+
n-Decylamine	+	+	+
n-Undecylamine	+	+	+
n-Tetradecylamine	+		+

[a] See footnote a in Table II for the description of assays.

mutagenic in *E. coli*. On the other hand, the strong carcinogen[(9)] nitrosomethylneopentylamine was not mutagenic in a number of microbial assays. Guttenplan's (column c) investigations showed that by lowering pH these two compounds could be converted into bacterial mutagens. Although it is not easily explainable how a simple change such as lowering pH could affect either the metabolism of these compounds or the mutant expression, it is important to note that a slight modification of testing methodology could markedly affect the end result. This indicates the importance of using multiple assays in a battery of test systems as more reliable than any one particular assay.

A large number of cyclic nitrosamines and their derivatives (Tables VI to X) were studied in a number of mutagenesis assays. The matrices of the comparative tables are more complete with this class of compounds. The concurrence among the mutagenicity assays is excellent. The tables are self-explanatory and some results deserve particular attention. 1-Nitrosopiperazine, a noncarcinogen (Table VI), was found to be nonmutagenic in a number of assays. Nitroso-3-pyrroline (Table VII), which was mutagenic in *Salmonella* only when activated by hamster S-9, was found to be mutagenic in *E. coli* and yeast. On the other hand, nitroso-1,2,3,6-tetrahydropyridine (containing an unsaturated ring and a homolog of nitrosopyrroline) was strongly mutagenic (Table VIII) in a number of test systems. Study with the substituted derivatives

Table V
Comparative Mutagenicity of Asymmetrical Nitrosamines[a]

Nitroso compound	Salmonella			Escherichia coli		Cultured hamster cells		Carcinogenicity
	a	b	c	a	c	a	b	
Methyl-ethylamine	0	0	+	+	+			+
n-Propylamine	+	+	+	+		+	+	+
n-Butylamine	+	+	+	+		+	+	+
Neopentylamine	0	0	+		0			+
Cyclohexylamine	+	+		+				+
Benzylamine		+		+				+
Aniline	0		+	+				+
Phenylethylamine	+	+				+	+	+
Dodecylamine		+				+	+	+

[a] See footnote a in Table II for the description of assays.

Table VI

Mutagenicity of Cyclic Nitrosamines: Effect of Ring Size[a]

Nitroso compound	Salmonella				Escherichia coli			Saccharomyces	Cultured hamster cells			Drosophila	SCE/CA	Carcinogenicity
	a	b	c	d	a	b	c		a	b	c			
Azetidine	+	+		+	+			+	+	+		+		+
Pyrrolidine	+	+	+	+	+			+	+	+	+	+		+
Piperidine	+	+		+	+	+		+	+			+	+	+
Piperazine	0			+	0			0	0					0
Morpholine	+	+	+	+	±	+		+	+	+				+
Hexamethyleneimine	+	+		+	+	+		+	+	+				+
Heptamethyleneimine	+	+		+	+	+		+	+	+				+
Octamethyleneimine	+	+		+	+			+						+
Dodecamethyleneimine	+	+		+	0				0	+				+
Dinitroso-piperazine	+	+		+	0	+			+	+	+			+
Homopiperazine	+	+			0	+		+						+

[a] See footnote a in Table II for the description of assays.

Table VII

Mutagenicity of Cyclic Nitrosamines: Substituted N-Nitroso-pyrrolidines[a]

Nitroso compound	Salmonella				Escherichia coli		Saccharo-myces	Cultured hamster cells			Drosophila	Carcinogenicity
	a	b	c	d	a	c		a	b	c		
Pyrrolidine	+	+	+	+	+	+	+	+	+	+	+	+
3-Pyrroline	0	+					+					+
3-Hydroxypyrrolidine	+				0							+
3,4-Dichloropyrrolidine	+				+		+					+
3,4-Dibromopyrrolidine	0				+	+						?
Proline	0				±	0	+				0	0
4-Hydroxyproline	0				±		0		0		0	0
2,5-Dimethylpyrrolidine	0						0					0
Nornicotine		+										+

[a] See footnote a in Table II for the description of assays.

Table VIII

Mutagenicity of Cyclic Nitrosamines: Substituted N-Nitrosopiperidines[a]

Nitroso compound	Salmonella				Escherichia coli	Saccharomyces	Cultured hamster cells		Drosophila	SCE/CA	Carcinogenicity
	a	b	c	d	a		a	b			
Piperidine	+	+		+	+	+	+	+	+	+	+
2-Methylpiperidine	+	+	+	+	+	+			+		+
3-Methylpiperidine	+					+			+	+	+
4-Methylpiperidine	+				+	+			+		+
2,6-Dimethylpiperidine	0				0	0			0	0	0
3,5-Dimethylpiperidine	+	+			+	+			+		+
2,2,6,6-Tetramethylpiperidine	0				0	0					0
1,2,3,6-Tetrahydropyridine	+	+			+	+			+		+
1,2,3,6-Tetrahydropyridine-3-methyl	+	+			+					+	+
1,2,3,4-Tetrahydropyridine	+				+						+
3-Hydroxypiperidine	+				0						+
4-Hydroxypiperidine	+				0	0			0		+
2,2,6,6-Tetramethyl-4-hydroxy-piperidine	0										0
4-Ketopiperidine	+	+	+	+	0				0	0	+
2,2,6,6-Tetramethyl-4-keto-piperidine	0										0

[a] See footnote a in Table II for the description of assays.

of cyclic nitrosamines (Tables VII to X) indicated similarities between their mutagenic and carcinogenic properties. Substitution with carboxy groups nullified their biological activity, while an enhancement was noted when substituted by halogens. A block in the alpha positions to the N-nitroso group eliminated their biological activity, probably because of steric hindrance. All of these observations suggest that these nitrosamines are activated into their ultimate mutagenic or carcinogenic form through similar metabolic pathways.

Results presented in this section emphasize the use of multiple genetic end points in genotoxicity determination. The correlations between genetic end points and carcinogenic properties are discussed in the next section. Study of these compounds clearly indicated the complexity of test methodologies and the precautions needed in interpreting test results.

4. RELATIONSHIP BETWEEN SHORT-TERM TEST RESULTS AND CARCINOGENICITY

4.1. Qualitative Relationships

After testing more than 150 N-nitroso compounds for both mutagenicity and carcinogenicity, it can be concluded that most of the carcinogens are mutagenic and most of the noncarcinogens are not mutagenic. This is similar to the conclusion of Ames and his colleagues following their survey of 300 compounds, belonging to a variety of chemical classes, including some N-nitroso compounds.[3] Nevertheless, within those broad conclusions lie considerable and important discrepancies that must modify any conclusion that the mechanisms of mutagenesis and carcinogenesis are always the same—or even similar.

Even the conduct of the mutagenesis test requires special attention, as demonstrated by some investigators[12] who obtained positive results and a considerable correspondence between mutagenic and carcinogenic activities, with a number of nitrosamines utilizing modification of "standard" microbial assays. The results demonstrate that superficial judgments of the outcome of short-term tests using standardized testing procedures can be misleading. (Simply stated, an investigator should not "screen" compounds; an investigator should carry out experiments.) Nevertheless, it is clear that there are several N-nitroso compounds that are carcinogenic and yet are not mutagenic to bacteria in anyone's hands and are often inactive in other short-term tests. This is also disappointingly true of many metabolites of nitrosamines that are isolated from whole animals or from organs or cells in culture. As discussed by Farrelly and Hecker,[24] their examination of the mutagenicity of several metabolites of cyclic nitrosamines has not shown any that could be considered part of the pathway of activation of these potent carcinogens. Although the parent

Table IX

Mutagenicity of Cyclic Nitrosamines: Substituted N-Nitrosopiperazines[a]

Nitroso compound	Salmonella			Escherichia coli			Saccharomyces	Cultured hamster cells		Drosophila	SCE/CA	Carcinogenicity
	a	b	c	a	b	c		a	b			
Piperazine	0			0	0		0	0	0			0
N-Methylpiperazine	+	+	+				+			+		+
3,5-Dimethylpiperazine		+										+
3,4,5-Trimethylpiperazine		+		+								+
4-Benzoyl-3,5-dimethyl piperazine	0						0					+
4-Acetyl-3,5-dimethyl piperazine		+										
2-Methyldinitrosopiperazine	+	+		0			+			+	0	+
2,5-Dimethyldinitrosopiperazine	+	+					+			+		+
2,6-Dimethyldinitrosopiperazine	+	+	+	0	0		+			+	0	+
2,3,5,6-Tetramethyldinitroso piperazine	0						0			0	0	0

[a] See footnote *a* in Table II for the description of assays.

Table X

Mutagenicity of Cyclic Nitrosamines: N-Nitrosomorpholines and Analogs[a]

Nitroso compound	Salmonella				Escherichia coli		Saccharo-myces	Cultured hamster cells		Drosophila	Carcinogenicity
	a	b	c	d	a	b		a	b		
Morpholine	+	+	+	+	±	+	+	+	+	+	+
2-Methylmorpholine		+	+		+						+
2,6-dimethylmorpholine	+	+			+			+	+	+	+
Thiomorpholine		0			±		+			0	+
Tetrahydro-1,3-oxazine		+			+						+
Oxazolidine		+			+						+
5-Methyloxazolidine		+			0						+
2-Methyloxazolidine		+			0						+
Phenmetrazine		+									0

[a] See footnote *a* in Table II for the description of assays.

cyclic nitrosamines are both carcinogenic and strongly mutagenic when activated by rat liver microsomes, no mutagenic metabolite has been demonstrated. It seems, therefore, that the pathways leading to carcinogenesis by these cyclic nitrosamines are very minor and undetected among the pathways that produce the major metabolites, or that the activated carcinogenic products are not mutagenic and possibly do not induce tumors by a mutagenic mechanism. In the absence of good experimental evidence, either of these possibilities is quite valid, along with others not yet considered.

Another interesting outcome of these comparative studies is discussed by many contributors to this volume, namely, the frequent observation that activation of nitrosamines by Syrian hamster liver S-9 fraction is more effective than activation by the comparable rat liver S-9 fraction. Attention was first drawn to this phenomenonn by Bartsch et al.[25] Thus, several nitrosamines that are nonmutagenic to *Salmonella* when activated by rat liver S-9 are active when the hamster liver fraction is substituted.[11] Furthermore, there is no strong relationship to the relative carcinogenic effectiveness of the particular compound in rats and hamsters. For example, a number of nitrosamines induce tumors in rats at much lower doses than are required for hamsters, although often in different organs in the two species. Nitroso-2,6-dimethylmorpholine is, by this measure, considerably more potent as a carcinogen in the rat than in the hamster,[26] yet it is much more mutagenic to *Salmonella* when activated by hamster liver S-9 than by rat liver S-9. In the phage induction assay of Elespuru,[15] hamster liver S-9 was used almost exclusively for activation since rat liver S-9 was ineffective. In the V79 cell mutagenesis assay of Huberman,[21] nitrosodimethylmorpholine was weakly active using rat liver (hepatocyte) activation. However, in the bacteriophage induction assay it was strongly positive. It appears, therefore, that the fact that the target organ of nitrosodimethylmorpholine in the rat is the esophagus, not the liver, might explain the low activity in the V79 cell assay. A reasonable conclusion from these disparate results is that the test systems have greatly different susceptibilities to the biologically active agent. Nitrosomethylethylamine, for example, is nonmutagenic to *Salmonella* (with rat or hamster S-9 activation), but strongly positive in the phage induction assay (with hamster liver S-9) and in V79 cells with rat hepatocyte activation; nitrosomethylethylamine induces hepatocellular carcinomas in rats.[27]

The *Salmonella* reversion assay is the one short-term assay in which a large number of N-nitroso compounds have been thoroughly studied. A substantial number of those for which chronic toxicity data in animals are available have also been examined in the phage induction assay. In the other assays described in this volume, selected groups of nitrosamines were used for studies in depth, and many of those were chosen because they have well-defined carcinogenic properties.

4.1.1. Nitrosoalkylamides

In the *Salmonella* assay, all of the nitrosoalkylamides were mutagenic and were directly acting. In every case, the addition of liver S-9 either had no effect or reduced the number of revertants produced by a given dose of nitrosamide. This was almost certainly due to removal of some of the nitrosamide from the system through reaction with a readily alkylated constituent of the biological mixture. Most of the nitrosoalkylamides were tested for carcinogenic activity by repeated application to mouse skin; some have also been tested by chronic administration by other routes.

In the mouse skin assay most, but not all, of the nitrosoalkylamides were carcinogenic. Nitrosomethylurethane, nitrosoethylurethane, and the nitrosated N-methylcarbamate insecticides were noncarcinogenic or weakly so on mouse skin,[28,29] although they were extremely potent mutagens. They were also very active in the phage induction assay,[15] and were positive at very low doses (1 to 10 μg). These nitrosoalkylcarbamates were quite effective carcinogens when given orally to rats, and gave rise mainly to carcinomas of the forestomach. However, they were of very similar carcinogenic potency by this route,[30,31] although their mutagenic potencies and their activities in the phage induction assay varied quite widely. For example, the number of revertants per micromole varied from approximately 1 million for methomyl to 1800 for nitrosoethylurethane. The cyclic nitrosoalkylcarbamates, nitrosooxazolidone and its 5-methyl derivative, were potent mutagens to *Salmonella* and were among the more potent carcinogens examined by mouse skin painting or by gavage to rats.[32] The reasons for the discrepancies between the ranking of mutagenic and carcinogenic potencies could include such factors as absorption, uptake by the cells, and stability of the compounds in the cellular milieu.

Only three nitrosoalkylguanidines were examined and among them there were also some discrepancies. Both nitrosomethylnitroguanidine and nitrosoethylnitroguanidine were potent mutagens and phage-inducing agents, while nitrosocimetidine was weaker in both short-term assays. Nitrosocimetidine, however, does not appear to be carcinogenic, either on mouse skin or by chronic administration in drinking water. By skin painting nitrosoethylnitroguanidine is considerably more carcinogenic than is the methyl compound.[28] This correlates well with the higher potency of nitrosoethylnitroguanidine compared with the methyl compound in the phage induction assay, but in *Salmonella* the methyl compound is more mutagenic than the ethyl compound.

The largest group of nitrosoalkylamides studied is the nitrosoalkylureas, which are in two groups, those with alkyl or aryl groups and those with substituted alkyl groups. Without exception these were directly acting mutagens to *Salmonella*, although some required very large doses to reveal significant mutagenic activity (nitrosotridecylurea and nitrosobenzylurea); nitrosophenylurea was not active in strain TA 1535, which was the strain of choice for all of the other N-nitroso compounds. In the phage induction assay four nitrosoalkylureas were inactive or very weakly active: nitrosoundecylurea, nitrosotridecylurea, nitrosoiso-butylurea, and nitrosohexylurea. The last was a surprise, since it was one of the more potent carcinogens in this group in the mouse skin painting study, whereas the other three inactive compounds were also very weak carcinogens in mouse skin.[33] The range of mutagenic potency of the nitrosoalkylamides was not as great as for the nitrosoalkylcarbamates, but there was no consistent relationship between mutagenicity and carcinogenic effectiveness in mouse skin. This was particularly true when the aromatic nitrosoureas are considered. None of them were carcinogenic in mouse skin, although nitrosophenylurea and nitrosobenzylurea were weak carcinogens by other routes of administration to rats[34,35]; nitrosophenylethylurea was a potent mutagen to *Salmonella*.

The substituted nitrosoalkylureas include nitrosofluorethyl-, nitrosohydroxyethyl-, nitroso-2-hydroxypropyl-, and nitroso-3-hydroxypropyl-urea. With the exception of the last compound, these were more potent mutagens than their unsubstituted parents by two orders of magnitude, although they were approximately equally potent carcinogens in mouse skin, including the weakly mutagenic nitroso-3-hydroxypropylurea.[32] Also, when administered to rats by gavage, nitrosohydroxyethylurea and nitroso-2-hydroxypropylurea were potent carcinogens (although not more potent than nitrosoethylurea, a much weaker mutagen), but differed greatly in the spectrum of organs in which they induced tumors.[32] On the other hand, nitrosohydroxyethylurea and nitroso-2-hydroxypropylurea had very similar carcinogenic effects in hamsters,[36] suggesting that some type of "activation" might play a role in carcinogenesis by even these "directly acting" carcinogens.

Among the directly acting N-nitroso compounds discussed above, which can be assumed to induce tumors and cause mutations by alkylation of DNA in target cells, there are very great differences in activity. There is, furthermore, a considerable disparity between mutagenic potency and carcinogenic potency, even when these activities are measured in the same systems under similar conditions. We conclude that there are so many possible factors influencing the extent of both types of biological activity, including physical factors such as liposolubility and transport across membranes, that it would be foolish to assume that measurement of mutagenic activity can be predictive of carcinogenicity, other than qualitatively to a considerable degree. This is particularly true considering the very complex nature of carcinogenesis, with its multitude of steps, mostly not known or understood. The role played by the biological and biochemical nature of the target cells in carcinogenesis is particularly vague, and this is even more apparently a stumbling block to understanding when we consider the nitrosamines, which have to be enzymically activated to display carcinogenic or mutagenic activity.

4.1.2. Nitrosamines

It is convenient to consider these in two groups, acyclic and cyclic, although the distinction between them is not clear and somewhat artificial. A very large number of nitrosamines have been tested in a number of laboratories, and there are numerous publications and reviews dealing with those studies, especially from the laboratories of Druckrey, Preussmann, Lijinsky, Schmähl, and Mohr. Those studies have had great influence on our understanding of carcinogenesis by N-nitroso compounds. However, for the purpose of this evaluation we will consider mainly the data in Chapter 10, from the laboratories of William Lijinsky,[9] because these studies in rats were conducted under quite standard conditions, using to a large extent equimolar doses of groups of nitrosamines, so that a pharmacological evaluation of effectiveness can be made. The evaluations of carcinogenic potency are not precise because of the complexity and variability of the biological system represented by a mammal. Nevertheless, the several dose–response studies of certain nitrosamines and the good reproducibility of some of the chronic tests suggest that there is an inverse relationship between carcinogenic potency and both the doses required to reach a particular end point in number of animals with tumors and also the age at which the animals die with tumors induced by the treatment. While the latter cannot be considered a true "latent period," since the time of appearance of the first microscopic tumor is not known, the time to death of half of the treated animals with tumors can be used as an index of relative carcinogenicity.

The simplest and most extensively studied nitrosamines, nitrosodimethylamine and nitrosodiethylamine, offer the greatest problems in understanding the relationship between the short-term assays, especially bacterial mutagenesis and carcinogenicity. Nitrosodimethylamine, about which there must be a thousand or more publications, is a very potent carcinogen and a strong hepatotoxin in rats. It is also a nitrosamine to which humans are undoubtedly exposed, albeit in usually low concentrations. In most strains of rat examined, nitrosodimethylamine induces hemangioendothelial sarcomas.[27] In F 344 rats very few tumors arising from parenchymal cells have been seen, except possibly at very low doses. The dilemma is that much of the study of mechanisms of action of nitrosodimethylamine has been carried out using liver parenchymal cells of the rat, which is not the cell neoplastically transformed by this carcinogen. On the other hand, nitrosodiethylamine, which is also not mutagenic to *Salmonella* in the plate test but is active following liquid preincubation (like nitrosodimethylamine), does induce hepatocellular carcinomas in rats, as well as hemangioendothelial sarcomas and esophageal tumors.[37] Both nitrosodimethylamine and nitrosodiethylamine are strongly mutagenic in the plate test when activated by hamster liver S-9, although the hamster is no more sensitive to the carcinogenic action of these nitrosamines than the rat. In the phage induction assay, nitrosodiethylamine was considerably more active than nitrosodimethylamine (using hamster liver S-9 for activation), which accords well with the much higher carcinogenic potency of the former in rats. The studies of Prival,[38] described in Chapter 8, have done much to explain the lack of mutagenicity of these simple nitrosodialkylamines in the plate test using rat liver S-9, and have explored the role of inhibitors and the characteristics of the microsomal enzymes from different species.

Some of the simple nitrosodialkylamines, exemplified by nitrosomethylethylamine and nitrosomethylneopentylamine, were not mutagenic to *Salmonella* when activated by rat liver S-9, whether in the plate test or using liquid preincubation. Nitrosomethylneopentylamine induces esophageal tumors in rats. Therefore, although it is a very potent carcinogen, it is understandable that rat liver enzymes might be incompetent to activate this compound to a bacterial mutagen (although rat liver S-9 is effective in activating many other nitrosamines that induce only tumors of the esophagus in rats). On the other hand, this is not an adequate explanation of the failure of rat liver S-9 to activate nitrosomethylethylamine to a bacterial mutagen, since this nitrosamine is a liver carcinogen of quite high potency in rats,[27] although it is, for unknown reasons, considerably less potent than either nitrosodimethylamine or nitrosodiethylamine, of which nitrosomethylethylamine can be considered a hybrid. In contrast with the results in *Salmonella*, nitrosomethylethylamine gave positive results in both the phage induction assay and in V79 cells.

In Table XI are listed those carcinogenic nitrosamines that gave negative results in the *Salmonella* mutagenesis assay with rat liver S-9 activation. The organ in which these nitrosamines induce tumors after chronic administration to rats is also given. Since the esophagus is the most common target of carcinogenic nitrosamines in the rat, it is not surprising that a large number of the nonmutagenic nitrosamines are esophageal carcinogens in the rat. A number of them have the bladder, the nasal cavity, or the nervous system as their targets in rats. The most interesting group is that, in addition to nitrosomethylethylamine, which induces liver tumors (usually hepatocellular carcinomas) in rats; it is comprised of seven compounds. Of these nitrosodiethanolamine is the most important, since, as a component of synthetic cutting oils and a trace contaminant of cosmetics (see Chapter 1), it is the nitrosamine to which human exposure is the greatest. The failure of compounds that induce liver tumors in rats to be activated to bacterial mutagens by rat liver S-9 is not unique to nitrosamines. Several commonly used drugs, including methapyrilene[39] and clofibrate, show the same behavior. These findings raise a perplexing question: Why is a substance that is activated by rat liver to products that transform liver cells to the

Table XI
Carcinogenic Nitrosamines Not Mutagenic to *Salmonella*
When Activated with Rat Liver S-9

Nitrosamine	Target organ in rats
Nitroso-	
Di-isopropylamine	Liver, nasal cavity
Diphenylamine	Bladder
Diethanolamine	Liver
Bis-(2-methoxyethyl)amine	Liver, esophagus
Bis-(2-hydroxypropyl)amine	Esophagus, nasal cavity
Bis-(oxopropyl)amine	Liver, colon
Methylethylamine	Liver, esophagus
Methyloxopropylamine	Esophagus, liver
Methyl-3-carboxypropylamine	Bladder
Methylneopentylamine	Esophagus
Methyl-n-tetradecylamine	Bladder
Methylcyclohexylamine	Esophagus
Methylaniline	Esophagus
Methyl-4-fluoroaniline	Esophagus
Methylbenzylamine	Esophagus
Phenylbenzylamine	Esophagus
Ethanolisopropanolamine	Liver, esophagus
Dihydroxypropylallylamine	Esophagus
Dihydroxypropylethanolamine	Liver
3-Pyrroline	Liver
Thiomorpholine	Esophagus
4-t-Butylpiperidine	Esophagus
3,5-Dimethylpiperazine	Thymus, nasal cavity
3,4,5-Trimethylpiperazine	Thymus, nasal cavity
4-Benzoyl-3,5-dimethylpiperazine	Forestomach
Methyldiethylurea	Nervous system
Ethyldimethylurea	Nervous system

neoplastic state not activated by the same liver enzymes to a bacterial mutagen? The findings raise the possibility that assays for mutagenicity are not necessarily good tests for carcinogenicity and suggest that, at least in some cases, carcinogenesis might not occur through a mutagenic (i.e., DNA-damaging) process.

Nitrosodiethanolamine has not been demonstrated to be mutagenic to *Salmonella* using hamster liver S-9, and it has, so far, been inactive in every short-term test to which it has been subjected, including phage induction. The mechanism of carcinogenic action of nitrosodiethanolamine is even more obscure than those of other nitrosamines. The closely analogous nitrosoethanolisopropanolamine, which is also a liver carcinogen in rats, induced some liver tumors as well as tumors of the lung and pancreas in hamsters, yet was nonmutagenic to *Salmonella* whether activated by rat or hamster liver S-9;

neither was it active in the phage induction assay. These negative results pose the same dilemma as does nitrosodiethanolamine.

Some of the rat liver carcinogens that are not activated to mutagens by rat liver S-9 are activated to mutagens very effectively by hamster liver S-9. These include nitrosobis(2-oxopropyl)amine, nitrosomethyl-2-oxopropylamine, and nitroso-3-pyrroline, of which the first two induce tumors of the liver (and pancreas) in hamsters[40,41]; the third compound has not been tested in hamsters. Nitrosobis(2-hydroxypropyl)amine, also not mutagenic when activated by rat liver S-9 and related closely to the bis-oxopropyl compound, is mutagenic when activated by hamster liver S-9 and also induces tumors of the liver and pancreas in hamsters. It is notable that these four compounds nonmutagenic to *Salmonella* with rat liver activation are also inactive in the phage induction assay with hamster liver S-9 activation. A majority of nitrosamines that have been examined are more mutagenic when activated by hamster liver S-9 than by rat liver S-9, to a large extent irrespective of the relative carcinogenic potencies of the compounds in rats and hamsters. This will be discussed later and suggests that the microsomal enzymes of hamster liver are more active in oxidizing nitrosamines to mutagenic agents than are those of rat liver. It could be that there is (or are) enzyme systems in hamsters that are unique to this species—although there is no good evidence for this—that are by coincidence good activators of nitrosamines. The liver is, of course, the main source of enzymes that detoxify foreign chemicals, and it would be unlikely that other organs are more active than the liver in converting nitrosamines to mutagens. There have, however, been some suggestions that this is so in a few cases.

Among cyclic nitrosamines, of which a very large number have been examined for both carcinogenic and mutagenic activity, the concordance between the two activities is much closer than with the acyclic nitrosamines. Only six cyclic nitrosamines were carcinogenic to rats and not mutagenic with rat liver S-9. Nitroso-3-pyrroline has already been mentioned as a liver carcinogen. Three of the remaining five compounds are mononitrosodimethylpiperazines, of which the 4-benzoyl derivative is very weakly carcinogenic. On the other hand, the latter is the only nitrosated piperazine tested that was active in the phage induction assay, although several of the noninducing compounds, especially the dinitrosopiperazine derivatives, were very potent carcinogens in rats. All of the cyclic nitrosamines nonmutagenic with rat liver S-9 activation were mutagenic when hamster liver S-9 was used, although the one compound that has been tested in hamsters as well as rats, nitroso-3,4,5-trimethylpiperazine, was a less effective carcinogen in the hamster than in the rat.

Few nitrosamines were mutagenic but not carcinogenic, and, in some cases, it might be that inadequate doses were given to rats to detect a possible weak carcinogenicity. It is always possible in a chronic toxicity test to administer a dose too small or to use too small a group of animals for significant numbers of

Table XII
Mutagens Not Carcinogenic

Nitroso-
  Diallylamine
  Di-*n*-octylamine
  Guvacoline
  Methoxymethylamine
  Phenmetrazine
  4-Cyclohexyl-piperidine
  Methylphenidate
  Methyl-4-nitroaniline

animals to die with tumors induced by the treatment. Therefore, it would be premature to draw conclusions or inferences from the data in Table XII, which lists those nitrosamines mutagenic with either rat or hamster liver S-9 activation but not shown to be carcinogenic in the tests so far peformed. Nitrosodiallylamine is the only one of this group that is a potent mutagen but is not carcinogenic to rats; it is very positive in the phage induction assay, also.

One of the more intriguing results of the cooperative studies described in this volume is the general observation that hamster liver S-9 is more effective in activating nitrosamines to bacterial mutagens than rat liver S-9. This has been discussed in previous publications and it is probably not restricted to nitrosamines.[25] The advantage of hamster liver enzymes is so pervasive that a number of nitrosamines not mutagenic with rat liver S-9 activation are quite active with hamster liver S-9, and they include both carcinogenic and noncarcinogenic compounds. The power of the *Salmonella* mutagenesis assay to detect carcinogenic nitrosamines is, therefore, very high when both rat and hamster liver S-9 are employed for activation.

In Table XIII are listed those nitrosamines that are not mutagenic to *Salmonella* with either rat or hamster liver S-9 activation (nitrosomethylethylamine is positive with Guttenplan's modification[12]). They include two compounds, nitrosodiphenylamine and nitrosophenylbenzylamine, that induce tumors (of the bladder and esophagus, respectively) in rats only at very high doses and that would be expected to be noncarcinogenic for chemical structural reasons. The aromatic compound nitrosomethylaniline and its fluoro derivative are esophageal carcinogens of reasonable potency in rats. Structural considerations would also lead one to the conclusion that these compounds are unlikely to be carcinogenic. On the other hand, there is no sound reason for the apparent failure of nitrosomethylethylamine and nitrosomethylneopentylamine to be activated to mutagens for *Salmonella* by rat or hamster liver S-9. Both compounds are readily metabolized by rat liver microsomes, as demonstrated by the

Table XIII
Carcinogenic Nitrosamines Not
Activated to Mutagens for *Salmonella*
by Either Rat or Hamster Liver S-9

Nitroso-
  Diphenylamine
  Diethanolamine
  Methylethylamine
  Methyl-3-carboxypropylamine
  Methylneopentylamine
  Methylaniline
  Methyl-4-fluoroaniline
  Phenylbenzylamine
  Ethanolisopropanolamine
  Thiomorpholine
  4-Benzoyl-3,5-dimethylpiperazine

identification of both possible aldehydes from each nitrosamine in the reaction mixtures.[42] Since oxidation of the more complex alkyl group in each case would release a methylating agent (as would oxidation of nitrosodimethylamine and all other nitrosomethylalkylamines), it is not clear why these compounds do not form a mutagenic product—unless, of course, formation of a methylating moiety is not the mechanism of mutagenesis by these compounds. It is not known whether such a methylating product is surely involved in carcinogenesis. The nitrosamino acid nitrosomethylcarboxypropylamine is a relatively weak bladder carcinogen.

Two of the remaining three compounds in Table XIII are both hydroxylated dialkylnitrosamines and they are carcinogens of only moderate potency. Nevertheless, both induce liver tumors on chronic administration to both rats and hamsters. They are both of very low toxicity. Nitrosodiethanolamine has been extensively studied and appears not to be metabolized by liver microsomes (J. G. Farrelly, unpublished observations). However, since it is highly unlikely to be a directly acting carcinogen, it is probable that it is activated by soluble (non-membrane-bound) enzymes in liver cells and that the product or products are the proximate carcinogens. The number of possibilities is limited and it would seem likely that one product would be an activated 2-hydroxyethyl species (the same product could be formed from nitrosoethanolisopropanolamine), which could be the proximate carcinogen. However, this is somewhat dissatisfying as an hypothesis for the carcinogenic action of nitrosodiethanolamine, since the same activated moiety would be expected to be formed from the directly acting nitroso-2-hydroxyethylurea, which has a totally different spectrum of carcinogenic activity and does not include the induction of liver tumors in rats.[32] The possibility must also be considered that microsomal activation is not necessarily important in carcinogenesis by all nitrosamines; instead, metabolism and activation by soluble enzymes might play an important role not previously considered.

4.2. Quantitative Relationships

Although the qualitative relationship between mutagenicity and carcinogenicity is quite good, there is no basis for assuming that any quantitative relationship between the two activities has been demonstrated, especially in the *Salmonella* assays. Few nitrosamines have been examined in the V79 assay and, while the concordance between mutagenicity and carcinogenicity is reasonably quantitative, too few of the compounds tested lie outside a small pocket of weakly mutagenic and weakly carcinogenic compounds. There are also notable exceptions that lie far from the straight line. Similarly, the correlation between phage induction and carcinogenic potency is quite good quantitatively for many nitrosamines and nitrosamides, but there are some outstanding exceptions, including whole groups of strong carcinogens that are undetected in this system, which make generalizations impossible and predictions unwise.

The lack of a good quantitative relationship between mutagenicity and carcinogenicity among the directly acting nitrosoalkylamides has been alluded to above. In *Salmonella*, nitrosomethylurea is a much more potent mutagen than nitrosoethylurea, although their carcinogenic potencies are not much different and the mutagenicity of the latter is similar to that of the very much weaker carcinogens nitroso-*n*-propylurea and nitrosoallylurea. The very potent mutagen nitroso-2-phenylethylurea was not detectably carcinogenic while the equally potent mutagen nitroso-2-fluorethylurea was a very potent carcinogen. Substitution of a hydroxyl group in the 2- position of nitrosoethylurea or nitroso-*n*-propylurea increased the mutagenic activity by one or two orders of magnitude but did not increase the carcinogenic potency of the ethyl compounds (which remained high), although it did increase greatly the carcinogenicity of the *n*-propyl compounds; the increase in carcinogenicity of the nitroso-*n*-propylurea substituted with hydroxyl in the 3-position was also great, although there was no increase in mutagenic potency.[32] There is a large disparity between the relative mutagenicities and carcinogenicities of nitrosomethylnitroguanidine and nitrosoethylnitroguanidine, the former being a more potent mutagen, while the ethyl compound is a much more potent carcinogen than the methyl compound; the structurally similar nitrosocimetidine, while strongly mutagenic, appears not to be carcinogenic. These results suggest that, even for the directly acting N-nitroso compounds, the mechanisms of mutagenesis and carcinogenesis are not the same, and that, if alkylation of DNA is involved in either or both of these biological actions, this alkylation has different outcomes in *Salmonella* and in mammalian epithelial cells.

The position is considerably clearer with the large group of cyclic nitrosamines that have been tested for both mutagenic and carcinogenic activity. If the exceptional nonmutagenic carcinogens in Table XIII are excluded, and we consider those nitrosamines that are mutagenic with hamster liver S-9 activation as well as with rat liver S-9, the parallel between mutagenicity and carcinogenicity is quite good. This suggests that the activated products of metabolism by the microsomes might be those responsible for both mutagenic and carcinogenic activity. Thus, nitrosooxazolidine, its 2- and 5-methyl derivatives, nitrosotetrahydrooxazine, nitrosomorpholine, and nitroso-2-methylmorpholine are all liver carcinogens in rats of approximately equal potency, and they are all of similar mutagenic potency. An exception in this series is nitroso-2,6-dimethylmorpholine, which is a much more potent carcinogen than nitrosomorpholine[43] but is less mutagenic. Among the dinitrosopiperazines, all of the methylated derivatives and dinitrosohomopiperazine are more carcinogenic and more strongly mutagenic than dinitrosopiperazine, although some are several times more mutagenic than their equally carcinogenic homologs. Among the nitrosopiperidines, the nitrosopyrrolidines, and the homologous unsubstituted cyclic nitrosamines, there is a good correspondence between carcinogenic and mutagenic potency, the more carcinogenic compounds being also the more mutagenic ones. An exception is the much greater mutagenicity of nitrosopyrrolidine than nitrosopyrroline, although the two compounds are equally carcinogenic to rat liver.[44]

Among the acyclic nitrosamines there are many discrepancies between carcinogenic and mutagenic potency, related perhaps to the very different target organs of these nitrosamines in the rat. Nitrosodiethylamine is little more mutagenic than nitrosodi-*n*-propylamine, although the former is a much more potent carcinogen. Similarly, nitrosodi-*n*-propylamine is about equally mutagenic with nitrosodi-*n*-butylamine and nitrosodi-iso-butylamine, although the latter two are less carcinogenic by one and two orders of magnitude, respectively.

The series of nitrosomethyl-*n*-alkylamines shows no correlation between mutagenic and carcinogenic potency, although it is believed in some quarters that they act through formation of a methylating agent that methylates DNA of the bacteria and of the mammalian target cells. The smallest member of the series, nitrosomethylethylamine, does not appear to be mutagenic to *Salmonella* in the standard assay. The next higher homologs containing 3, 4, 5, and 6 carbon atoms in the chain are all very potent carcinogens, but are not very potent mutagens, while the higher homologs are increasingly weaker carcinogens, but much more potent mutagens. Some of the members of this series are liver carcinogens in rats, but that seems to have little bearing on the rat-liver-microsome-mediated mutagenicity of the compounds, which shows a characteristic pattern regardless of the target organ in the rat, which can be lung, liver, esophagus, or bladder. The same can be said of the substituted nitrosomethylethylamines and nitrosomethyl-*n*-propylamines, which are often very potent carcinogens but relatively weak mutagens. An exception is nitrosomethyl-2-phenylethylamine, which is both a very potent mutagen and a potent carcinogen, although it induces only tumors of the esophagus in rats, not liver tumors.

Among the series of hydroxylated *n*-propylnitrosamines and the associated oxo compounds, only nitrosoisopropanolethanolamine was nonmutagenic. The remainder were weakly mutagenic with rat liver S-9 activation, excepting those containing an allyl group, which were strong mutagens. However, all were potent mutagens with hamster liver S-9 for activation. They ranged in carcinogenic potency from very weak to highly potent, and the organs of the rat in which tumors were induced included liver, lung, esophagus, and nasal cavity, but these bore no quantitative relation to their mutagenic potencies. Although almost all of these compounds were more mutagenic when activated by hamster liver S-9 than by rat liver S-9, all were not more potent carcinogens in the hamster than in the rat, although many were.

The conclusion of these comparative studies is that short-term tests, especially bacterial mutagenesis, are a good qualitative index of carcinogenic activity of N-nitroso compounds but are poorly correlated quantitatively and are not informative about mechanisms of carcinogenesis by this group of compounds.

5. PRECAUTIONS IN INTERPRETATION AND EXTRAPOLATION

If the large body of information gathered within these chapters can be considered examples of in-depth analyses of structurally related compounds, a number of precautions can now be expressed. First, if the question is simply "Is a compound mutagenic?" then no single assay can be completely relied upon to provide an answer. If the question posed becomes "Is the compound mutagenic in higher organisms, given a response or lack of response in prokaryotic systems?" the answer is that one has to carry out the experiment with the specific system. Predictions and extrapolations from one organism to others are subject to error because of deficiencies inherent in the "mutagenic determinations."

The results summarized throughout this chapter, especially those dealing with the microbial responses, point to the difficulty in attempting to force a standardized protocol upon even a group of structurally related molecules. Minor changes in the approach or modifications of the "standard" procedure can lead to false positive (or negative) responses. Thus, each compound in each system must be considered in its own right and "prescreening" and "screening" become steps in an evolution of thought, not the end point in an evaluation.

In the extrapolation from one organism to another, consideration must be given not only to the commonality of the chemical structure of the material, but to the structure and function of the target; to the metabolism of the chemical toxicant in question; to the genetic constitution of the test system and its control mechanisms; and to the effect, or perhaps repair, of the damage, if and when perpetrated upon the cell. Each biological end point might be a measure of that end point alone, and attempts to generalize and/or extrapolate are subject to error because of gaps in our knowledge of the metabolism and fate of the damaging agent and the mechanism of action upon the target molecule. Thus, the "correlation-concordance" between the end point of mutagenesis in one organism versus the end point of carcinogenesis in another organism becomes simply another example.

However, the evidence that a chemical can damage the genetic material even in the simplest organism should be sufficient to call for further study. The qualitative relationship between the two end points of mutagenesis and carcinogenesis is, obviously, not infallible. There are many false negatives in the *Salmonella* assay, particularly using rat liver activation. For N-nitroso compounds the predictive value of mutagenic effects appears to be more a function of the class of compound or type of substitution as modified by the increased sensitivity of the microbial tests for that specific class (i.e., the "false positives" noted in Table XII). Given the knowledge that a response, either positive or negative, can sometimes vary with slight modifications of assay conditions, and that the extent of any response is influenced by a large number of experimental parameters including the genetic constitution of the test system, quantitative extrapolation from one end point to another seems an improbable expectation.

ACKNOWLEDGMENT. Research sponsored by the Office of Health and Environmental Research, U.S. Department of Energy, under contract W-7405-eng-26 with the Union Carbide Corporation.

REFERENCES

1. C. E. Nix, B. Brewen, R. Wilkerson, W. Lijinsky, and J. L. Epler, Effects of N-nitrosopiperidine substitutions in mutagenicity in *Drosphila melanogaster, Mutat. Res. 67*, 27–38 (1979).
2. C. E. Nix, B. Brewen, W. Lijinsky, and J. L. Epler, Effects of methylation and ring size on mutagenicity of cyclic nitrosamines in *Drosophila melanogaster, Mutat. Res. 73*, 93–100 (1980).
3. J. McCann, E. Choi, E. Yamasaki, and B. N. Ames, Detection of carcinogens as mutagens in the *Salmonella*/microsome test: Assay of 300 chemicals, *Proc. Natl. Acad. Sci. USA 72*, 5135–5139 (1975).
4. J. Ashby and J. A. Styles, Does carcinogenic potency correlate with mutagenic potency in the Ames assay? *Nature (London) 271*, 452–455 (1978).
5. B. N. Ames, J. McCann, and E. Yamasaki, Methods for detecting carcinogens and mutagens with the *Salmonella*/mammalian-microsome mutagenicity test, *Mutat. Res. 31*, 347–364 (1975).

6. J. L. Epler, The use of short-term tests in the isolation and identification of chemical mutagens in complex mixtures, in: *Chemical Mutagens: Principles and Methods for Their Detection* (F. J. deSerres and A. Hollaender, eds.), pp. 239–270, Plenum Press, New York (1980).
7. J. L. Epler, W. Winton, A. A. Hardigree, and F. W. Larimer, The appropriate use of genetic toxicology in industry, in: *The Scientific Basis of Toxicity Assessment* (H. Witschi, ed.), pp. 45–59, Elsevier/North-Holland Biomedical Press, Amsterdam (1980).
8. F. J. deSerres, The utility of short-term tests for mutagenicity, *Mutat. Res.* 38, 1–2 (1976).
9. W. Lijinsky, Structure-activity relations in carcinogenesis by N-nitroso compounds, in: *Topics in Chemical Mutagenesis* (F. J. deSerres, series ed.), Vol. 1, *Genotoxicology of N-Nitroso Compounds* (T. K. Rao, W. Lijinsky, and J. L. Epler, eds.), Plenum Press, New York, pp. 189–232 (1983).
10. T. K. Rao, Structural basis for the mutagenic activity of N-nitrosamines in the *Salmonella* histidine reversion assay, in: *Topics in Chemical Mutagenesis* (F. J. deSerres, series ed.), Vol. 1, *Genotoxicology of N-Nitroso Compounds* (T. K. Rao, W. Lijinsky, and J. L. Epler, eds.), Plenum Press, New York, pp. 45–58 (1983).
11. A. W. Andrews and W. Lijinsky, N-Nitrosamine mutagenicity using the *Salmonella* mammalian-microsome mutagenicity assay, in: *Topics in Chemical Mutagenesis* (F. J. deSerres, series ed.), Vol. 1, *Genotoxicology of N-Nitroso Compounds* (T. K. Rao, W. Lijinsky, and J. L. Epler, eds.), Plenum Press, New York, pp. 13–44 (1983).
12. J. B. Guttenplan, Effects of pH and structure on the mutagenic activity of N-nitroso compounds, in: *Topics in Chemical Mutagenesis* (F. J. deSerres, series ed.), Vol. 1, *Genotoxicology of N-Nitroso Compounds* (T. K. Rao, W. Lijinsky, and J. L. Epler, eds.), Plenum Press, New York, pp. 59–90 (1983).
13. E. Zeiger and A. T. Sheldon, The mutagenicity of N-nitrosopiperidines for *Salmonella typhimurium*, *Mutat. Res.* 57, 85–89 (1975).
14. E. Zeiger and A. T. Sheldon, The mutagenicity of heterocyclic N-nitrosamines for *Salmonella typhimurium*, *Mutat. Res.* 57, 1–10 (1978).
15. R. K. Elespuru, Induction of bacteriophage lambda by N-nitroso compounds, in: *Topics in Chemical Mutagenesis* (F. J. deSerres, series ed.), Vol. 1, *Genotoxicology of N-Nitroso Compounds* (T. K. Rao, W. Lijinsky, and J. L. Epler, eds.), Plenum Press, New York, pp. 91–118 (1983).
16. R. K. Elespuru and W. Lijinsky, Mutagenicity of cyclic nitrosamines in *Escherichia coli* following activation with rat liver microsomes, *Cancer Res.* 36, 4099–4101 (1976).
17. T. K. Rao, B. E. Allen, W. Winton, W. Lijinsky, and J. L. Epler, Nitrosamine-induced mutagenesis in *Escherichia coli* K-12 (343/113). 1. Mutagenic properties of certain aliphatic nitrosamines, *Mutat. Res.* 89, 209–215 (1981).
18. T. K. Rao, J. T. Cox, B. E. Allen, J. L. Epler, and W. Lijinsky, Mutagenicity of N-nitrosopyrrolidine derivatives in *Salmonella* (Ames) *Eschcerichia coli* K-12 (343/113) assays, *Mutat. Res.* 89, 35–43 (1981).
19. F. W. Larimer, D. W. Ramey, W. Lijinsky, and J. L. Epler, Mutagenicity of methylated N-nitrosopiperidines in *Saccharomyces cerevisiae*, *Mutat. Res.* 57, 155–161 (1978).
20. F. W. Larimer, A. A. Hardigree, W. Lijinsky, and J. L. Epler, Mutagenicity of N-nitrosopiperazine derivatives in *Saccharomyces cerevisiae*, *Mutat. Res.* 77, 143–148 (1980).
21. C. A. Jones and E. Huberman, The relationship between the carcinogenicity and mutagenicity of nitrosamines in a hepatocyte-mediated mutagenicity assay, in: *Topics in Chemical Mutagenesis* (F. J. deSerres, series ed.), Vol. 1, *Genotoxicology of N-Nitroso Compounds* (T. K. Rao, W. Lijinsky, and J. L. Epler, eds.), Plenum Press, New York, pp. 119–128 (1983).
22. T. Ho, J. R. San Sebastian, and A. W. Hsie, Mutagenic activity of nitrosamines in mammalian cells. Study with the CHO/HGPRT and human leukocyte SCE assays, in: *Topics in Chemical Mutagenesis* (F. J. deSerres, series ed.), Vol. 1, *Genotoxicology of N-Nitroso Compounds* (T. K. Rao, W. Lijinsky, and J. L. Epler, eds.), Plenum Press, New York, pp. 129–148 (1983).

23. J. Althoff, C. Grandjean, and B. Gold, Diallylnitrosamine: A potent respiratory carcinogen in Syrian golden hamsters, *J. Natl. Cancer Inst. 59*, 1569–1571 (1977).
24. J. G. Farrelly and L. I. Hecker, The relationship between metabolism and mutagenicity of two cyclic nitrosamines, in: *Topics in Chemical Mutagenesis* (F. J. deSerres, series ed.), Vol. 1, *Genotoxicology of N-Nitroso Compounds* (T. K. Rao, W. Lijinsky, and J. L. Epler, eds.), Plenum Press, New York, pp. 169–188 (1983).
25. H. Bartsch, C. Malaveille, and R. Montesano, The predictive value of tissue-mediated mutagenicity assays to assess the carcinogenic risk of chemicals, in: *Screening Tests in Chemical Carcinogenesis*, IARC Scientific Publications No. 12, pp. 467–486, International Agency for Research on Cancer, Lyon, France (1976).
26. W. Lijinsky, M. D. Reuber, and H. M. Reznik-Schüller, Contrasting carcinogenic effects of nitroso-2,6-dimethylmorpholine given by gavage to F344 rats and Syrian golden hamsters, *Cancer Lett. 16*, 281–286 (1982).
27. W. Lijinsky and M. D. Reuber, Comparative carcinogenicity of some aliphatic nitrosamines in Fischer rats, *Cancer Lett. 14*, 297–302 (1981).
28. W. Lijinsky, Comparison of the carcinogenic effectiveness in mouse skin of methyl- and ethyl-nitrosourea, nitrosourethane and nitrosonitroguanidine and the effect of deuterium labeling, *Carcinogenesis 3*, 1289–1291 (1982).
29. W. Lijinsky and D. Schmähl, Carcinogenesis by nitroso derivatives of methylcarbamate insecticides and other nitrosamides in rats and mice, in: *Environmental Aspects of N-Nitroso Compounds*, IARC Publications No. 19, pp. 495–501, International Agency for Research on Cancer, Lyon, France (1978).
30. W. Lijinsky and H. W. Taylor, Carcinogenesis in Sprague–Dawley rats of N-nitroso-N-alkylcarbamate esters, *Cancer Lett. 1*, 275–279 (1976).
31. W. Lijinsky and D. Schmähl, Carcinogenicity of N-nitroso derivatives of N-methylcarbamate insecticides in rats, *Ecotoxicol. Environ. Safety 2*, 413–419 (1978).
32. W. Lijinsky and M. D. Reuber, Carcinogenicity of hydroxylated alkylnitrosoureas and of nitrosooxazolidones by mouse skin painting and by gavage in rats, *Cancer Res. 43*, 214–221 (1983).
33. W. Lijinsky and C. Winter, Skin tumors induced by painting nitrosoalkylureas on mouse skin, *Cancer Res. Clin. Oncol. 102*, 13–20 (1981).
34. R. Preussmann, H. Druckrey, and J. Bucheler, Carcinogene Wirkung von Phenyl-Nitrosoharnstoff, *Z. Krebsforsch. 71*, 63–74 (1968).
35. S. Ivankovic, Carcinogene Wirkung von N-Benzyl-N-Nitrosoharnstoff (BzNH) an BD-Ratten, *Z. Krebsforsch, 91*, 63–67 (1978).
36. W. Lijinsky, Dose–response studies with nitrosamines and species differences, in: *Banbury Report 12: Nitrosamines and Human Cancer*, pp. 257–269, Cold Spring Harbor Laboratory, Cold Spring Harbor, New York (1982).
37. W. Lijinsky, M. D. Reuber, and C. W. Riggs, Dose–response studies in rats with nitrosodiethylamine, *Cancer Res, 41*, 4997–5003 (1981).
38. M. J. Prival and V. D. Mitchell, Dimethylnitrosamine demethylase and the mutagenicity of dimethylnitrosamine: Effects of rodent liver fractions and dimethylsulfoxide, in: *Topics in Chemical Mutagenesis* (F. J. deSerres, series ed.), Vol. 1, *Genotoxicology of N-Nitroso Compounds* (T. K. Rao, W. Lijinsky, and J. L. Epler, eds.), Plenum Press, New York, pp. 149–166 (1983).
39. A. W. Andrews, J. A. Fornwald, and W. Lijinsky, Nitrosation and mutagenicity of some amine drugs, *Toxicol. Appl. Pharm. 52*, 237–244 (1980).
40. P. Pour, L. Wallcave, R. Gingell, D. Nagel, T. Lawson, S. Salmasi, and S. Tines, Carcinogenic effect of N-nitroso(2-hydroxypropyl)-(2-oxopropyl)amine, a postulated proximate pancreatic carcinogen in Syrian hamsters, *Cancer Res. 39*, 3828–3833 (1979).

41. P. Pour, R. Gingell, R. Langenbach, D. Nagel, C. Grandjean, T. Lawson, and S. Salmasi, Carcinogenicity of N-nitrosomethyl(2-oxopropyl)-amine in Syrian hamsters, *Cancer Res. 40*, 3585-3590 (1980).
42. J. G. Farrelly and M. L. Stewart, The metabolism of a series of methylalkylnitrosamines, *Carcinogenesis, 3*, 1299-1302 (1982).
43. W. Lijinsky and M. D. Reuber, Comparative carcinogenicity by nitrosomorpholines, nitrosooxazolidines and nitrosotetrahydrooxazine in rats, *Carcinogenesis, 3*, 911-915 (1982).
44. H. Garcia and W. Lijinsky, Tumorigenicity of five cyclic nitrosamines in MRC rats, *Z. Krebsforsch. 77*, 257-261 (1972).

Index

Accelerators of nitrosation, 4
Activated N-nitrosamines, 73
Agricultural chemicals, 3, 6
Aldehydes, 4
Aliphatic nitrosamines, 203
Alkanolamines, 4
Alkanesulfonates, 131
Alkonium ion, 85, 86
Alkylated nucleic acids, 201
Alkylation of DNA, 61, 189
Alkylating agent, 2, 60, 113, 114, 201
Alkylating agent, formation of a reactive, 39
Alkyldiazonium ions, 61, 84
O^6-Alkylguanine, 61, 113, 202
O^6-Alkylguanine, repair activity, 63
7-Alkylguanine, 71, 202
Alkyliodides and bromides, 61
Alkylnitrites, 3
Alkylsulfonates, 61, 131
Alkylureas, 6
Allantoin, 6
α-Acetoxy nitrosamines, 169
α-1-Antitrypsin, 159
Alpha carbon, susceptibility to activation, 219
Alpha-hydroxylation, 167, 172, 184
Alpha-hydroxynitrosamines, 73, 167
Alpha oxidation, 180, 214, 219, 224
  inhibition of, 38
Alpha tocopherol, 4
Amine
  secondary, 1, 5, 6
  tertiary, 3, 5

γ-Aminobutyrate, 173
ε-Aminocaproic acid, 172
Aminopyrine, 6
Ampicillin, 97
Ampicillin resistance, 47
Analgesic, 5
Arginine, 6
Arylating agent, 86
Aryl groups, 249
Arg-locus, 236
Ascorbic acid, 4
Asymmetrical oxygenated nitrosamines, 210

Bacon, nitrosamines in, 112
Bacterial mutagenesis tests, 5
Bacterial reduction, 5
Bacteriophage induction assay, 92
Bacteriophage lambda, 91
Beer, nitrosamines in, 3
Benzidine, 163
Benzopyrene, 65
Beta carbon, in nitrosomethylethylamine, 206
Betel nuts, 223
5-Bromo-2'-deoxyuridine, 135
1, 4-Butanediol, 168, 170, 173, 175, 179

Canavanine resistance, 236
Carbamate insecticides, 100
Carbonyl compounds, 4
Carcinogenic damage, 234

265

Carcinogenic potency, 123
Carcinogenic mechanisms, 183
Carcinogenesis, two stage, 199
Carcinogenesis, and structural symmetry, 55
Carcinogenic risk, 5, 6, 8
Carbonyl, 6
Carbonium ion, 168
  primary, 86
  secondary, 86
Carboxynitrosopiperidines, 137
Cellular macromolecules, 2, 207
Cheese, 5
Chinese hamster cells, (V79), 120, 236
Chinese hamster ovary cells, 130, 236
Chloral, 4
Chlordiazepoxide, 6
Chlorpheniramine, 6
Chlorpromazine, 6
Chromatid breaks, 233
Chromosomal aberations, 233, 236
Chromosomal breaks, 236
Cigarette smoke, 212
Clofibrate, 252
Cosmetic preparations, 4
Croton oil, 199
Cured meat and fish, 5
Cyclic nitrosamines, 16, 38
Cyclophosphamide, 136
Cytochrome-linked mixed function oxidases, 45
Cytosolic fraction, 155, 157, 162

Dealkylation, oxidative, 45
Deuterium substitution, 206, 211
Dialkylnitrosamines, 102, 168
Dialkylsulfates, 61
Diazohydroxide, 168
Diazonium ions, 85, 86, 169
Diethanolamine, 4
Diethylamine, 2
Dimethylamine, 2, 4, 5, 172
Dimethyldodecylamine, 6
Dimethyldodecylamine-N-oxide, 6
Dimethylnitrosamine, binding to protein, 162
Dimethylnitrosamine demethylase I, 62, 65, 67, 149, 152, 155, 157, 160, 161, 162, 163
Dimethylnitrosamine demethylase II, 152
Dimethylsulfoxide, 150, 159, 163
Dinitroso-2, 5-dimethylpiperazine, 19, 38, 217

Dinitroso-2, 6-dimethylpiperazine, 19, 135, 137, 138, 217, 218
Dinitrosohomopiperazine, 19, 133, 217, 258
Dinitroso-2-methylpiperazine, 19, 137, 138, 217
Dinitrosopiperazine, 19, 38, 49, 137, 138, 217, 218, 258
Diphenhydramine, 6
Dipyrone, 6
Disulfiram, 6
DNA
  alkylation of, 47, 250, 257
  binding to protein, 162
  damaging agent, 97
  repair, 233
Dose-response curve, 34
Dried milk, 4
Drosophila melanogaster, 236
Drosophila X-linked recessive lethal system, 130, 236
Drugs, 3, 6

Environmental mixtures and effluents, 234
Epoxide formation, 39, 223
Esophageal tumors, 251
Esophagus, methylation in, 203
Ethers, 208
Ethoxylating agent, 203
Ethylating agents, 114
Ethylation of DNA, 40
Ethyldiazonium ions, 84
O^6-Ethylguanine, 82, 112, 114
Ethylmethanesulfonate, 133
Excision repair deficient mutants, 236
Excision repair system, 47, 48

Fenuron, 6
Filamentation, 91
Flue gases, 3
Fluorine, substitution in nitrosamines, 207
Food preservatives, 129
Formaldehyde, 4, 152

β-Galactosidase, 92, 97
gal locus, 263
Gene mutations, 233, 235
Genetic toxicants, 235
Genotoxic/genotoxicity, 203
Genotoxicology of nitrosamine, 235

Index

Genotoxic properties, 233, 235
Glutathione, 4
Guanine O^6-methyltransferase, 113
Guvacine, 223

Haloethanes, 131
Hamster liver, S9, 34, 38, 96, 163, 254
Hemangioendothelial sarcomas, 251
Hepatocellular carcinomas, 172
Hepatocytes, 120
Hepatotoxins, 251
Heterocyclic nitrosamines, 131, 167
1, 6-Hexanediol, 173
HGPRT, see Hypoxanthine-guanine phosphoribosyl transferase
Human leukocyte culture system, 130, 137
Hydrochlorothiazide, 6
Hydrophilic, 211
4-Hydroxybutanal, 175, 179
γ-Hydroxybutyrate, 168, 170, 172, 173, 175, 179
2-Hydroxyethyl species, 256
α-Hydroxylation, 167, 172, 184
6-Hydroxy-n-hexanal, 180
α-Hydroxynitrosamine, 73, 167
α-Hydroxnitrosopyrrolidine, see Nitroso-2-hydroxypyrrolidine
β-Hydroxynitrosopyrrolidine, see Nitroso-3-hydroxypyrrolidine
2-Hydroxytetrahydrofuran, 168, 169, 171, 172, 173, 175, 179, 184
Hypoxanthine-guanine phosphoribosyl transferase (HGPRT) locus, 130

Inhibition of nitrosation, 4
Initiators, 199
Insecticides, nitroso-derivatives of, 6, 106

Kinetic assay of induction, 97
Kupffer cells, 172

Lambda repressor, 91, 92, 113
Lambda lysogen, 92
Leukemia of the thymus, 218
Leukocytes, human, 236
Lex A mutation, 94
Lipophilic, 211

Lipophobicity, 53
Lipopolysaccharide barrier, 47
Liposolubility, 250
Liquid preincubation assay, 13
Liquid suspension assay, 13
Liver necrosis, 1
Liver parenchymal cells, 251
Lucanthone, 6
Lymphoma of the thymus, 218

Mammalian cell systems, 236
Mammalian enzymes, 234
Mammalian stomach, 2
Meat-curing mixtures, 129
Metabolic activation, 234
Metabolism of N-nitrosamines, 235
Metal ions, 4
Methapyrilene, 6, 252
Methomyl, 249
3-Methyladenine glycosylase, 113
Methylating agent, 206, 256, 258
3-Methylcholanthrene, 65
Methyldiazonium ion, 84
Methylmethanesulfonate, 64
Methylphenidate, 223
7-Methylguanine, 70, 71, 78, 79
O^6-Methylguanine, 70, 71, 78, 79, 80
Methylguanidine, 6
Methyl groups, oxidizable, 206
Methyl substitution, 38
Microsomal fraction, 155, 157, 162
Microsomes
  metabolism by, 161
  supernatant, 173, 178, 184
Milk, 129
Mitosis, 225
Molecular configuration and conformation, 212
Monomethylnitrosamine, 61
Monuron, 6
Morpholine, 4, 6, 214
  analogs, 109
  derivatives, 103
Mouse liver, S9, 150
Mouse skin, 189
Mouse skin assay, 249
Multiplex CHO system, 131
Mutagenesis
  and aliphatic nitrosamines, 54, 73
  and chemical ring size, 49

Mutagenesis, (cont'd)
  and chemical ring substitution, 51
  DNA repair in, 48
  and halogen substitution, 52
  nitrosamines as model compounds in, 56
  NMU-induced, 81
  and structural symmetry, 55
Mutagenic intermediates, 120
Mutagenicity, chemical structure and, 34, 38, 48
Mutagenic potency, 74, 122
  and chemical structure, 13
Mutants, canavanine-resistant, 46
Mutations, reversion and forward, 236

Nad locus, 236
Neurospora conidia, 46
Nitrilotriacetic acid, 6
Nitrite, 4, 5
Nitrite, fish and meat cured with, 5
Nitrogen oxides, 3
Nitrosamides, 28, 60
Nitrosamines
  activated, 184
  acyclic, 24, 40, 239
  in air, 4
  aromatic, 24
  asymmetric, oxygenated, 210
  cyclic, 48, 49, 102, 168, 172, 180, 240, 245
  cyclic alkylated, 52
  dialkyl, 48, 168
  mutagenic activity of, 184
  nonmutagenic, 253
  α-substituted, 86
  β-substituted carcinogenic, 85
N-Nitrosamines, 63, 64, 120, 131, 172, 239
Nitrosating agents, 2
N-Nitroso-
  acetoxydibenzylamine, 24, 39, 203
  acetoxydimethylamine, 203
  4-acetyl-3, 5-dimethylpiperazine, 19, 38, 219
  aldicarb, 104
  allylethanolamine, 28, 211
  allylhydroxypropylamine, 28, 211
  allylurea, 107, 201, 257
  n-amylurea, 201
  anabasine, 220
  azetidine, 16, 38, 49, 103, 212
  baygon, 106
  4-benzoyl-3, 5-dimethylpiperazine, 19, 38, 219, 254

N-Nitroso-, (cont'd)
  benzylurea, 41, 201, 250
  bis-(2-chloroethyl)amine, 54, 208, 209
  bis-(2-chloropropyl)amine, 54, 209
  bis-(2,2-diethoxyethyl)amine, 209
  bis-(2-ethoxyethyl)amine, 208
  bis-(2-cyanoethyl)amine, 85, 208
  bis-(2-cyanomethyl)amine, 208, 209
  bis-(2-hydroxypropyl)amine, 4, 24, 40, 54, 163, 207, 209, 210, 254
  bis-(2-methoxyethyl)amine, 54, 208
  bis-(2-oxopropyl)amine, 24, 40, 210, 254
  bis-(2-trifluoroethyl)amine, 209
  4-*t*-butylpiperidine, 19, 39, 83, 86, 219, 221
  n-butylurea, 201
  buxten, 104
  carbaryl, 6, 104, 199
  carbofuran, 104
  3-chloropiperidine, 52, 224
  4-chloropiperidine, 52, 137, 224
  cimetidine, 28, 249, 257
  4-cyclohexylpiperidine, 19, 39, 105, 109, 222
  cyclohexylurea, 201
  diallylamine, 24, 40, 54, 121, 209, 211, 239, 254
  dibenzylamine, 39, 207
  3, 4-dibromopiperidine, 52, 105, 224
  3, 4-dibromopyrrolidine, 52, 221
  di-sec-butylamine, 86
  di-*n*-butylamine, 24, 40, 102, 163, 205, 258
  3, 4-dichloropiperidine, 52, 105, 137, 224
  3, 4-dichloropyrrolidine, 221
  diethanolamine, 4, 24, 40, 41, 54, 86, 203, 207, 208, 209, 210, 252, 253, 256
  diethylamine, 2, 24, 40, 63, 81, 85, 94, 101, 108, 121, 133, 163, 201, 203, 204, 205, 207, 251, 258
  2, 3-dihydroxypropylallylamine, 28, 41
  dihydroxypropylethanolamine, 28, 41, 211
  dihydroxypropyl-2-oxopropylamine, 28, 211
  di-isobutylamine, 24, 40, 258
  di-isopropanolamine, *see* bis(2-hydroxypropyl)amine
  di-isopropylamine, 4, 86, 203
  dimethylamine, 1, 4, 24, 40, 45, 46, 61, 63, 74, 85, 94, 101, 108, 121, 132, 133, 149, 152, 155, 161, 189, 201, 204, 205, 251, 256
  2, 6-dimethylmorpholine, 19, 215, 218, 248, 258
  2, 6-dimethylmorpholine-*cis* and *trans*, 103, 109, 215

Index

N-Nitroso-, *(cont'd)*
2, 4-dimethyl-1, 3-oxazolidine, 16
4, 4-dimethyl-1, 3-oxazolidine, 16
3, 5-dimethylpiperazine, 19, 218, 219
2, 6-dimethylpiperidine, 52, 86, 137, 221
3, 5-dimethypiperidine, 19, 39, 51
3, 5-dimethylpiperidine-*cis* and *trans*, 19
2, 5-dimethylpyrrolidine, 133, 219
di-*n*-octylamine, 24, 40, 54, 205
diphenylamine, 3, 24, 39, 203, 255
di-*n*-propylamine, 24, 40, 86, 201, 208, 209, 258
dodecamethyleneimine, 19, 38, 214
3, 4-epoxypiperidine, 19, 38, 223
ethanoloxopropylamine, 28, 210
ethanolhydroxypropylamine, 28, 41, 86, 210, 253, 256, 259
ethyldimethylurea, 34, 42
ethylethoxyamine, 203
2-ethyl-5-methyl-1, 3-oxazolidine, 16
ethylnitroguanidine, 28, 41, 68, 75, 77, 85, 199, 249, 257
2-ethyl-4-methyl-1, 3-tetrahydrooxazine, 19
2-ethyl-1, 3-tetrahydrooxazine, 19
ethylurea, 41, 42, 78, 82, 199, 250, 257
ethylurethane, 41, 199, 249
fluroethylurea, 42, 201, 250, 257
guvacine, 53
guvacoline, 19, 39, 53, 105, 137, 223
heptamethyleneimine, 19, 38, 49, 213
hexamethyleneimine, 19, 49, 168, 171, 173, 180, 185, 213
n-hexylurea, 41, 107, 201, 250
α-hydroxydimethylamine, 61
hydroxyethylurea, 42, 201, 250, 256
2-hydroxyhexamethyleneimine, 180
3-hydroxyhexamethyleneimine, 169, 181
4-hydroxyhexamethyleneimine, 169, 181, 183, 185
3-hydroxypiperidine, 137, 180, 224
4-hydroxypiperidine, 105, 180, 224
hydroxyproline, 3, 220
hydroxypropyloxopropylamine, 28
2-hydroxypropylurea, 201, 250
3-hydroxypropylurea, 34, 201, 250
2-hydroxypyrrolidine, 175, 180
3-hydroxypyrrolidine, 168, 172, 180, 220
iminodiacetaldehyde, 209
iminodiacetic acid, 3, 208, 209
iminodiacetonitrile, *see* bis-(2-cyanomethyl)amine

N-Nitroso-, *(cont'd)*
iminodipropionitrile, *see* bis-(2-cyanoethyl)amine
isobutylurea, 41, 250
isonipecotic acid, 53, 223
2-isopropyl-4, 4-dimethyl-1, 3-oxazolidine, 16
isopropylurea, 201
landrin, 104
methomyl, 104
methylacetonitrile, 209
methylallylamine, 209
methylaniline, 24, 39, 83, 86, 202, 206, 255
methylbenzylamine, 24, 39, 206
methyl-*n*-butylamine, 28, 40, 124, 201, 204
methyl-tert-butylamine, 202
methyl-4-carboxybenzylamine, 24, 39
methyl-2-carboxyethylamine, 205
methyl-3-carboxypropylamine, 28, 41, 205, 256
methyl-4-chlorobenzylamine, 24
methyl-4-cyanobenzylamine, 24
methylcyclohexylamine, 24, 39, 202, 207
methl-*n*-decylamine, 28, 205
methyldiethylurea, 34, 42, 201
methyldihydroxypropylamine, 28, 41, 210
methyl-*n*-dodecylamine, 28, 205
methylethylamine, 28, 40, 54, 55, 85, 108, 204, 205, 206, 239, 248, 252, 255, 258
methyl-4-fluoroaniline, 24, 39, 207, 255
methyl-4-fluorobenzylamine, 24
methyl-*n*-heptylamine, 28, 204
methyl-*n*-hexylamine, 28, 204
methyl-2-hydroxypropylamine, 28, 41, 210
methyl-isopropylamine, 54
methylmethoxyamine, 28, 110, 203
methyl-4-methoxybenzylamine, 24
methyl-4-methylbenzylamine, 24
2-methylmorpholine, 19, 216, 258
methylneopentylamine, 28, 40, 54, 55, 85, 206, 240, 252, 255
methyl-4-nitroaniline, 24, 39, 207
methyl-4-nitrobenzylamine, 24
methylnitroguanidine, 2, 28, 64, 68, 75, 97, 111, 199, 249, 257
methyl-*n*-nonylamine, 28, 205
methyl-*n*-octadecylamine, 28
methyl-*n*-octylamine, 28, 204
2-methyl-1, 3-oxazolidine, 16, 216, 258
5-methyl-1, 3-oxazolidine, 4, 16, 216, 258
5-methyloxazolidone, 249
methyl-2-oxopropylamine, 28, 41, 210, 254

N-Nitroso-, *(cont'd)*
  methyl-*n*-pentylamine, 28, 40, 204
  methylphenidate, 16, 39, 223
  methylphenylcarbamate, 104
  methyl-2-phenylethylamine, 24, 39, 55, 124, 206, 259
  N-methylpiperazine, 3, 133, 218
  2-methylpiperidine, 51, 105, 221
  3-methylpiperidine, 51, 137, 221
  4-methylpiperidine, 51, 105, 221
  3-methyl-4-piperidone, 19, 39, 225
  methyl-*n*-propylamine, 28, 40, 85, 163, 204, 210
  methyl-*n*-tetradecylamine, 28, 40, 205
  3-methyl-Δ^3-tetrahydropyridine, 19, 38
  methyl-*n*-tridecylamine, 28, 40
  methyl-trifluoroethylamine, 206, 209
  methyl-*n*-undecylamine, 28, 205
  methylurea, 41, 63, 64, 79, 111, 199, 257
  methylurethane, 2, 199, 249
  morpholine, 4, 19, 38, 49, 103, 133, 214, 215, 258
  nipecotic acid, 53, 137, 223
  nornicotine, 16, 220
  octamethyleneimine, 19, 38, 49, 103, 213
  1, 3-oxazolidine, 4, 16, 38, 103, 215, 216, 258
  oxazolidone, 249
  oxopropylallylamine, 28, 211
  phenmetrazine, *see* 2-phenyl-3-methylmorpholine
  phenylbenzylamine, 24, 39, 203, 207, 255
  phenylethylurea, 41, 201, 250, 257
  2-phenyl-3-methylmorpholine, 19, 133, 216
  4-phenylpiperidine, 19, 39, 53, 104, 109, 219, 221, 222
  phenylurea, 34, 201
  pipecolic acid, 53, 223
  piperazine, 38, 45, 103, 121, 130, 137, 138, 217, 218, 240
  piperidine, 19, 38, 39, 49, 53, 86, 104, 108, 123, 130, 137, 168, 172, 180, 213, 220, 221, 258
  4-piperidone, 19, 39, 105, 137, 224
  proline, 3, 5, 53, 220
  propylnitroguanidine, 68, 75, 77, 85
  n-propylurea, 42, 201, 257
  pyrrolidine, 5, 16, 38, 48, 49, 53, 101, 103, 133, 162, 168, 170, 177, 184, 212, 214, 219, 220, 258
  pyrroline, 16, 38, 49, 221, 240, 254, 258
  sarcosine, 205, 209

N-Nitroso-, *(cont'd)*
  tetrahydro-1, 3-oxazine, 4, 19, 38, 215, 258
  1, 2, 3, 6-tetrahydropyridine, 19, 38, 39, 49, 105, 222, 240
  1, 2, 3, 4-tetrahydropyridine, 19, 38, 105, 222
  2, 2, 6, 6-tetramethylpiperidine, 52, 221
  thiazolidine, 16, 38
  thiomorpholine, 19, 38, 215
  tridecylurea, 41, 201, 250
  triethylurea, 34, 42
  trihydroxydipropylamine, 28, 211
  2, 2, 5-trimethyl-1, 3-oxazolidine, 16
  2, 4, 4-trimethyl-1, 3-oxazolidine, 16
  3, 4, 5-trimethylpiperazine, 19, 218, 254
  trimethylurea, 34, 42
  undecylurea, 201, 250
N-Nitrosoalkylamides, 2
N-Nitrosoalkylcarbamates, 100
N-Nitrosoalkylguanidines, 100
Nitrosoalkylureas, 34, 100–102
N-Nitroso compounds, 61, 62, 63
  ethylating, 75, 78
  formation in the stomach, 5
  human exposure to, 6
  methylating, 75, 78
  propylating, 75, 78
N-Nitroso dialkylamines, 83
N-Nitrosodimethylamine, poisoning by, 1, 60
N-Nitroso-N-ethyl compounds, 60, 63, 80
N-Nitroso-N-methyl compounds, 60, 63, 78, 80
Nitrosotrialkylureas, 34
Nitrous acid, 1, 3, 4
Nucleophiles, 3

Open-chain nitrosamines-symmetrical/unsymmetrical, 207
Organotropy, of nitrosamines, 172
α-oxidation, 180, 214, 219, 224
α-oxidation, inhibition of, 38
β-oxidation, 172, 205, 216
Oxidative dealkylation, 45
Oxygen, alkylation, 61
Oxygenated nitrosopiperidines, 225
Oxytetracycline, 4, 6
Ouabain, 121, 123, 236

Pancreatic duct, 210, 215
Permeability mutation, *envA*, 93

Peroxyacyl nitrites and nitrates, 3
Pesticides, 41
Phage induction, 252
Phage induction assay, 234
Phenacetin, 163
Phenmetrazine, 5, 216
Phenols, 4
Phenyldiazonium ions, 207
Piperazine, 6
Piperidine, 6
Piperidine substitution, 104
Piperine, 6
Plate incorporation assay, 13
Platinum (II) chloramines, 131
Point mutation, 236
Polycyclic aromatic hydrocarbons, 120, 131, 199
Prokaryotic systems, 259
Promutagens/carcinogens, 46
Prophage induction assay, 92
Proteins, nitrosamine metabolizing, 82
Proximate carcinogen, 256
2-Pyrrolidone, 172

Quinacrine, 61
Quinolines, 131

Rat liver S9, 34, 38, 96, 150, 163, 255
recA protein, 91
Reticulo-endothelial tumors, 172
Rubber factories, 214
Rubber industry, 3

S-8 fraction, 178
Saccharomyces cerevisiae, 236
Saccharomyces gene conversion test, 130
Saliva, 5
Salmonella, 182
Salmonella histidine reversion assay, 46, 130, 131, 235
Salmonella mammalian-microsome mutagenicity assay, 13, 34, 234

Sister chromatid exchange, 130, 135, 137, 236
Smog, 3
Sodium nitrite, 2
Somatic mutation theory, 233
Soybean trypsin inhibitor, 159
Specificity, species and organ, 45
Steric factors, 38
Steric hindrance, 39, 245
Stomach, 5
Structure, chemical
  and biological activity, 46
  and mutagenic activity, 13
Substituted aliphatic nitrosamines, 207
4-Substituted nitrosopiperidines, 39
Succinic semialdehyde, 172
Sulfhydryl groups, 72
Synthetic cutting oils, 4

Tannins, 4
Thiocyanate, 3
6-Thioguanine, 121, 123, 130, 236
Thiram, 6
[^{3}H]Thymidine, 135
Tobacco smoke, 129, 213, 220
α-Tocopherol, 4
Tolazamide, 6
Tolbutamide, 6
Trace impurity, 34
Transitional cell carcinomas of the urinary bladder, 204
Transnitrosation, 5, 218
Transplacental effect, 5
Transport across membranes, 250
Tribenzylamine, 3
Triethanolamine, 4
Trypsin, 159
Tumor promotors, 61
Tyr locus, 236

Unsaturation, effect on carcinogenesis, 222
Urea, 172